Familienunternehmen und KMU

Edited by
A. Hack, Berne
A. Calabrò, Witten/Herdecke
H. Frank, Vienna
F. W. Kellermanns, Tennessee
T. Zellweger, St. Gallen

W0037763

Both Family Firms and Small and Medium Sized Enterprises (SME) feature a number of distinct behaviors and characteristics which could provide them with a competitive advantage in the market but could also lead to certain risks. The scientific series at hand presents research which provides an empirical and theoretical contribution to the investigation on these specific characteristics and their impact on business practice. The overall aim of this series is to advance the development of theory in the areas of family firm and SME management.

Edited by

Professor Dr. Andreas Hack
University of Berne

Professor Dr. Andrea Calabrò
University of Witten/Herdecke

Professor Dr. Hermann Frank
Vienna University of Economics and
Business

Professor Franz W. Kellermanns, Ph.D.
University of Tennessee

Professor Dr. Thomas Zellweger
University of St. Gallen

Nadine Kammerlander

Organizational Adaptation to Discontinuous Technological Change

The Effects of Family Influence and Organizational Identity

Foreword by Prof. Dr. Björn Ivens

 Springer Gabler

Nadine Kammerlander
St.Gallen, Switzerland

Dissertation University of Bamberg, 2012

ISBN 978-3-658-01314-1   ISBN 978-3-658-01315-8 (eBook)
DOI 10.1007/978-3-658-01315-8

The Deutsche Nationalbibliothek lists this publication in the Deutsche Nationalbibliografie; detailed bibliographic data are available in the Internet at http://dnb.d-nb.de.

Library of Congress Control Number: 2013932869

Springer Gabler
© Springer Fachmedien Wiesbaden 2013

Printed on acid-free paper

Springer Gabler is a brand of Springer DE.
Springer DE is part of Springer Science+Business Media.
www.springer-gabler.de

Foreword

Adaptation to discontinuous technological changes is a pivotal yet not insurmountable challenge for incumbent organizations. While previous literature has made substantial advances in explaining why established organizations vary in how they respond to such breakthroughs, multiple facets of organizational adaptation still remain to be discovered. One important, yet so far under-explored, clue to better understanding of the adaptation to discontinuous technologies is the social context of owners and managers. In particular, the role of family owners and family managers in the context of discontinuous change is under-investigated.

In her dissertation, Nadine Kammerlander examines how family ownership and identity affect adaptation to discontinuous technologies. She thereby contributes to available literature on organizational adaptation to discontinuous change as well as research on family businesses and organizational identity. Her work follows a three-step process: First, she develops a conceptual integration of the related, yet so far distinct literature streams on organizational adaptation and family influence. She then provides empirical evidence based on multiple case studies in three German industries, illustrating how family firms vary in their adaptation behavior. In a last step, based on empirical evidence from 14 interpretative case studies in the German publishing industry, she induces a mid-range model that formalizes the interplay between organizational identity, organizational adaptation and family influence.

The results from Nadine Kammerlander's thesis are relevant for both, researchers and practitioners. Particularly owners and managers of family firms can derive valuable advices how to master situations of discontinuous technological changes. I thus wish that this thesis will attract the broad readership that it deserves.

Bamberg, December 2012

Prof. Dr. Björn Ivens

Preface

The following thesis is a result of a cooperation from 2010 to 2012 between the Otto-Friedrich University, Bamberg, the Friedrich-Alexander University Erlangen, Nuremberg, as well as the IMD in Lausanne. Numerous people, whom I wish to thank in the following, supported me in one way or another and helped me to achieve success with this "project," which was certainly one of the most exciting, educational and enjoyable ones of my career.

Foremost, I wish to thank Prof. Dr. Björn Ivens for his mentorship of this thesis, for his constant openness to new ideas and his unwavering optimism throughout its development. I owe my gratitude to Prof. Dr. Albrecht Enders for numerous constructive discussions and suggestions whether in Lausanne, San Antonio, or at the Munich Sailing School. A special thank you to Dr. Andreas König, who animated my interest in "technological discontinuity", who, in numerous conversations emphasized what is paramount in scientific papers and publications, and whose creativity and erudition will remain a guiding influence for me.

Thank you to Prof. Dr. Christoph Bläsi for his collaboration and support in the execution of case studies in German publishing, especially for his assistance in establishing contact to interviewees and his valuable insights on "the industry" and the operations of publishing houses.

Furthermore, I wish to thank my former diploma graduates, Barbara Schillo, Kilian Bornschein, Matthias Eck and Benjamin Giebler, as well as both my Master's graduates at the University of Mainz, Kristina Rubel and Ruth Speil, for accompanying me to numerous interviews and for serving as coding partners.

A lot of gratitude goes to the numerous companies, experts and management teams, who devoted their time in anonymous interviews, and without whose candid participation in these interviews a significant part of this thesis would not have been possible.

I also wish to thank various (anonymous) reviewers and conference participants (e.g., FERC and Doctoral Consortium 2011 in Grand Rapids and in Montreal in 2012; EIASM, 2011 in Witten; AOM 2011 in San Antonio; The annual convention of the Internationale Buchwissenschaftliche Gesellschaft, Mainz 2011; AOM 2012 in Boston), for commenting on the preliminary versions of some excerpts of this thesis and for their countless constructive remarks.

Moreover, I wish to thank my (internal and external) fellow PhD students at Bamberg University, my McKinsey Fellow colleagues in Munich, as well as the team at the Center for Family Business at the St. Gallen University for their diverse field and non-field discussions, diversions and practical tips.

In conclusion, I wish to thank my family, especially my husband, Friedrich, my son, Jonathan, and my parents for their infinite patience, their constant encouragement and their invigorating spontaneity.

St. Gallen, Oktober 2012

Nadine Kammerlander

Table of Contents

List of Figures

All Figures displayed in this thesis are own illustrations of the author unless otherwise noted.

List of Tables

List of Abbreviations

BS   business system

CD   compact disc

CD-ROM compact disk read only memory

CEO   chief executive officer

DAX   Deutscher Aktienindex (German stock market index)

DT   discontinuous technology

DVD   digital versatile disc

e.g.,   exemplum gratia: for example

E-N   exclusive/environment-centric (type of organizational identity)

E-S   exclusive/self-centric (type of organizational identity)

epub   electronic publication: free and open ebook standard

et al.   et alii: and others

EUR   EURO

FIBER Family control and influence, Identification of family members with the firm, Binding social ties, Emotional attachment of family members, Renewal of family bonds to the firm through dynastic succession (scale to measure socioemotional wealth)

F-PEC Family Influence—Power, Experience, Culture (scale to measure family influence)

FS   family system

i.e.   id est: that is

I-N   inclusive/environment-centric (type of organizational identity)

I-S   inclusive/self-centric (type of organizational identity)

IT   information technology

n/a   not applicable

P	proposition
p.	page
pdf	portable document format
R&D	Research & Development
S&P	Standard & Poors
SEW	socioemotional wealth
TV	television
US	United States
VP	vice president

1 Introduction

1.1 State of Research

Understanding how and why incumbent organizations do or do not adapt to discontinuous technological changes is one of the most intensively studied topics in organizational science and is also highly relevant for practitioners (e.g., Chesbrough, 2001; Christensen, 1997; Hill & Rothaermel, 2003). This is particularly true because such breakthroughs impose immense challenges on organizations in that they deviate substantially from established performance trajectories (Christensen & Bower, 1996), render established standards and knowledge obsolete (Anderson & Tushman, 1990), and necessitate new sets of competencies within the affected organizations (Anderson & Tushman, 1990).

Discontinuous (frequently also called radical or disruptive[1]) innovations differ from established technologies in the perceived value of the product (Bowman & Ambrosini, 2000), in the underlying processes for value creation (Christensen & Bower, 1996; Hulin & Roznowski, 1985), in how the value is captured (Bowman & Ambrosini, 2003), or in a combination of any of the above (König, 2009). Well known and frequently cited examples of discontinuous technologies include the innovation of quartz as opposed to mechanical watches (Glasmeier, 1991), electric powered vehicles as opposed to those relying on internal combustion engines (Christensen, 1997), internet telephony as opposed to analog/digital telephony (Benner, 2008), digital as opposed to analog photography (Benner, 2010), and hard disk drives as opposed to former, physically larger storage media (Christensen, 1997).

Adequate adaptation to such technological discontinuities that occur on an irregular basis across industries (Anderson & Tushman, 1990) is crucial, because such disruptions have the potential to entirely change the market structure and result in a redistribution of market shares, frequently at the cost of the 'incumbents'—the established players in a given industry. Many examples show how new entrants, after having successfully introduced the novel technology, become the new market leaders, whereas the established market leaders, due to missing or inadequate adaptation, lose their leading positions and/or retreat into market niches—a phenomenon that is frequently labeled 'incumbent inertia.' The failure of Borders and rise of Amazon in the retail book industry is a famous recent example of market leadership change induced by technological discontinuities. According to Christensen and

---

[1] For a more detailed differentiation between the various terms, see Chapter 2.1.1

Bower (1996), the computer industry has experienced several discontinuous technological changes throughout its history, with market leadership being transferred from Digital Equipment and Data General (incumbents in the minicomputer market) to Commodore and Tandy (new players in the desktop market, together with IBM), and ultimately to Compaq, Zenith, Toshiba and Sharp (new players in the portable computer market).

Given the abundance of anecdotal empirical evidence on incumbent inertia, it is not surprising that most academic studies so far have particularly documented the predisposition of incumbent firms to "[resist] strategic renewal outside the frame of current strategy" (Huff, Huff, & Thomas, 1992: 56) and, in turn, to respond sluggishly to radical breakthroughs (Miller & Friesen, 1980). In those studies, scholars have built on various technological lenses such as bounded rationality (March & Simon, 1958), population ecology (Hannan & Freeman, 1984), and theory of power and politics (Pfeffer & Salancik, 1978) to explain this phenomenon of 'organizational inertia.' They have also identified various economic, strategic, organizational (see Hill & Rothaermel, 2003 for a summary), cognitive, and emotional (Kaplan, 2011; Tripsas & Gavetti, 2000) factors that impede adaptation to discontinuous technological changes.

When investigating the organizational adaptation of incumbents, scholars mainly rely on four distinct dimensions of organizational adaptation that are closely connected to adaptation performance: (1) speed—the rate at which organizations recognize and interpret technological discontinuities and implement competitive responses (Miller & Friesen, 1980; Szymanski, Troy, & Bharadwaj, 1995); (2) intensity—the degree in which established organizations commit resources to the exploration of discontinuous technologies rather than to reinvest in existing technologies and capabilities (Christensen & Bower, 1996; Gilbert & Newbery, 1984); (3) stamina—the extent in which incumbents continue to commit resources to a new technology after an initial investment, even when (temporary) setbacks occur (Block & MacMillan, 1985); and (4) routine flexibility—the degree in which organizations replace established routines with new, non-paradigmatic routines (Feldman & Pentland, 2003; Tripsas & Gavetti, 2000).

Criticizing the unilateral view of incumbent players as a homogeneous group which reacts inadequately to discontinuous technological changes, various researchers have started to challenge the portrayal of incumbent inertia as inevitable and have shown that established organizations vary—often significantly—in the manner in which they adopt radical innovations (Mitchell, 1989). For instance, Gilbert (2005) observed that, contrary to the predictions of disruptive innovation theory (Christensen & Bower, 1996), some incumbents aggressively adopt technological discontinuities. Given the importance of these observations, researchers have begun to study factors that might cause an established organization to resist the prevailing pattern and to overcome organizational paralysis (Teece, Pisano, & Shuen, 1997). Such factors include, for instance, external influence (Gilbert, 2005) and structural ambidexterity (Tushman & O'Reilly, 1996).

1.2   Research Gap and Relevance

Although research on the variance of organizations' responses to discontinuous technological change has provided a wealth of insights, the behavioral heterogeneity of incumbents faced with technological discontinuities cannot yet be fully explained. For example, it remains unclear why NCR, an incumbent of the mechanical cash register industry, successfully adapted to the emergence of electronic products, while Smith Corona, a former leader in the US typewriting industry, failed to adapt to this discontinuity and ultimately went bankrupt in 1995 (Danneels, 2011; Hill & Rothaermel, 2003; Rosenbloom, 2000).

Most previous studies aiming to explain adaptation behavior have focused on managerial behavior and incentives (e.g., Christensen & Bower, 1996; Tripsas & Gavetti, 2000) as well as institutional influence affecting organizational behavior (DiMaggio & Powell, 1983). Intriguingly, so far almost no attention has been devoted to examining how the social contexts of major shareholders in general, and family influence in particular determine organizational adaptation. Only recently, Benner (2007) provided evidence on how shareholders can hinder organizational adaptation to discontinuous technological change through negative stock market reactions to announced adoption strategies. While previous studies in the inertia literature indeed include companies that are influenced by family owners (e.g., Gilbert, 2005), the implications of family influence on adaptation to discontinuous technological change have never been explicitly and systematically studied. Thus, the following research question constitutes a significant gap in the current literature and guides the theorizing and the empirical research conducted within the dissertation project and presented in this thesis.

4

> *How does family influence affect the adaptation patterns of incumbent organizations to discontinuous technological change?*

Answering the research question as to how family influence affects adaptation to discontinuous change has both theoretical and practical relevance. The outlined research gap is astonishing because owners and investors in general (Benner, 2007), and family owners in particular, play important roles in shaping the strategic activities of organizations (Berrone, Cruz, Gómez-Mejía, & Larraza-Kintana, 2010; Fiss & Zajac, 2004; La Porta, Lopez-de-Silanes, & Shleifer, 1999). Hence, owners can be expected to substantially influence their respective companies' adaptation patterns when faced with discontinuous technological change.

To a greater extent than other owners, family owners shape their organization's interpretation, decision-making, and resource allocation processes (Berrone et al., 2010; Gómez-Mejía, Núnez-Nickel, & Gutierrez, 2001) and also affect the organization's embeddedness in its environment (Berrone, Cruz, & Gómez-Mejía, forthcoming), its culture (Dyer, 1986; Miller & Le Breton-Miller, 2006), and its identity (Arregle, Hitt, Sirmon, & Very, 2007; Gómez-Mejía, Takács Haynes, Núnez-Nickel, Jacobson, & Moyano-Fuentes, 2007) through what scholars label 'family influence.' Family influenced businesses differ from non-family influenced ones with regards to their capabilities and motivation. The intermingling of ownership and management as frequently observed in family influenced businesses and the mostly existing independence from outsiders (e.g., Berrone et al., forthcoming) affect family firms' capabilities: The specific structures present in family firms allow these firms to react differently to changes compared to non-family influenced businesses. For instance, the owner-manager duality enables fast decision making in organizations. Regarding the motivation for any strategic moves, non-economic and, in particular, transgenerational intentions[2] play a crucial role for family businesses (e.g., Gómez-Mejía et al., 2007). The idiosyncratic goals and intentions of family influenced businesses substantially affect their interpretation and decision-making processes. Prior research has provided evidence that this line of argument is not only applicable to small, owner-managed firms, but, to a substantial degree, also holds true for larger firms with family influence,

---

[2] 'Transgenerational intentions' refer to an owner-family's desire to be able to hand over a prosperous family influenced business to the next generation (Gómez-Mejía et al., 2007).

including public firms with a substantial share of family ownership (Berrone et al., 2010). For instance, abundant empirical evidence has shown how family influence affects strategic behavior such as participation in alliances (Gómez-Mejía et al., 2007), diversification (Gómez-Mejía, Makri, & Larraza-Kintana, 2010), acquisition (Miller, Le Breton-Miller, & Lester, 2010), and performance under institutional pressure (Berrone et al., 2010) of both, small and large family influenced businesses.

Understanding the implications of family influence on organizational adaptation is also crucial from a practical perspective, particularly because of the economic weight of family influence worldwide. In fact, scholars estimate that between 50% and 96% of all companies around the globe are in the hand of families, depending on the respective measurement used (Klein, 2008). For example, in Germany, more than 80% of the companies are family businesses, contributing roughly 60% of all job positions in Germany, and approximately 50% of total revenues created by the German corporate sector[3] (Gottschalk & Keese, 2011). Although many family businesses are small or medium-sized, family influence is also prevalent amongst larger companies. For instance, according to an analysis of Anderson and Reeb (2003), 35% of the S&P 500 firms were family-owned at the time of their study.

Due to the increasing amount of practically orientated literature on discontinuous technological change (e.g., Christensen & Raynor, 2003) as well as the strong media coverage of companies that failed in their adaptation (e.g., Borders, Ritz, Kodak), managers and owners of organizations have become increasingly aware of path-breaking technological shifts and have begun to search for remedies to avoid incumbent inertia. However, only when having a comprehensive understanding of all stakeholders promoting or hindering adaptation processes and knowledge about which specific barriers and enablers are relevant to their respective organization, managers and owners are likely to be prepared for a successful change process.

Based on the considerations outlined above, integrating the owner perspective into research on discontinuous technological change is a promising approach: It will provide a more fine-grained understanding of observed adaptation patterns and their antecedents and

---

[3] Those numbers are based on a sample called 'Mannheimer Unternehmenspanel (MUP).' This dataset is built on data provided by Creditreform, who systematically scan public registers, newspapers, public announcements, annual reports and balance sheet disclosures and whose data, in turn, represent roughly 80% of all corporate activities in Germany. The MUP sample includes firms of any legal structure (such as 'Einzelunternehmen,' 'Personengesellschaften,' and 'Kapitalgesellschaften') except those governed as 'e.V.' or 'eG.' Moreover, any banks and other firms of the public sector, agricultural businesses, interest clubs, religious associations, exterritorial organizations, and private households were excluded from the MUP sample. In sum, the sample contained 2.8 million businesses, employing 26.2 million individuals.

thus allow for better predictions, making the above-mentioned research question relevant from a theoretical perspective. Moreover, knowledge of the implications of family influence on organizational adaptation will help managers and owners to better respond to technological discontinuities.

1.3 Research Objective

In the realm of describing and analyzing the effect of family influence on organizational adaptation to discontinuous technological change, this doctoral thesis has three main research objectives which are illustrated in Figure 1). My first objective (number 1 in the Figure) is to provide an intimate understanding of how various elements of family influence affect determinants of organizational adaptation. Since family influence can induce a variety of beneficial and detrimental behaviors, and thus constitutes a multi-faceted research phenomenon, it is important to identify and knit nuanced linkages between family business literature and research on organizational adaptation. I will therefore build on extant literature to bridge the gap between the two literature streams and deduce propositions as to how manifestations of family influence affect barriers to and enablers of adaptation to discontinuous technological change, and how these relationships will, in turn, entail characteristic adaptation patterns. To the best of my knowledge, this is the first scholarly effort to integrate the distinct, yet related research streams on family businesses and organizational adaptation to discontinuous technological change and thus this theoretical integration constitutes a ponderous part of this thesis.

The second research objective (number 2 in Figure 1) is to empirically study how and why family influence affects various well established ingredients for adaptation performance. In particular, I will investigate how considerations regarding 'socioemotional wealth,' an umbrella construct that encompasses all non-economic elements of family owners' utility functions (Gómez-Mejía et al., 2007), will shape organizational adaptation.

As I will show, direct influence by owners can only explain part of an organizations' behavioral variation when adapting to discontinuous changes. During my first empirical analysis, 'organizational identity,' which is shaped yet not fully determined by family influence, emerged as an important predictor of organizations' response patterns. Hence, the third research objective of my doctoral thesis addresses two questions: How can firms with varying levels of family influence (number 3a in Figure 1) be categorized into different organizational identities?; and: How does variety in organizational identity affect organizations' response patterns to discontinuous technological change (number 3b in Figure 1)?

Figure 1: Research Objectives

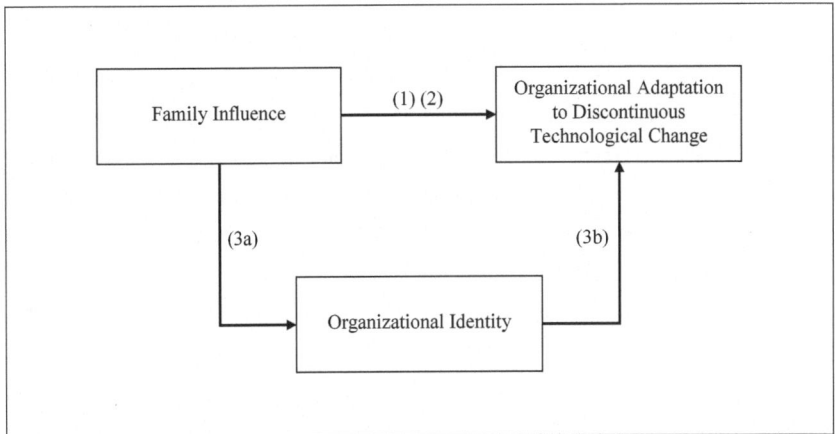

1.4 Epistemological Position

This thesis is rooted in a moderate epistemological position that combines elements of positivism with a more social constructivist notion. Adhering to the advice of Gioia and Pitre (1990), this thesis consists of parallel studies, thereby applying a multiparadigm approach that allows for reciprocal enrichment of the different perspectives. In the following I will outline the importance of epistemology, discuss some of the most disputed epistemological positions in social science, and classify this doctoral thesis in order to build a thorough base for methodological legitimization as discussed in Chapter 1.5.

Advancing theory, which refers to "an ordered set of assertions about a generic behavior or structure assumed to hold throughout a significantly broad range of specific instances" (Sutherland, 1975: 9), is at the core of any solid scientific work (Brannick & Coghlan, 2007). However, there is little consensus among social science researchers what differentiates strong theory from weak theory (Sutton & Staw, 1995). One important root cause of this controversy lies in the fact that different assumptions about ontology—the "nature of [...] phenomena"—entail divergent epistemological positions—i.e. perspectives on "the nature of knowledge about [the] phenomena [relevant for theory]"—which in turn result in dissent about appropriate methodologies, that denote "the nature of ways of studying those phenomena" (Gioia & Pitre, 1990: 585). Indeed, one long standing, ongoing dispute within the academic community refers to epistemological positions to be obtained in social science (Gioia & Pitre, 1990: 585).

One of the most important epistemological positions, positivism, can be traced back to the French philosopher Auguste Comte (1798—1857) who was the first to explicitly formalize that "[r]easoning and observation, duly combined, are the means of [...] knowledge," whereby knowledge refers to the "invariable relations of succession and resemblance" (Comte, 1853: 26). Based on the core assumption "that scientific truth reflects an independent external reality" (Suddaby, 2006: 633), positivism encourages researchers to deduct propositions from pre-existing literature, focus on quantitative, empirical testing of predicted causal relationships, and build theory on positive results (Gioia & Pitre, 1990). Positivism, also labeled 'functionalism' (Morgan & Smircich, 1980) within the social science perspective, flourished particularly well in the 1960s and 1970s (Morgan & Smircich, 1980). Although today positivism is still highly appreciated amongst academics, critical voices of scholars in varying research fields such as Popper's advocacy for falsification instead of verification (Popper, 1935) have steadily gained in popularity.

In the second half of the 20th century, social science scholars like Glaser and Strauss (1967) as well as Gioia and Pitre (1990) criticized the epistemological perspective of positivism as well as Popper's principles of falsification, as falling short of building rich, new theory and of adequately including processes such as interpretation, meaning-construction, and conflict. As a consequence, Glaser and Strauss (1967), who were particularly influenced by pragmatism and early symbolic interactionism (Suddaby, 2006), developed the concept of 'grounded theory' for building novel, rich, and relevant theory, which substantially relies on the interpretive processes of the researcher and hence clearly contradicts positivistic assumptions. As Suddaby (2006: 634) noted, Glaser and Strauss "offered a compromise between extreme empiricism and complete relativism." Subsequently, from the 1980s, research based on an interpretive perspective started to flourish (Morgan & Smircich, 1980), and now co-exists with positivism and several epistemological tonalities.

Social science can either be advanced by theory building or theory testing which follow different target courses as Sutton and Staw describe: "[Quantitative] data describe *which* empirical patterns were observed and theory explains *why* empirical patterns were observed or are expected to be observed" (1995: 374). This doctoral thesis can be classified as theory building research, thus rendering positivistic (here: deduction of propositions) and interpretive/constructivist (here: grounded theory based on multiple case studies) epistemological positions appropriate. Firstly, I apply a positivistic perspective by logically deducting propositions on the effect of family influence on organizational adaptation to discontinuous technological change from the extant literature (Chapter 3). Next, I adopt a constructivist/interpretive[4] perspective and investigate the same phenomenon by drawing on qualitative, empirical data to build grounded theory (Chapter 4). Since my research question includes the investigation of interpretation processes—e.g., 'how do family firm owners interpret discontinuous technological changes and why do they do so?'—an interpretive approach is particularly appropriate (Yin, 1994). The combined approach allows me to scrutinize whether different perspectives yield similar results (Gioia & Pitre, 1990). To investigate how organizational identity affects organizational adaptation to discontinuous technological change, I build exclusively on empirical data from multiple case studies to induce propositions, thus applying an interpretive approach (Chapter 5). I refrain from a positivistic examination of this part of the research question, because the relevant dimensions of organizational identity still remain under investigated and, to a large degree, undefined (Corley et al., 2006). Therefore, the current status of the nascent field of research on organizational identity is inadequate for logically deducting relevant propositions on the targeted relationship.

1.5   Overview of Methodology

In line with the epistemological positions described above, I apply two distinct approaches of theory building: (1) logical deduction, and (2) qualitative, empirical work, based on multi-case studies (Eisenhardt, 1989). I thereby aim to build a mid-range theory (Merton, 1957), that finds a balance within the inevitable tradeoff between generality, simplicity, and accuracy, inherent to any theorizing (Sutton & Staw, 1995).

(1) The starting point for logical deduction of propositions is general knowledge about a defined research area, from where a chain of analytical arguments leads to

---

[4] In line with other scholars (e.g., Schwandt, 1994) I use the terms 'constructivism' and 'interpretivism' synonymously.

propositions for specific contexts (Ketokivi & Mantere, 2010). Whetten (1989) summarizes four elements of deductive theory building: First, the concise definition of concepts that are comprehensive and parsimonious at the same time; second, the causal relations between those concepts; third, a profound explanation as to why the proposed linkages exist; and fourth, a clarification of the contextual boundaries of the assumed propositions. Following this logic of deduction, I first partition the specific research setting my work focuses on and define the constructs that I later use for theorizing such as 'discontinuous technology' or 'family influence' (Chapter 2.1.1 and 2.2.1). In a second step I conduct a comprehensive literature review on the two topics of interest—family influence, organizational adaptation to discontinuous technological change—which provides me with a general, thorough knowledge of the two research areas (Chapter 2.3). In a third step (Chapter 3), I bridge the two related yet distinct research fields by identifying logical arguments that coalesce the constructs of both literature streams while, in parallel, explaining the underlying causal relationships for these linkages in detail. Furthermore, I discuss the critical assumptions and boundary conditions.

(2) Multi-case studies are an important instrument for building grounded theory (Yin, 1994). While some researchers envisage theory building from cases to be a purely inductive approach (Glaser & Strauss, 1967), other scholars emphasize the interplay between induction and deduction involved in case study research due to the iterative interpretation of data and extant literature (Strauss & Corbin, 1998).

My case-based research process mostly followed the eight-step process suggested by Eisenhardt (1989): First, I defined my research questions. Second, I selected cases based on theoretical sampling—in contrast to convenience and/or probabilistic sampling—that best allowed for comparing the individual cases and, hence, applying replication logic. Third and fourth, I engaged in the data collection process, thereby relying on several sources of information (including interviews, observations, primary and secondary documents, and website information) and involving multiple investigators (students with varying educational backgrounds). At the same time I started engaging in the preliminary data analysis process, which provided additional flexibility, e.g., to schedule further interviews or collect supplementary sets of data. Fifth, I systematically conducted single and cross-case analyses, applying well-established methods such as axial coding (Charmaz, 2006; Strauss & Corbin, 1998) and frequency counts (Krippendorff, 2004). Sixth, I searched for patterns across the spectrum of cases and derived formal propositions. I deviated from Eisenhardt's seventh process step—enfolding and comparing relevant literature—by encompassing extant research work in parallel, beginning already in step one, rather than envisaging it as a distinct and subsequent step in the process. This upfront inclusion of relevant literature is in line with calls of various scholars (e.g., Suddaby, 2006). Lastly, I concluded my findings and presented formalized propositions as suggested by Eisenhardt. When analyzing the data (steps 4-6), I applied van Maanen's (1979a) approach, separating between two distinct types of data: first order data encompasses 'facts' provided by the interviewees and gained through observations and primary and secondary documents. Second order data, however, comprises concepts that encompass the researcher's organization and explanation of this first level data. The analyst thereby needs to account for the specific context in which the first order data was gained.

Irrespective of the chosen epistemological positions and methodological approaches, there is a void for any scientific work to meet requirements regarding relevance, validity and reliability (Popper, 1994; Yin, 1994). Relevance addresses the necessity that the new theory substantially advances extant knowledge by providing improved predictions of observed phenomena. Validity consists of three distinct components: construct, internal, and external validity. Construct validity refers to the question whether the theoretical constructs upon which the propositions are built can indeed be measured by the practical operationalizations utilized. Internal validity refers to the foundation of the causal relations of the theory and the 'logic' of the arguments. It also refers to the existence of alternative explanations. External validity refers to the domain, in which the propositions are valid. High reliability suggests that in case of repetition of parts of the study (e.g., reproducing the data collection or data analysis), the results would remain unchanged.

I assume relevance to be fulfilled for each part of this thesis due to the more nuanced picture of understanding derived from my theorizing. Throughout my empirical research, I adhered to Yin's (1994) tactics to maximize validity and reliability. In this vein, I drew on multiple sources such as interviews, observations, and archival data to strengthen construct validity because triangulation of data minimizes the risk of overly subjective judgments. Moreover, I established a chain of evidence during the data collection process and presented my preliminary results to the key informants in order to get their feedback (e.g., electronically or as an oral presentation at a conference). Furthermore, I referred to constructs already utilized in prior research (e.g., F-PEC scale for family influence (Astrachan, Klein, & Smyrnios, 2002); Chapter 4.4), wherever possible. In cases where, due to a lack of prior measurement, such reference to prior research was non realizable (e.g., Chapter 5), I clearly defined the constructs I aimed to measure and explicitly outlined the exact assessment of the variables and the underlying rationale.

12

To enhance internal validity, I built on various techniques such as pattern matching, explanation building, consideration of rival explanations, and the use of logic models, thereby utilizing computer-assisted tools (NVivo, Excel) where appropriate. Potential rival explanations are discussed in detail in the respective chapters on discussion and limitations. Involving multiple researchers, each analyzing the data independently in a first step, further enhanced internal validity. Building on multiple cases and thus adopting replication logic allowed me to enhance external validity.

To address potential concerns regarding reliability, I set up an internet-based comprehensive case study database[5] containing all raw and processed materials as well as any other documents relevant for the studies included in this thesis. This database also contains a detailed case study protocol that was iteratively updated throughout the study.

1.6 Structure of Thesis

This doctoral thesis is structured as follows (see also Figure 2):

---

[5] Realized by services of http://hidrive.strato.com .

Figure 2: Structure of this Thesis

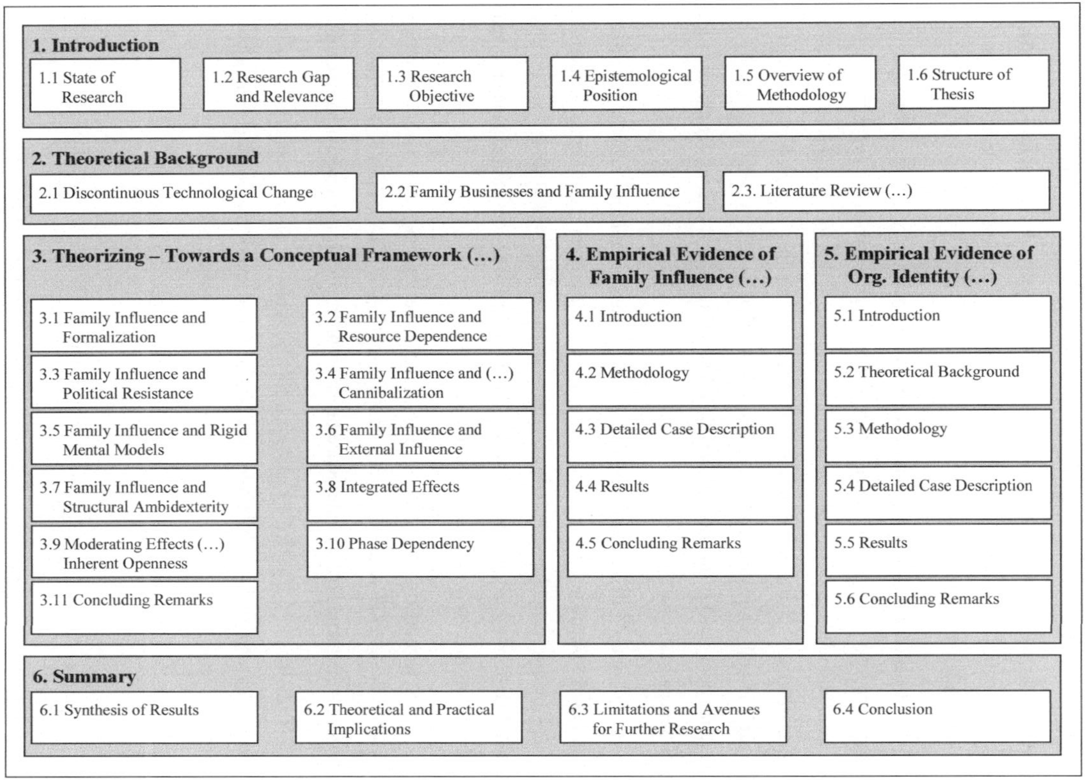

1. Introduction

- 1.1 State of Research
- 1.2 Research Gap and Relevance
- 1.3 Research Objective
- 1.4 Epistemological Position
- 1.5 Overview of Methodology
- 1.6 Structure of Thesis

2. Theoretical Background

- 2.1 Discontinuous Technological Change
- 2.2 Family Businesses and Family Influence
- 2.3 Literature Review (...)

3. Theorizing – Towards a Conceptual Framework (...)

- 3.1 Family Influence and Formalization
- 3.2 Family Influence and Resource Dependence
- 3.3 Family Influence and Political Resistance
- 3.4 Family Influence and (...) Cannibalization
- 3.5 Family Influence and Rigid Mental Models
- 3.6 Family Influence and External Influence
- 3.7 Family Influence and Structural Ambidexterity
- 3.8 Integrated Effects
- 3.9 Moderating Effects (...) Inherent Openness
- 3.10 Phase Dependency
- 3.11 Concluding Remarks

4. Empirical Evidence of Family Influence (...)

- 4.1 Introduction
- 4.2 Methodology
- 4.3 Detailed Case Description
- 4.4 Results
- 4.5 Concluding Remarks

5. Empirical Evidence of Org. Identity (...)

- 5.1 Introduction
- 5.2 Theoretical Background
- 5.3 Methodology
- 5.4 Detailed Case Description
- 5.5 Results
- 5.6 Concluding Remarks

6. Summary

- 6.1 Synthesis of Results
- 6.2 Theoretical and Practical Implications
- 6.3 Limitations and Avenues for Further Research
- 6.4 Conclusion

To provide comprehensive understanding of the core concepts upon which this work builds, Chapter 2 will provide a detailed literature review of organizational adaptation to discontinuous technological change, family influence, and also the current status in literature of integrating these two research streams.

Chapter 3 will rely on extant literature to conceptually develop propositions as to how manifestations of family influence affect barriers to and enablers of organizational adaptation to discontinuous technological change. The purpose of this theorizing is to provide a nuanced understanding of the various linkages between family influence and determinants of organizational adaptation and a detailed explanation as to how and why family influenced organizations react differently to discontinuous technological change compared to non-family influenced businesses.

Chapter 4 describes an empirical study, based on multiple case studies, that provides fact-based evidence of the reactions of firms with varying levels of family influence to recent discontinuous technological changes. Building on the findings of these studies, propositions are induced on the adaptation patterns of family influenced companies, measured against four dimensions that are well established in literature.

Subsequently, Chapter 5 will investigate the influence of variation in organizational identity on the adaptation patterns of 14 firms with varying levels of family influence, that are currently threatened by the emergence of electronic media as a discontinuous technology.

Ultimately, Chapter 6 concludes this thesis with a short summary and discussion of the results and their contribution and implications, exposition of the limitations, and an outlook for further related research.   illuminates this structure.

2 Theoretical Background

2.1 Discontinuous Technological Change

2.1.1 Definition and Characteristics

Despite the ubiquity of the phenomenon of discontinuous technological change in the academic literature of the last decades, there is still a lack of definition for this term (Chesbrough, 2001). Even more importantly, scholars refer to similar yet somewhat different terms such as 'disruptive,' 'radical,' 'non-paradigmatic,' or 'breakthrough' when discussing related phenomena without clarifying the similarities and differences of the concepts they are building on (Suarez & Rogelio, 2005). Table 1 provides an overview of the definitions used in some articles that have been frequently cited by academics and that are particularly relevant for this thesis.

Table 1: Overview of Definitions Used in Prior Research

Authors	Year	Term	Definition	Antonym
Miller & Friesen	1980	Major change *synonymous for:* serious change, dramatic change	A change, for instance "in the external environment caused by competitor strategies, technological obsolescence, economic booms or recessions, etc." (p. 596)	Small, piecemeal change
Abernathy & Clark	1985	Revolutionary innovation	An "innovation that disrupts and renders established technical and production competence obsolete, yet is applied to existing markets and customers" (p.12)	Regular and niche innovation
Foster	1985	Technological discontinuity	Technological opportunities with "superior performance improvement potential" (p. 132), i.e. more remote "technical limits" (p.128), lying on a different "S-curve" as compared to the old technology	Evolutionary progress (along established S-curve)

Authors	Year	Term	Definition	Antonym
Tushman & Anderson	1986	Product/process discontinuities/major technological shifts	"Technical advance so significant that no increase in scale, efficiency, or design can make older technologies competitive with the new technology" (p. 441)	Incremental improvement
Anderson & Tushman	1990	Technological breakthrough, *synonymous for:* technological discontinuity, breakthrough innovation	"innovations that dramatically advance an industry's price vs. performance frontier" (p. 604); innovations that "command a decisive cost or quality advantage and […] strike not at the margins of the profits and the outputs of the existing firms, but at their foundations, their very lives" (Schumpeter, 1942:84) (p. 606)	(Continuous) incremental innovation
Henderson & Clark	1990	Architectural innovation	"reconfiguration of an established system to link together existing components in a new way" (p. 12)	Radical, incremental or modular innovation
Christensen	1997	Disruptive technology	Disruptive technologies' are characterized by "worse product performance, at least in the near-term," offering "very different value proposition," often "cheaper, simpler, smaller […] more conveniently to use," thus being attractive for "few fringe (and generally new) customers " (p xviii)	Sustaining technology
Chesbrough	2001	Technological change	None (highlights a classification along 3 dimensions: technical complexity, external linkages, institutional environment)	None

Authors	Year	Term	Definition	Antonym
Hill & Rothaermel	2003	Radical technological innovation	"involves methods and materials that are novel to incumbents [in order to achieve a commercial or industrial objective]" and can be based either on an "entirely different knowledge base or [stem] from the recombination of parts of the incumbents' established knowledge." (p. 258)	Incremental technological innovation
Gilbert	2005	Discontinuous change	"external changes that require internal adaptation along a path that is nonlinear relative to a firm's traditional innovation trajectory" (p.742)	None
Benner	2007	Radical technological change	A change that "shifts the underlying base of technological knowledge in an industry's products and promises dramatic improvement in price and performance, possibly resulting in product substitution" (p. 704)	None
Kaplan & Tripsas	2008	Discontinuous change/technological discontinuity	None (referring to Anderson & Tushman's models of evolutionary cycles)	None
Tripsas	2009	Identity-challenging technology	"technologies that deviate from the expectations associated with an organization's identity" (p. 442)	None
König, Schulte, & Enders	Forthcoming	Non-paradigmatic change	"a distinctively novel model or pattern for resolving selected problems in a collectively shared cognitive domain" (p. 5)	Paradigmatic change

It seems important to begin the definitional discussion by emphasizing the difference between 'inventions' and 'innovations' as first described by Schumpeter (1934). 'Inventions' refer to "the discovery of new knowledge" (Hill & Rothaermel, 2003: 258) and as such are mainly relevant for the R&D departments of organizations. 'Innovations,' however, go one step beyond and constitute "attempts to commercialize an invention" (Hill & Rothaermel, 2003: 258). As such, their importance and impact is not constrained to organizations' R&D departments but they also substantially affect, for instance, production, sales and marketing as well as the purchasing departments. Put differently, innovation refers to a change in technology, whereby 'technology' depicts the "process by which an organization transforms labor, capital, materials, and information into products and services of greater value" (Christensen, 1997: xvi). Thus, I will henceforth use the terms 'innovation' and 'technological change' synonymously.

Common to most of the definitions given in this context is the comparison of disruptive/discontinuous/radical as opposed to sustaining/continuous/incremental/ evolutionary technological change. Any input or output functions, such as knowledge accumulation, resource requirements, or product performance of the latter type typically shows what mathematicians call 'continuously differentiable behavior'[6]. On the contrary, the characteristic curves of discontinuous technological changes exhibit sharp bends and steps: Knowledge, resources, or product features that are crucial for the old technology become irrelevant whilst new knowledge, resources, or product features become necessary. Whereas some authors, for instance Christensen, become very specific about detailed characteristics of these changes such as the development of product performance criteria over time, others such as Kaplan remain rather vague when defining the nature and characteristics of their focal construct.

In this thesis, I build on a combination of Gilbert's (2005) and Hill and Rothaermel's (2003) view envisaging discontinuous technological changes as:

Definition 1: Discontinuous Technological Changes

'Innovations triggered by external players, that involve methods and/or materials that are novel to incumbents and that require internal adaptation along a path nonlinear to the firm's traditional trajectory.'

---

[6] Given an appropriate level of granularity.

By drawing on this definition, I aim to highlight three aspects:

1) The phenomenon I am investigating is triggered *externally* (as opposed to internally)

2) via a new *technology* (as opposed to political turmoil, general economic crises,…); and

3) adoption of such a new technology requires *non-paradigmatic changes* of internal resource commitments and routines.

This definition allows me to sufficiently focus the core construct in order to build reliable and specific theory, whilst at the same time being broad enough to allow for generalization and empirical evidence within multiple settings.

More specifically, the definition used here includes product and process innovations (Anderson & Tushman, 1990) as well as innovations referring to business models, i.e. the way of capturing value, and any combinationary cases (König, 2009). The same holds true for Henderson and Clark's (1990) differentiation between architectural innovations (referring to reconfiguring existing components in a novel way) as opposed to radical innovations (referring to innovations based on substantially new components requiring a new architecture), both of which are discontinuous technological changes as defined here. Furthermore this definition comprises, but is not restricted to, 'identity-challenging' innovations that contravene the organizational members' shared beliefs about the nature of the central, distinctive, and enduring elements of the organization (Tripsas, 2009). This thesis focuses on competence-destroying rather than competence-enhancing innovations (Anderson & Tushman, 1990).

A static analysis of adaptation to discontinuous change holds little promise; yet, any study of adaptation to discontinuous change needs to consider the temporal evolution of such changes. In their ground-breaking articles (1990; 1986), Anderson and Tushman propose a cyclical model of continuous and discontinuous technological changes as depicted in Figure 3. At a given point in time, a technological discontinuity emerges entailing an 'era of ferment'. In this phase, characterized by high uncertainty, many potential designs co-exist while the residual fit with the established technology steadily decreases. This phase ends with the emergence of a 'dominant design' that heralds the 'era of incremental change,' a phase that is dominated by incremental innovations and standardization. To come full circle, a new technological discontinuity triggers the next evolutionary cycle[7].

Figure 3: Evolutionary Cycle of Technological Changes

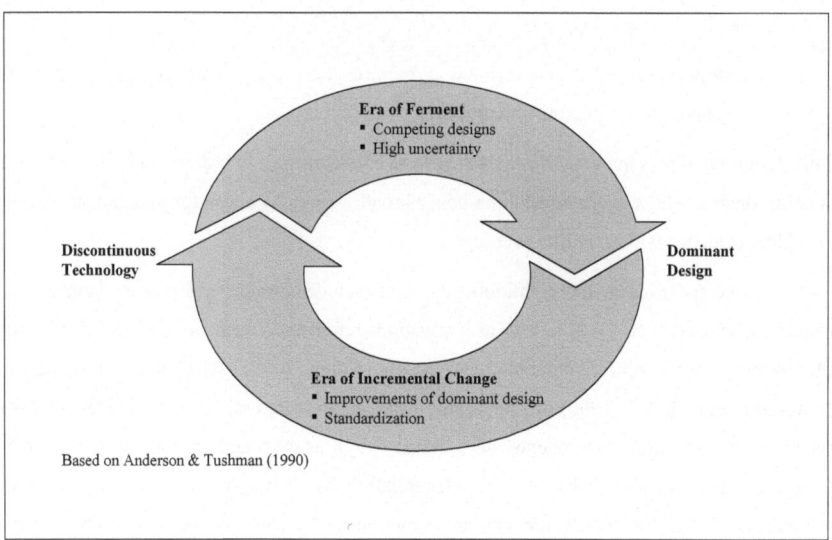

In their intriguing study on the hard disk industry, Christensen and Bower (1996) investigated how technological discontinuities come to maturity, i.e., in terms of Anderson and Tushman (1990), how a dominant design emerges. Disruptive innovations—as they are labeled by Christensen and Bower (1996)—are often cheaper, easier, or more convenient to use. Christensen and Bower (1996) describe the nature and evolution of disruptive innovations as follows: The technologies are first commercialized by new entrants, in many cases spin-offs of established incumbents, and initially only attract fringe customer groups due to their lower performance regarding traditional performance criteria (such as areal recording density in the case of hard disks). It is, however, important to note, that product performance and customer requirements on product performance often deviate for mature technologies. In other words,

---

[7] The discontinuous technological changes selected for this work are situated on different phases of this evolutionary cycle: Electronic toys and games were in a phase of incremental innovation at the time of this study. In the retailing industry, a dominant design for online businesses had already been established for most segments (except perishable food), whereas the publishing industry was amidst an era of ferment. The metering industry anticipated the emergence of smart meters as a discontinuous technological change in the close future.

the 'old' technologies already fulfill or even over-fulfill customer demands and sustaining innovations do not contribute to better customer satisfaction. Disruptive, initially inferior, technologies, however, typically improve their performance rapidly, thus meeting the customers' performance requirements after a certain time span. At the same time, they are superior regarding new performance criteria, e.g., size, price, or convenience. Thus, after a given period of time, established customers will also begin to switch from the old technology to the new technology, i.e. from purchasing goods or services from the incumbent players to the new entrants (compare Christensen & Bower, 1996).

The above described evolution refers to successful discontinuous technological changes. As stated, for instance by Hill and Rothaermel (2003), however, it is difficult, if not impossible to assess the future success of a technological discontinuity *ex-ante*.

2.1.2 Organizational Adaptation to Discontinuous Technological Change

Although established companies vary in the ways in which they adapt to discontinuous technologies, they typically find it difficult to change internal processes and structures radically and, in turn, tend to respond sluggishly to discontinuous technological change (Christensen & Bower, 1996; Hill & Rothaermel, 2003; Miller & Friesen, 1980). This phenomenon has often been referred to as incumbent 'inertia' (Gilbert, 2005; Hannan & Freeman, 1984). In this respect, researchers have observed four specific dimensions of incumbent inertia that affect whether, when, and how incumbents adapt to discontinuous technologies: speed, intensity, stamina, and routine flexibility (Christensen & Bower, 1996; Gilbert, 2005). Scholars have identified a broad range of adaptation strategies, including but not limited to adoption[8], re-investment in current technologies, or retrenchment into niches (Ford & Baucus, 1987; Zammuto & Cameron, 1985). The following discussion of adaptation dimensions is tailored to the case of 'adoption' as this reaction type is specifically challenging for incumbent companies and is at the core of the innovator's dilemma (Christensen, 1997).

---

[8] There is a fundamental difference between the terms 'adapt' and 'adopt.' According to Oxford Dictionaries, 'adapt' means "become adjusted to new conditions," whereas 'adopt' means to "choose to take up or follow an idea." Hence, adoption of discontinuous technologies refers to companies actively striving to commercialize the innovation. Adaptation to discontinuous change is more broadly defined, including adoption of the innovation but also non-adoption strategies such as retrenchment into niches (Ford & Baucus, 1987). For the sake of parsimony, the conceptual model of this thesis will focus on the link between family influence and adoption. The empirical parts, however, include cases of non-adopters and consequently, I will discuss other adaptation strategies wherever required.

2.1.2.1 Dimensions of Organizational Adaptation to Discontinuous Technological Change

Table 2 summarizes the definitions of and the previous scholarly work conducted on the four above-mentioned dimensions of organizational adaptation to discontinuous change. These dimensions will be discussed in more detail in the following subchapters.

Table 2: Dimensions of Organizational Adaptation

Dimension	Definition	Relevant Literature
Speed	Inverse of time lag between emergence of technological discontinuity and adaptation of the firm. Time is the sum of a) time until awareness, b) time for interpretation, c) time for decision-making, d) time for implementation	Gatignon, Robertson, & Fein, 1997; Golder & Tellis, 1993; Szymanski, Troy, & Bharadwaj, 1995; Tripsas & Gavetti, 2000; Tushman & Anderson, 1986
Intensity	Time-averaged amount of resources (financial, manpower, operational) committed to the technological discontinuity, as opposed to re-investment in the existing technology	Christensen & Bower, 1996; Gilbert & Newbery, 1984; Gilbert, 2005; March, 1991; Reinganum, 1983
Stamina	Degree of continued investment in the technological discontinuity as opposed to abandonment of the investment after initial setback	Block & MacMillan, 1985. Implicitly: Christensen, 1997; Gilbert, 2006; Suarez & Lanzolla, 2000; Tripsas & Gavetti, 2000
Routine flexibility	Degree to which established routines within the organization, i.e. processes, systems, structures, are replaced by new, non-paradigmatic ones	Feldman & Pentland, 2003; Gilbert, 2005; March & Simon, 1958; Tripsas & Gavetti, 2000

2.1.2.1.1 Speed

'Speed of adaptation' denotes the speed at which organizations recognize and interpret technological discontinuities, decide on the respective domain creation strategies (Ford &

Baucus, 1987), and ultimately implement competitive responses based on the discontinuous technology (Bower, 1986; Burgelman, 1983; Weick, 1995a).

Figure 4 depicts the individual components of adaptation speed. First, organizational members need to notify the emergence of the new technology, which is frequently developed and first commercialized by new entrants outside the organizations' established networks (Rothaermel, 2002). Although scholars have emphasized the importance of early awareness as a competitive advantage (Chen, 1996), this section of the speed chain appears to be the least critical part when adapting to technological discontinuities. As Christensen and Bower (1996) highlight, not only are incumbent players frequently early aware of new discontinuous technologies; in many cases they even have prototypes ready but struggle with the further steps of adaptation such as interpretation, decision-making, and implementation. Second, organizations need to interpret these technological changes as relevant for their own business, despite potential performance disadvantages in early phases of the technological cycle (Christensen & Bower, 1996). Third, managers need to make the decision to adapt to the new technology despite potential uncertainties concerning the future market, and communicate those adaptation strategies throughout the organization. Fourth and last, the organization needs to implement, i.e., develop, produce, and sell products based on the new technology.

Researchers have found that established companies tend to react to technological discontinuities more slowly than new entrants (Gatignon, Robertson, & Fein, 1997; Golder & Tellis, 1993; Szymanski et al., 1995; Tripsas & Gavetti, 2000). Tushman and Anderson (1986) used the examples of the airline and cement industries to show how failure to timely adopt the new technology ultimately resulted in the loss of market dominance of incumbent players in those industries.

Despite this abundant empirical and theoretical evidence of late adaptation as a major impediment for successful adaptation, there is no *a priori* association between incumbents' early adoption of discontinuous technologies and adoption success (Lieberman & Montgomery, 1988; Suarez & Lanzolla, 2007). Suarez and Lanzolla (2007) theorize that the existence of a 'first mover advantage' depends on several firm- and industry-dependent factors such as the ability to keep up with steady product innovations, efficiency of patents, customer buying and switching behavior, and pace of market development. When "vintage effects" (Suarez & Lanzolla, 2007: 383) occur, late entrants might even enjoy competitive advantages as compared to first movers.

24

Not all incumbent players adopt a discontinuous technology. Instead, some of them—deliberately or unintentionally—retreat into niches, and engage in what Ford and Baucus (1987) describe as domain defense, domain offense, domain consolidation, or passive reactions. These organizations 'stop' at some point along the adaptation chain, mostly because they interpret the technological change as irrelevant, inferior, incommensurate with the incumbents' conceptions, or too resource or knowledge intensive. The following theorizing will focus on incumbents with positive speed (i.e. those adopting the new technology), the empirical part, however, will discuss several cases of businesses reacting passively (and thus at a 'zero' speed).

Figure 4: Speed of Adaptation

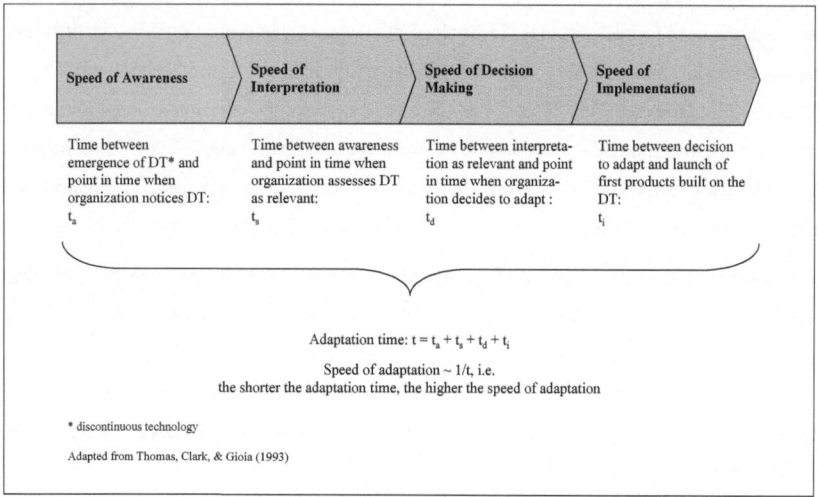

2.1.2.1.2 Intensity

'Adoption intensity' refers to the degree to which established organizations commit resources to the exploration of discontinuous technologies, as opposed to reinvesting in existing technologies and capabilities (Gilbert & Newbery, 1984). The term 'resources' thereby refers to financial investments ($R_{financial}$ in figure 5), manpower (including managerial attention; $R_{manpower}$ in Figure 5), and equipment utilization ($R_{equipment}$ in figure 5). Low resource commitment can either occur because managers decide to commit only few or no resources to the discontinuous technology during the decision-making phase or because, after an initial resource commitment, resources are instantaneously re-allocated from the discontinuous technology back to the established technology due to 'firefighting' issues (Christensen & Bower, 1996).

Resources committed to the discontinuous technology can stem from two different sources: First, managers can decide to re-allocate internal resources, originally planned for the continuous improvement ($R_{continuous}$ in figure 5) of the old technology to the new, discontinuous technology ($R_{discontinuous}$ in figure 5). In other words, the stock of available resources (R_{stock} in figure 5) becomes re-distributed (see option 1 in figure 5). Second, the organization can acquire new resources (R_{new} in figure 5) from external sources, such as new bank loans and/or raising capital to acquire new financial resources, hiring or joining alliances to rise manpower, or purchasing new tools to augment operational resources, and allocate these to the new, discontinuous technology (see option 2 in figure 5). Alternatively, any combination of redistribution and raising of new resources is also considerable (see option 3 in figure 5).

Although the absolute amount of resources invested in the development and commercialization of the discontinuous technology is an important dimension of adoption, the relative intensity of adoption also constitutes a meaningful number. The relative intensity ($I_{relative}$ in figure 5) denotes the ratio of resources invested in the discontinuous technology as opposed to the amount of resources invested in the incremental improvements of the established technology. Referring to this number allows for better comparison of the resource commitments of organizations of varying sizes. figure 5 summarizes the individual components of resource intensity as well as the different investment options in discontinuous technologies.

Figure 5: Resource Investment in Discontinuous Technology

Formal description of resource investment (R) in discontinuous technology

Initial situation

$$R_{stock} = R_{financial} + R_{manpower} + R_{equipment} = R_{continuous}$$

Option 1: Redistribution

$$R_{stock} = R_{continuous} + R_{discontinuous}$$

Option 2: New resources

$$R_{stock} = R_{continuous} \quad \text{and} \quad R_{new} = R_{discontinuous}$$

Option 3: Redistribution & new resources

$$R_{stock} + R_{new} = R_{continuous} + R_{discontinuous} \text{ with } R_{discontinuous} > R_{new}$$

Relative intensity of adaptation

$$I_{relative} = \frac{R_{discontinuous}}{R_{continuous}}$$

R: Absolute amount of resources
I: Resource distribution ratio

Resource investment patterns do not necessarily remain constant over time (e.g., Bowman & Hurry, 1993): for instance, some organizations may enter the market with intense resource commitment which decreases over time. Other organizations may begin cautiously with little investment that becomes augmented as soon as the first successes are realized, and still others might hold their level of intensity constant. To avoid exuberant complexity, when referring to 'intensity of adoption,' I draw on the average intensity for those time spans, in which the level of intensity committed to the discontinuous technology is above zero.

Scholars have found that incumbents systematically under-invest in discontinuous technologies (March, 1991; Reinganum, 1983), which was, for instance, the case in the disk-drive (Christensen & Bower, 1996) and the newspaper industries (Gilbert, 2005). Somewhat similar to 'speed of adaptation' (Chapter 2.1.2.1.1), however, the relationship between intensity of adoption and success of adoption is non-linear: contextual factors such as the complexity of the new technology determine the amount of resource investment required for a successful adoption and increased levels of flexibility can render lower resource investment beneficial, particularly in times of high uncertainty (Steensma & Fairbank, 1999). Despite this non-linearity, researchers agree that investment intensity is an important precursor to adoption success, as some threshold level of resource commitment is required to successfully embrace a new technology (Tushman & Anderson, 1986). The underlying rationale is that adoption of discontinuous technologies per definition requires novel knowledge, new capabilities, and the implementation of non-paradigmatic routines (Gilbert, 2005; Hill & Rothaermel, 2003), which, in turn, can only be acquired by drawing on adequate resources.

2.1.2.1.3 Stamina

'Stamina' describes the degree to which established companies continuously reinvest in a new technology following initial investments (Block & MacMillan, 1985). Anderson and Tushman's cyclic model of technologies (1990) highlights that the development of a technological discontinuity to a new dominant design is a process—of varying and unclear duration and outcome ex-ante—rather than a punctual event. Thus, organizations may experience a time lag of uncertain length between first investments into the new technology and first commercial success. For example, first attempts to commercialize trolley suitcases occurred as early as the 1970s, but because of resentment on the part of various retailers, who rejected the sale of the product, large scale production and sales did not start before the 1990s (Gunkel, 2011).

Furthermore, anecdotal evidence shows that discontinuous technologies often emerge in 'waves' that include an initial hype, temporary setbacks, and ultimately after some time, the re-emergence and ultimately the dominance of that technology. One such example is the emergence of internet-based business models such as online retailing, pushed by the internet boom in the mid- to end-1990s, set back by the burst of the dotcom-bubble in 2000, and ultimately re-emerging from around the mid-2000s. Video games which disrupted the traditional gaming market and started to emerge in the 1970s, also experienced a severe setback after a few successful years: In 1983, the entire US video gaming market collapsed temporarily with revenues shrinking by more than 90% before re-flourishing nationally and internationally from the end of the 1980s onwards (Sheff, 1999).

As a consequence of these temporal patterns, incumbent players need to keep up their commitment—at least at a certain level—for an extended period of time, particularly if initial adoption initiatives have failed (Gilbert, 2005). These patterns also imply that the process of interpretation, decision-making, and implementation, as displayed in

Figure 4, frequently does not end after some implementation efforts, but rather constitutes an iterative process as displayed in figure 6. During implementation, particularly if setbacks have occurred or sales and profit have not gained momentum for a long period of time, organizational members will re-engage in the interpretation and decision-making process. Built on new information and insights available, they will re-assess the relevance of the discontinuous technology for their respective organizations, and once more decide whether, and if yes, how many resources they will commit to the new technology. These decisions lay the ground for further implementation activities.

Although rarely explicitly addressed in prior research, evidence for the crucial role of 'stamina' for the adoption of discontinuous technologies has been provided by many qualitative studies (Christensen, 1997; Gilbert, 2006; Tripsas & Gavetti, 2000). Suarez and Lanzolla (2007) support the notion that non-recurring investment is insufficient when technologies change rapidly. Instead of a one-time investment, "keep[ing] up with the evolution of knowledge within the industry" (Suarez & Lanzolla, 2007: 382) constitutes a key success factor. Despite the importance of 'stamina,' however, there is also a flip side of the coin, particularly in cases where stamina leads to escalation of commitment (Brockner, 1992; Staw, Sandelands, & Dutton, 1981). As highlighted, for instance, by Hill and Rothaermel, it is difficult, ex ante, to identify "whether and when a radical technological innovation will become a commercial success" (2003: 258). Frequently cited examples of such failed discontinuous technological changes are the cases of supersonic Concorde and the Iridium satellite telephone system (Santos, Doz, & Williamson, 2004).

Figure 6: Fine-grained Adaptation Process—Iterative Cycles

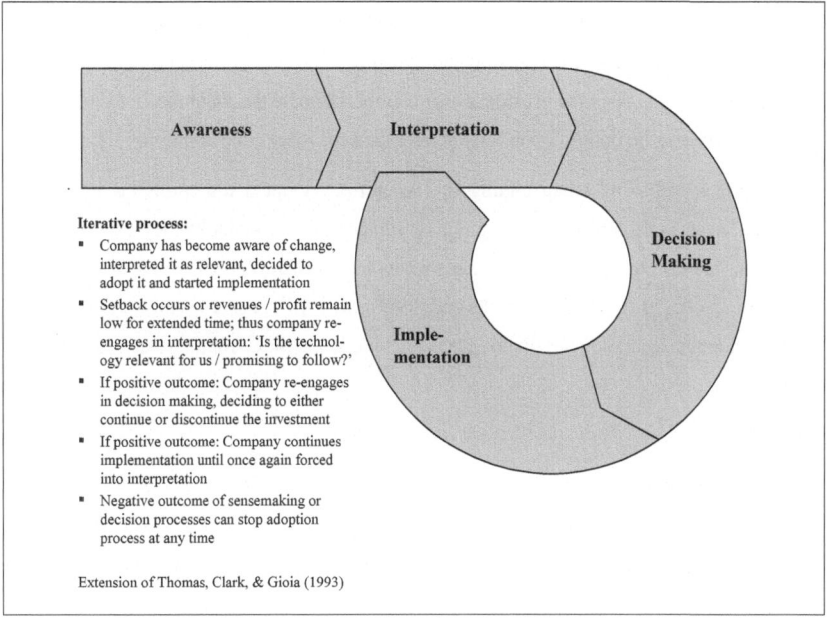

Iterative process:

- Company has become aware of change, interpreted it as relevant, decided to adopt it and started implementation
- Setback occurs or revenues / profit remain low for extended time; thus company re-engages in interpretation: 'Is the technology relevant for us / promising to follow?'
- If positive outcome: Company re-engages in decision making, deciding to either continue or discontinue the investment
- If positive outcome: Company continues implementation until once again forced into interpretation
- Negative outcome of sensemaking or decision processes can stop adoption process at any time

Extension of Thomas, Clark, & Gioia (1993)

2.1.2.1.4 Routine Flexibility

The term 'adoption of flexible routines' denotes the degree to which organizations reconfigure their structures, systems, and internal processes when implementing discontinuous technologies (Bockmühl, König, Enders, Hungenberg, & Puck, 2011; Feldman & Pentland, 2003; Leonard-Barton, 1992; March & Simon, 1958; Nelson & Winter, 1982; Tripsas & Gavetti, 2000). Routines can thereby be defined as "repetitive, recognizable patterns of interdependent actions, carried out by multiple actors" (Feldman & Pentland, 2003: 95). They are frequently perceived as "one product [... and] a key component of organizational learning" (Argote, 1999; Feldman & Pentland, 2003: 97-98) and, are therefore tightly linked to knowledge and capabilities (Feldman & Pentland, 2003). As knowledge and capabilities required to adopt a discontinuous technology significantly deviate from established competencies (Anderson & Tushman, 1990), incumbents need to adopt non-paradigmatic routines to successfully commercialize the discontinuous technology.

Timely and continuous commitment of a certain amount of resources is therefore a necessary yet insufficient condition for successful adaptation to discontinuous change; it is at least equally important, *how* the company appoints these resources during the implementation phase. In other words, the way organizational members handle the input resources to create a product or service as an output is a crucial determinant for successful adoption.

More specifically, when aiming to successfully adopt a discontinuous innovation, incumbents need to reconsider, for instance, the materials and methods they utilize (Hill & Rothaermel, 2003), the performance dimensions and criteria upon which their product evaluations are based (Christensen & Bower, 1996), and the revenue channels adequate for the discontinuous technology (Gilbert, 2005). Several of these routines are illustrated in Figure 7.

Figure 7: Adaptation of Routines

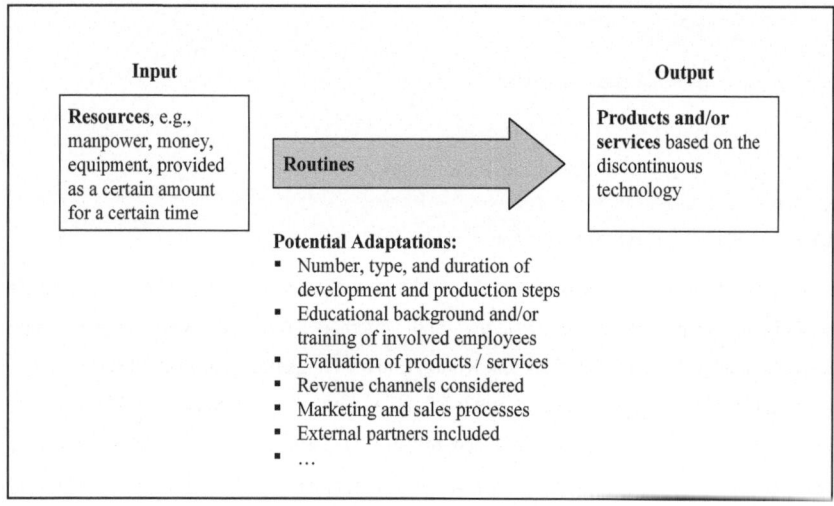

Incumbents frequently tend to 'cram' the technologies (Christensen & Raynor, 2003), which means that they apply the same routines for both technologies. For instance, incumbent newspapers reacting to the emergence of online news frequently ignored the altered customer preferences and performance criteria for online as opposed to offline media. Thus, they neglected the customers' void for concise and timely news enhanced by interactive and multimedia features such as bulletin boards, news alerts, and video messages and rather stuck to established criteria such as elaborate full-length articles. Gilbert (2005) empirically showed that US newspapers indeed reacted with 'routine rigidity' to this discontinuous technology and 'reproduced' their offline content online, ultimately resulting in inferior performance.

Research on organizational inertia shows that routine flexibility is a crucial determinant of adoption performance. In other words, when incumbents implement a discontinuous technology and thereby fail to adequately adjust their processes and routines, they are unable to take full advantage of the innovation (Gilbert, 2006).

Regarding the interrelated nature of the four dimensions of organizational adaptation to discontinuous technological change described, Gilbert (2005) identified an intriguing 'paradox:' he provided empirical evidence, that threat framing releases 'resource rigidity,' i.e. it enhances the intensity of resources committed to the new technology, whilst at the same time aggravating 'routine rigidity.' This shows that the determining antecedents can have contradicting effects on each of the four dimensions and an integrated discussion of all four dimensions is required for a full understanding of the phenomenon.

2.1.2.2 Determinants of Organizational Adaptation to Discontinuous Technological Change

Following the publication of Schumpeter's revelatory studies on the phenomenon of incumbent failure, a multitude of studies have sought to identify the causal mechanisms underlying organizational inertia on various levels, such as individual (Barley, 1990; Karasek, 1979; Noble, 1984), group (Burgelman, 1983), organizational (Chattopadhyay, Glick, & Huber, 2001; Christensen & Bower, 1996; Henderson & Clark, 1990), population (Hannan & Freeman, 1977, 1984), and institutional (Abrahamson & Fombrun, 1994; Debruyne & Reibstein, 2005; DiMaggio & Powell, 1983; Kraatz, 1998) levels. As implied by the various summaries of the abundant discontinuous technological change literature (Chesbrough, 2001; Hill & Rothaermel, 2003), several sets of interconnected barriers to the adoption of paradigm-contradicting technologies can be discerned: economic and strategic disincentives, organizational factors, and cognitive-emotional barriers. Researchers exploring these sources of organizational paralysis identify a dilemma imminent to any organization: to succeed in stable circumstances, decision makers and their organizations are bound to adopt certain practices of interpretation, decision-making and strategic action (Daft & Weick, 1984;

Thomas & McDaniel, 1990). However, many of these practices cause organizations to fail when confronted with discontinuous technological change.

More recently, scholars have also identified factors fostering the adaptability of organizations, such as external influence and organizational ambidexterity (Gilbert, 2005). In the following subchapters, I will outline the most relevant barriers to, and enablers of, adaptation to discontinuous change. Thereby I will focus on the individual and organizational level, thereby disregarding important institutional factors such as isomorphism[9] (Abrahamson & Fombrun, 1994; DiMaggio & Powell, 1983; Haveman, 1993). Moreover, this overview will not discuss absorptive capacity in detail: Absorptive capacity (Cohen & Levinthal, 1990) denotes the degree in which organizations and, in particular, their leaders, can recognize, understand, and use ambiguous environmental information (Cohen & Levinthal, 1989; Lichtenthaler, 2009) and it is thus an important antecedent for organizational adaptation. However, as Hill and Rothaermel (2003) note, absorptive capacity can even turn into an impediment to change if it is tied to the old rather than the new knowledge, thus there is no linear relationship between the level of absorptive capacity available in an organization and this organization's adoption of discontinuous technologies.

2.1.2.2.1 Formalization

'Formalization' refers to the tendency of incumbent organizations to establish standardized and stable forms, roles, procedures, conventions, and structures (Hannan & Freeman, 1984; Levitt & March, 1988). This phenomenon and its implications have been intensively studied by organization theorists (e.g., Levitt & March, 1988) and population ecologists (e.g., Hannan & Freeman, 1977), predominantly on an organizational level.

The underlying rationale for organizations to establish standardized and stable structures and processes is at least two-fold (Hill & Rothaermel, 2003): First, such structures enable efficient organizational learning in undisturbed environments (Levitt & March, 1988) by (Hill & Rothaermel, 2003: 260) "reducing the costs of information acquisition and utilization and for coping with bounded rationality" (Arrow, 1974; Cyert & March, 1963; Hill & Rothaermel, 2003; Nelson & Winter, 1982; Simon, 1955). Second, as population

---

[9] These authors propose that organizations continuously compare themselves with similar organizations and imitate their strategic moves, particularly those of market leaders. Such behavior is even more pronounced in times of uncertainty. This phenomenon, called mimetic isomorphism, ultimately results in collective inertia within an industry, since market leaders are particularly prone to resisting adoption of the new technology (König, 2009).

ecologists propose, established structures and processes augment an organization's reliability and accountability, two characteristics that are highly valued by the organization's environment (Hannan & Freeman, 1984).

In order to permanently anchor these processes and structures within the organization, incumbents typically resort to high levels of formalization of processes, structures and bureaucracies (Hill & Rothaermel, 2003). Such structurized processes include but are not limited to the methods of looking for new business opportunities, evaluating them, developing new products, and measuring the success of products.

In times of discontinuous change, however, such organizational behavior entails 'structural inertia' (Hannan & Freeman, 1984) and the companies' former core competencies can turn into core rigidities (Leonard-Barton, 1992), since the established structures and processes imply "that organizations respond relatively slowly to the occurrence of threats and opportunities in their environments" (Hannan & Freeman, 1984: 151).

First, formalization can impede early awareness of technological discontinuities because of the organizations' "limited search and selective knowledge acquisition activities" (Hill & Rothaermel, 2003: 260). Organizations which use formalized search processes may overlook technological discontinuities (Danneels, 2002; Day, 1994; Govindarajan & Kopalle, 2006; Howell & Shea, 2001; Tellis, 2006). Second, formalized processes may negatively affect the interpretation of technological discontinuities (e.g., Arrow, 1974; Cyert & March, 1963; Nelson & Winter, 1982) as relevant business opportunity because responsible organizational members underrate discontinuous strategic alternatives that are inconsistent with their time-honored business paradigms. Such 'underrating' ultimately prolongs the time span until the discontinuous innovation is perceived as 'relevant' and decreases the amount of resources committed to the exploration of the new technology. For instance, market leaders in the minicomputer industry underrated future success of the emerging desktop computers due to the reduced storage capacity (Christensen & Bower, 1996), and hence did not react timely. Third, bureaucracy hampers fast decision-making and implementation, particularly in case of scarce or ambiguous information which is typical for many discontinuous technologies (Tushman & Anderson, 1986). Overall, formalization has a negative effect on the speed and intensity of adaptation to discontinuous technological change.

A similar line of argument can be applied to the recurring interpretation and decision-making processes as described in 2.1.2.1.3, since low levels of formalization facilitate decisions to continue investing in discontinuous technology, specifically after initial setbacks.

The underlying rationale here is that formalized processes typically imply quantitative, exact measures of success for each strategic activity of the company. Since discontinuous technologies emerge over an extended period of time (Anderson & Tushman, 1990), during which they often produce scarce financial success (Utterback & Abernathy, 1975), efficient, formalized evaluation schemes will assess the respective strategic investments as 'failure,' which in turn encourages decision makers to abandon the activities related to the discontinuous technology. For instance, many of the early e-commerce adopters resigned from that market when profitability targets had not yet been reached around the turn of the century. However, Jeff Bezos, CEO of today's market leader Amazon, deliberately decided to refrain from traditional measures of assessing performance during an investment phase of several years, and thus maintained the investment in online retailing (Schneeberger, 2000).

Finally, formalization does not only refer to structures and processes related to searching for information, interpretation, and decision-making, but to any structures and processes "around which organizations are constructed and through which they operate" (Levitt & March, 1988: 326), thus also affecting the implementation phase. As a consequence, formalization might also exacerbate routine flexibility (Weiss & Ilgen, 1985) by forestalling that implementation of the novel technology is executed in a non-paradigmatic way for which no standards initially exist.

In sum, high levels of formalization decrease the speed, intensity, stamina, and flexibility of organizations' adoption of discontinuous technologies.

2.1.2.2.2 Resource Dependence

Resource dependence (Pfeffer & Salancik, 1978) constitutes a further theory through which scholars have examined the barriers to technology adoption on an organizational level. Christensen (1997) was the first to apply the principles of resource dependence theory (Pfeffer & Salancik, 1978) to disruptive innovations, thereby identifying the 'innovator's dilemma.'

Resource dependence theory is based on the assumption that controlling resources that are critical and important to others, such as financial capital, physical assets, or information, equals power and entails relational interdependencies between organizations and individuals. Control of resources thereby refers to regulation, possession, ownership, access, and actual control of use of the respective resources (Pfeffer & Salancik, 1978). In the words of Pfeffer and Salancik (1978: 43): "It is the fact of the organization's dependence on the environment that makes the external constraint and control of organizational behavior both possible and almost inevitable."

As discussed in chapter 2.1.2.1, adoption of a discontinuous technology requires the decision to invest a certain amount of resources into the commercialization of the innovation plus to maintain this investment for a certain period of time. Resource dependence theory predicts that external stakeholders such as investors and customers substantially affect these investment decisions regarding discontinuous technologies due to the power these constituents inherently exert (Bower & Christensen, 1995; Christensen & Bower, 1996).

Investors providing capital to incumbents require predictable profit margins and high certainty regarding market growth outlooks (Benner, 2007). As discontinuous technological changes are connected to high levels of uncertainty and mostly unprofitable in early stages (Anderson & Tushman, 1990), investors are unlikely to provide financial support for investments in discontinuous technologies (Benner, 2007). Moreover, stockholders are unlikely to endorse resource re-allocations from continuous improvement of the established technology to commercialization of the discontinuous technology, as this potentially diminishes the return on their dividends in the short-term. In this context, Benner (2008, 2010) provided empirical evidence on how financial analysts first decried activities based on digital as opposed to analog photography. As a consequence of this resource dependence on capital providers, incumbents have strong incentives to focus on established segments with predictable growth and profit margins, to invest in sustaining innovations, and to avoid any activities based on the discontinuous technology (Christensen & Bower, 1996; Gilbert, 2005). Ultimately, resource dependence on capital providers will negatively affect the speed of the attention to and interpretation of the technology and ultimately the speed of adaptation. In case of adoption, the resource dependence on external capital providers allays the intensity of resource commitment that can only be overcome by threat perception (Gilbert, 2005). Moreover, due to their resource dependence incumbents are more likely to abandon a new technology after temporary setbacks. This is because capital markets are too impatient to await growth and profits from new technologies (Benner, 2010; Bushee, 2001; Christensen, 1997) and consequently sanction failure, which many companies are required to disclose in their quarterly reports. Ultimately, capital markets exert pressure on the organization to discontinue 'failed innovations' (Benner, 2007, 2010).

Besides the influence exerted by capital providers, customers also influence how companies commit resources to discontinuous technologies (Christensen & Bower, 1996).

Companies typically generate the majority of their profits from their most demanding, high-end customers, who are generally reluctant to disruptive innovations, i.e., discontinuous innovations that do not perform well initially in terms of the traditionally valued metrics, but offer new performance attributes (for example, by being simpler, smaller, more convenient, or cheaper). In continuous circumstances, this focusing on high-end customers rather than competing for the low-end customer base pays off because high-end segments carry higher profit margins and revenue volumes.

Extant customers, seeking sustaining rather than discontinuous innovations, will probably object to any redistribution of firm resources in favor of the new, discontinuous and at the cost of the old, established technology as discussed in 2.1.2.1.2 (options 1 and 3), resulting in lower resource commitment to the radical innovation. Contrary to providers of financial capital, however, extant customers will not per se oppose the commercialization of the discontinuous technology, as long as the focal company continues to develop and sell their favored products. However, high-end customers are likely to start to oppose the discontinuous change as soon as incumbents begin to alter their business routines in a non-paradigmatic way[10], i.e. change the 'value network' and the 'innovation ecosystems' (Adner & Kapoor, 2010; Christensen, 1997). That is the case, for instance, when incumbents begin to withdraw resources from the development of the old technology and commit them to the new technology instead. Therefore, extant customers will impede the implementation of radically new routines, subsequently diminishing routine flexibility.

A similar line of argument holds true for other external stakeholders. Following the work of Ghemawat (1991), Rosenbloom and Christensen (1998), and Sull, Tedlow, and Rosenbloom (1997), Hill and Rothaermel explicitly amplify the notion of resource dependence to the "value network of suppliers [...] complementary product providers, communities, and so on, to which the firm has made strategic commitment" (2003: 261). Similar to customers, those external stakeholders will be unaffected by adoption activities as long as routines are not altered radically. However, when incumbents start to radically change their business routines and, for instance, consider replacing established suppliers by new contractors, the long established partners of the firm will rebel against those plans and work against them, thereby weakening the company's flexibility.

---

[10] The underlying rationale for this violation is that organizations face difficulties when simultaneously executing processes in radically different ways. As research on ambidextrous structures has shown, this inability can be overcome by establishing decoupled structures. This enabler of change is discussed in chapter 2.1.2.3.2.

Summarized, resource dependence on external providers of capital decreases the speed, intensity, and stamina of organizations' adoption of discontinuous technologies. Resource dependence on other external stakeholders such as suppliers decreases resource investment and routine flexibility when implementing new innovations.

2.1.2.2.3 Political Resistance

Theory of power (Cyert & March, 1963; Pfeffer, 1992) cannot only be applied to inter-firm relationships but also within organizations, relating to individual levels, specifically to middle managers, in order to further explain inertia (Hill & Rothaermel, 2003). Middle managers de facto control the use of scarce resources within the organization and thereby exert high levels of control (Burgelman, 1983; Pfeffer & Salancik, 1978). Theory of power within the organization can explain an important barrier to adoption—political resistance of managers.

While a 'truce' regarding the distribution of power and control is established among organizational members in periods of stability (Cyert & March, 1963; Hill & Rothaermel, 2003), organizational adaptation distorts those political equilibriums (Hannan & Freeman, 1977). As a consequence, managers engage in political resistance activities, which can be defined as "observable, but often covert actions by which [managers] enhance their power to influence a decision" (Eisenhardt & Bourgeois Iii, 1988: 738).

There are multiple reasons to assume that middle managers generally dismiss the adoption of discontinuous technologies (Lüscher & Lewis, 2008): First of all, redistribution of power and influence implies that some managers' positions will change for the worse. Typically, this affects those managers that are responsible for the established technology and thus had a substantial amount of power in the established organization. For instance, the shift from printed to digital content in the publishing industry deprived the formerly powerful print production managers of their responsibility and power; managers with IT skills, in turn, frequently climbed up the hierarchy very quickly. As a consequence, managers fearing disempowerment will rebel against the adoption of the discontinuous technology.

Furthermore, scholars assume that middle managers typically behave as "purposive managers" (Bower, 1986: 73). As such, they tend to seek monetary enrichment and political influence. Accordingly, to minimize the risk of jeopardizing their own goals, middle managers prefer continuous innovations to discontinuous innovations because the latter, by definition, carry a high risk of failure (Bower, 1986). Moreover, such decision makers tend to change positions frequently within their respective organizations. Consequently, they prefer to support projects that quickly lead to measurable financial success (Christensen & Raynor, 2003), while, in most cases, the adoption of new technologies is profitable only in the medium or long-term (Bower, 1986).

The tendency to engage in political resistance is reinforced by middle managers' incentive systems that typically mirror the preferences of external stakeholders who, as described, prefer predictable over non-predictable returns. Consequently, incentive systems for middle managers are typically aligned to short-term instead of long-term firm goals and enhance middle managers' political resistance to the adoption of discontinuous changes.

As a result of political resistance, middle managers typically support continuous innovations (Christensen & Bower, 1996) and stall the interpretation and decision-making processes, resulting in a decreased speed (Hill & Rothaermel, 2003; Pfeffer, 1992). As middle managers substantially influence resource allocation processes (Bower, 1970; Rouleau, 2005), their disincentives to investing in technological discontinuities will furthermore lower resource intensity. Moreover, the middle managers' motivation and incentives to implement discontinuous technologies will be low, thus decreasing routine flexibility.

The level of 'centralized authority' within the organization plays a crucial, yet ambiguous role in the extent to which political resistance can develop and result in hampered adoption. Eisenhardt and Bourgeois show that political resistance emerges because autocratic CEOs "preserve their power through use of politics" (1988: 765). Somewhat to the contrary, Hill and Rothaermel, following Pfeffer's (1992) argumentation, conclude that "absent strong leadership" (Hill & Rothaermel, 2003: 261) provides the ground for political contentions.

In sum, political resistance is an important inhibitor of technological adoption that decreases the response speed, resource investment, and routine flexibility.

2.1.2.2.4 Avoidance of Cannibalization

Avoidance of cannibalization (Chandy & Tellis, 1998; Reinganum, 1983) denotes managers' emotional attachment to the business and/or its assets and their unwillingness to shed extant resources. The unwillingness to cannibalize provides an explanatory approach in understanding incumbent inertia on an organizational and individual level by illustrating organizations' and organizational members' incentives to invest in established rather than discontinuous technologies.

Not only organizational (chapter 2.1.2.2.1) and strategic (chapter 2.1.2.2.2) incentives motivate incumbent players to invest in continuous rather than discontinuous innovations but also economic considerations (Gilbert & Newbery, 1982; Reinganum, 1983). Particularly due to market entry barriers, continuous technologies expand incumbents' bases for market dominance (Henderson, 1993; Hill & Rothaermel, 2003), whereas discontinuous innovations challenge their leading positions in their existing businesses. Consequently, "incumbent position reinvestment" (Gilbert, 2005: 747) theory and game-theoretical evidence (Reinganum, 1983) predict that incumbents typically will try to solidify their established business positions by re-investing in established technology, especially if they are quasi-monopolists (Gilbert & Newbery, 1984). Incumbents' scarce investments in discontinuous technologies can be explained by fear of cannibalizing sales of extant products (Gilbert & Newbery, 1984). For instance, players in the music industry were reluctant to invest in the digital music business as they feared that customers would see digitized music as a substitute for CDs, tapes, and long-playing records and, as a consequence, each unit of sales in the low profit digital market segment would diminish the number of sales units in the high margin traditional segment.

Chandy and Tellis (1998) expand this notion of fear of cannibalization by a psychological factor and by extension to the individual level by emphasizing that decision makers in incumbent organizations are afraid of cannibalizing sales in their established businesses. These authors define willingness to cannibalize as the "extent to which [organizations and their respective members are] prepared to reduce the actual or potential value of [their] investments. It is an attitudinal trait of the key decision makers of the firm and resides in the culture, or shared values, and beliefs, of the firm" (Chandy & Tellis, 1998: 475, referring to Deshpande & Webster, 1989).

The attempts to avoid cannibalization are reinforced in circumstances of 'residual fit' (Gilbert, 2006), when decision makers face the difficult task of evaluating the marginal effects of leaving an established business to enter into a new business. The established business often functions as a 'cash cow' in that it is an important source of internal funding, whereas the prospects of the new business are often unknown, particularly in terms of market predictability (Siggelkow, 2001). As a result, incumbents try to erect barriers to entry and they tend to put off their response until a new entrant actually threatens their controlling position (Reinganum, 1983).

Both the economic and psychological facets of avoidance of cannibalization hamper quick adoption by stalling the interpretation and decision-making processes. As Hill and Rothaermel (2003: 259-260) note: "The problem is akin to opening Pandora's box: once the box of new technology has been opened, even by an incumbent, the technology may trigger changes that alter the structure of the industry and lead to the demise of the incumbent." Moreover, avoidance of cannibalization, in particular it psychological component, lowers the intensity of resource commitment (Hill & Rothaermel, 2003) and routine flexibility as decision makers are reluctant to re-distribute resources and/or give up established routines.

In sum, political resistance has a negative effect on an organization's speed and intensity of technology adoption as well as its flexibility of routines.

2.1.2.2.5 Rigid Mental Models

Theory of bounded rationality (March & Simon, 1958) provides a fruitful base from which to investigate non-rational sources of inertia on an individual level by introducing the concept of rigid mental models. Mental models, also termed cognitive frames (Kaplan, 2011), stem from the need to interpret the environment and take decisions despite inevitable limits of human cognition (March & Simon, 1958) and they "consist of concepts and relationships an individual uses to understand various situations or environments" (Barr, Stimpert, & Huff, 1992: 16, referring to Weick & Bougon, 1986). Such models are based on "historical experience as opposed to current knowledge" and frequently constitute a "set of beliefs" that are shared among organizational members (Tripsas & Gavetti, 2000: 1148).

Numerous researchers have pointed out that rigid mental models and limited cognitive abilities of managers to appropriately identify and interpret environmental changes are a reason for inadequate responses to discontinuous change (Ford & Baucus, 1987; Kaplan, 2011; Kaplan & Tripsas, 2008; Staw et al., 1981; Weick, 1995b). For instance, Tripsas and Gavetti (2000) showed in a longitudinal case study how the photographic camera and film manufacturer Polaroid, despite valuable capabilities regarding digital imaging, was unable to successfully commercialize this new technology due to rigid cognitive frames shared by its top management. In a similar vein, Kaplan (2008b) provided empirical evidence on how firms in the telecommunication sector adapted differently to fiber-optics due to variances in their respective mental models. Besides evidence from academic studies, abundant historical anecdotes from various social, technological and scientific domains illustrate how rigid mental models can entail flawed assessments of change. For instance, in 1895, the British scientist and president of the British Royal Society, Lord Kelvin, believed that machines that were heavier than air, would be unable to fly (Mannermaa, 2004). Somewhat similar, in 1943, Thomas Watson, chairman of IBM, assessed the world market for computers to be less than ten (Remenyi, 2002).

As prior work shows, incumbent firms are prone to overlook discontinuous external developments because individuals in established firms develop mental models, or 'frames,' that cause them to focus their screening efforts on developments in the realm of their existing businesses, customers, and competitors (Levitt & March, 1988). In periods of continuous change, this approach is reasonable because it helps decision makers to reduce the apparent complexity of a firm's environment and to focus on the most important issues. However, in periods of discontinuous technological change, these established screening patterns tend to let companies overlook new customers and new competitors that enter the market from outside the traditional reference group (Kaplan, 2008a). Furthermore, when confronted with a discontinuous technology, the application of established criteria during the search for, and interpretation of, environmental change results in the detection of irrelevant issues, which in turn leads managers to make maladaptive response decisions (Barr et al., 1992; Garud & Rappa, 1994; Tripsas & Gavetti, 2000). Therefore, rigid mental models prolong the adaptation time due to delayed awareness and lengthened interpretation. As Kaplan (2008b) and Tripsas (2000) proposed, rigid mental models also limit the amount of resources invested in a new technology and decrease the flexibility of routines (Barr et al., 1992).

Lastly, it is important to note that it is not only the rigidity of cognitive frames that determines strategic flexibility but also their congruence (Kaplan, 2008b). As Kaplan (2008b) shows, incongruent frames entail 'framing contests' and engage middle managers in highly political discussions. These ongoing debates hamper decision making and ultimately contribute to a further decreased speed of adoption.

Summarized, rigid mental model has a negative effect on adoption speed, intensity, and flexibility.

2.1.2.3 Enablers of the Adoption of Discontinuous Technologies

Scholars have also increasingly questioned why, contrary to the prevailing view, some incumbents do not respond to discontinuous technological change with inertia (Mitchell,

1989). This research has identified two salient enablers[11] of change that can work as 'dynamic capabilities' (Teece et al., 1997) to overcome inertia: external influence and structural ambidexterity.

2.1.2.3.1 External Influence

'External influence' refers to the incorporation of outsiders into scanning, interpretation, decision-making processes, and implementation. External actors—particularly those from outside established industry boundaries—are able to provide unbiased assessments of ongoing trends and thus might be able to offer valuable input on the potential future of a field and on adequate adaptation strategies, and they can act as catalysts of discontinuous technological change (Gilbert, 2005; König et al., forthcoming; Vasudeva & Anand, 2011). One particular 'complementary asset' of outsiders is that they are less biased by traditional mental schemata (McDonald, Khanna, & Westphal, 2008). Thus, outsiders improve the absorptive capacity (Cohen & Levinthal, 1990) of incumbents not only by supplying new knowledge but also by enabling decision makers to interpret this knowledge appropriately. Depending on the phase, in which outsiders are brought in, they speed up the adaptation process by shortening the interpretation time, provide insights on the amount of required resources, and/or help to overcome outdated routines.

The commercialization of telephones is one intriguing, frequently cited example to illustrate how industry outsiders are, in some cases, better able to foresee the potential of discontinuous technologies as compared to industry insiders who are bound to their traditional mental schemata (Hounshell, 1975; Kaplan & Tripsas, 2008): Elisha Gray, an expert working in the telegraph business was among the first to develop phone technology. However, he later dismissed the idea, assessing telephony as a 'scientific toy.' Ultimately, it was an industry outsider, Graham Bell, who recognized the potential of the new technology and activated its commercialization.

In sum, external influence has a positive effect on the speed of adoption of a discontinuous technology as well as the flexibility of routines to implement the new innovation.

---

[11] Scholars have identified further structural and organizational enablers, such as high turnover rates in the top management teams (Tripsas & Gavetti, 2000), internal rivalry, product champions, and future market orientation (Chandy & Tellis, 1998) which should be mentioned here in the interests of completeness.

2.1.2.3.2 Structural Ambidexterity

'Structural ambidexterity' denotes the development and commercialization of discontinuous technologies in "highly differentiated, internally inconsistent organizational architectures" (Tushman, Smith, Westerman, & O'Reilly, 2010: 1336). Such loosely coupled architectures are independent at a lower team level but integrated at a senior team level and they foster experimentation and improvisation (O'Reilly & Tushman, 2008; Tushman & O'Reilly, 1996). Ambidexterity research (Raisch & Birkinshaw, 2008) provides evidence that such designs allow organizations to "recombine and reconfigure assets and organizational structures as markets and technologies change" (Teece, 2006: 38) and subsequently relax barriers to adoption, such as an emphasis on exploitative efficiency relative to explorative effectiveness or a mismatch between the organizational routines and customer demands of new markets (March, 1991). Moreover, structural ambidexterity helps to overcome avoidance of cannibalization, political resistance, and, to some degree, rigid mental models. Consequentially, structural ambidexterity increases both the resource intensity and routine flexibility[12]. For instance, Spiegel, an established German player in the print magazine sector, successfully adapted to online news by building up a largely independent business unit called 'Spiegel Online.'

Table 3 summarizes the barriers to and enablers of change described above.

---

[12] Structural ambidexterity only weakly influences speed of adoption (mostly speed of implementation), as such decoupled units only affect the implementation, not the earlier phases of the process.

	Formalization	Resource Dependence	Political Resistance	Avoidance of Cannibalization	Rigid Mental Models	External Influence	Ambidextrous Structures
Underlying theory	Organizational Science, Population Ecology	Theory of Power and Politics	Theory of Power and Politics	Economics, Psychology	Theory of Bounded Rationality	Network Theory	e.g., Organizational Learning, Organizational Design[13]
Important authors	Hannan, Freeman, Levitt, March	Pfeffer, Salancik, Christensen, Bower, Benner	Pfeffer, Salancik, Eisenhardt, Bourgeois	Gilbert, Newbery, Reinganum, Chandy, Tellis	March, Simon, Kaplan, Eggers, Tripsas	Gilbert, Geletkanycz, Hambrick, Podolny	March, Tushman, O'Reilly, Christensen
Predominant level of analysis	Organization/ population of organizations	Organization	Individuals	Organization/ individuals	Individuals	Organization/ group/ individuals	Organization
Assumed effects on							
Speed	decreases speed of awareness, interpretation and decision-making	decreases speed of interpretation	decreases speed of interpretation and decision-making	decreases speed of interpretation	decreases speed of awareness and interpretation	increases speed of awareness and interpretation	no effect
Intensity	decreases intensity	decreases intensity	decreases intensity	decreases intensity	decreases intensity	increases intensity	increases intensity
Stamina	decreases stamina	decreases stamina	no effect	no effect	no effect	no effect	no effect
Routine flexibility	decreases flexibility	decreases flexibility	decreases flexibility	decreases flexibility	decreases flexibility	increases flexibility	increases flexibility

Table 3: Determinants of Organizational Adaptation

---

[13] Raisch and Birkinshaw (2008) refer to contributions of organizational learning, technological innovation, strategic management, organizational design, and organizational adaptation to the field of organizational ambidexterity. Tushman and O'Reilly (1996) emphasize the role of population ecology.

2.2 Family Businesses and Family Influence

2.2.1 Definition of Family Businesses and Family Influence

Research on family businesses has long emphasized that the behavior of firms influenced by families differs from the behavior of other firms (e.g., Chua, Chrisman, & Sharma, 1999; Habbershon & Williams, 1999). In this thesis, I adopt a system view of family businesses (Distelberg & Sorenson, 2009; Habbershon, Williams, & MacMillan, 2003; Tagiuri & Davis, 1996) to define 'family influence' as the overlap between the 'family system' and the 'business system' in a profit-seeking organization.[14] The family system is formed by the "individual [members of one or a few families], along with their interactions with one another within their immediate environment. [Family members] share common goals and resources, and [their] interactions revolve around shared goals and resource management" (Distelberg & Sorenson, 2009: 67). The business system is formed by "individuals who are employed by the business or who share common goals, values, and commitment to the whole [and it is enacted by the] interdependence and interactions of these individuals within their business environment"(Distelberg & Sorenson, 2009: 67). The more a company's strategic interpretation, decision-making, and actions become affected by the attributes, interests, values, and cultures of one or a few families, the larger the overlap between the family system and the business system (Stafford, Duncan, Dane, & Winter, 1999).

The family system influences the business system through both formal and informal mechanisms. Formal mechanisms include family ownership and family involvement in management. These conditions are a necessary albeit insufficient condition for family influence (Chua et al., 1999). Informal mechanisms include, for instance, language and narratives that over time become shared by organizational members across hierarchical levels (Sirmon & Hitt, 2003) and idiosyncratic approaches to resolving conflicts (Astrachan et al., 2002). Informal mechanisms are important as they help to align the values, goals, and identity of a family with those of the business, thereby triggering the development of a

---

[14]In the attempt to describe the essence of family businesses, researchers have employed a variety of terms that are similar, yet not identical, to 'family influence,' such as 'family involvement' (Chua et al., 1999), 'family control' (Mishra & McConaughy, 1999), and 'familiness' (Habbershon & Williams, 1999). I adhere to the term 'family influence' because it best reflects the active role that family members take in shaping the behavior of an organization (in contrast to mere 'involvement'), while it simultaneously denotes the intangible aspects stemming from the overlap of the family and the business systems, such as family traditions, culture, and identification (in contrast to 'family control').

'family business culture' (Astrachan et al., 2002).[15] Summarized, I define family influence as

Definition 2: Family Influence

> *'The overlap between the family system and the business system in a profit-seeking organization caused by formal (e.g., management positions) and informal (e.g., cultural) mechanisms.'*

Figure 8 illustrates possible extents of overlap between the family and the business system. Firms with either no family owners or family owners (e.g., people owning 5% of shares[16]) who have no intention to shape and influence the company in the present or future, are categorized as firms with no family influence, as shown in row 1. With increasing overlapping of the family and the business system, family influence increases (row 2). Row 3 depicts the special case of conflict-ridden, dysfunctional families (Gersick, Davis, McCollom Hampton, & Lansberg, 1997). I argue that family influence in this extreme case is evanescent, as no comprehensive family system exists; instead there might be overlaps between individual family systems, i.e. the power and intentions of selected family members and the business system. However, due to potentially diverging interests and goals of the individual family systems overlapping with the business system, the net effect of the overlap might even be non-existent.

---

[15] At this point, I have deliberately omitted the number of effected successions as a further antecedent for family influence, given the equivocal views in the literature. Whereas Astrachan et al. (2002) assume logarithmically increasing family influence with cumulating successions due to augmented experience, Gómez-Mejía et al. (2007) and several other researchers claim that family attachments to the business and family influence are at their peak for founder-owners and diminish for subsequent generations.
[16] Threshold value of family ownership often used in family business literature to distinguish publically listed family from non-family businesses, e.g., Chrisman and Patel (2012).

47

Figure 8: Family and Business Systems

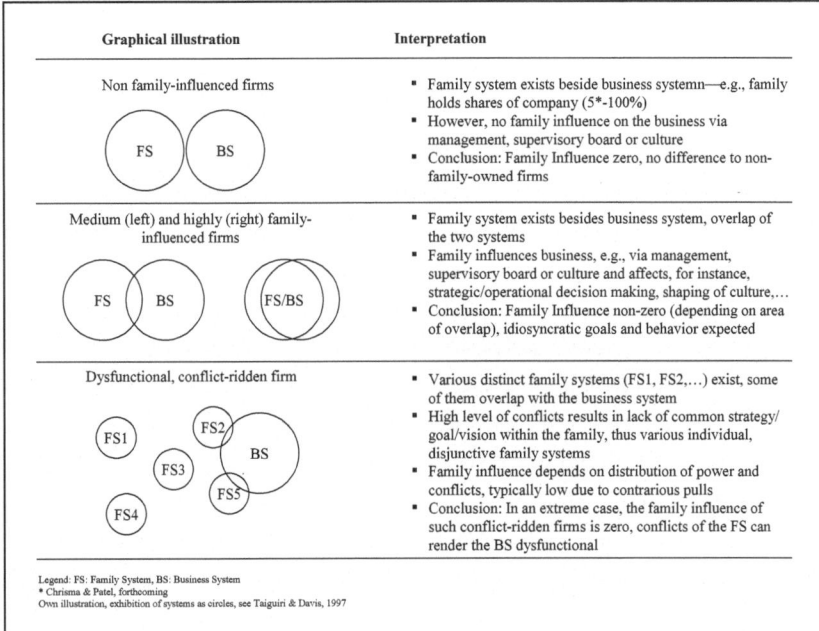

Given that formal and informal mechanisms can lead to varying levels of overlap between the family and the business systems—from zero overlap to full intersection—my definition implies that family influence is a continuous dimension ranging from low to high, in which all companies can be arrayed. This is crucial to my theorizing, as it allows me to differentiate between various levels of family effects within family-owned businesses while also including non-family-owned businesses. I am therefore able to avoid restricting my theorizing to the dichotomous differentiation between 'family firms' and 'non-family firms,' which has been criticized as being somewhat over-simplistic (Astrachan et al., 2002). In addition, this conceptualization of family influence as an overlap of the family and business systems provides an exhaustive, yet granular, foundation for my theorizing: it focuses not only on the components of a family firm, such as the percentage of ownership held by one family, but also on the essence of family influence, that is the family's common vision, its desire to pass on the business to future generations, and its willingness to commit to the business. As Chua et al. highlight, encompassing the "essence" (1999: 19) of family influence is pivotal to any theoretical definition of family influenced firms, as this essence comprises those characteristics that render family influenced business systematically different from other types of organizations.

2.2.2 Characteristics of Family Influenced Businesses

A fundamental tenet of family business theory is that family influence engenders idiosyncratic firm propensities and attributes (e.g., Chrisman, Chua, & Kellermanns, 2009; Habbershon & Williams, 1999). An abundant body of family business research has thus worked to systematically describe and categorize the goals, capabilities, and motivations of family influenced firms. Among the various taxonomies that family business researchers have developed to systematize manifestations of family influence, two approaches are salient and will be further illustrated in the following subchapters: First, Miller and Le Breton Miller's '4Cs'-framework has been particularly well received (Chrisman, Kellermanns, Chan, & Liano, 2010), arguably because of its exhaustiveness and because it is corroborated by a rich body of theoretical and empirical family business research. It encompasses four distinct firm-internal and -external characteristics of family influenced firms (Miller & Le Breton-Miller, 2005a). Second, Gómez-Mejía's umbrella construct of socioemotional wealth (Gómez-Mejía et al., 2007) that encompasses all non-economic goals of family owners' utility functions has recently emerged as an important concept to explain family firm behavior from a prospect theory view.

Other, widely accepted frameworks are Carney's (2005) '3P' model, which encompasses parsimony, personalism, and particularism as three important family firm propensities, and Astrachan et al.'s (2002) F-PEC scale, which is comprised of power, experience, and culture as characteristics of family influenced businesses. I refrained from adopting Carney's model, as it focuses on the effect of family influence on governance alone rather than on all organizational levels and functions, which is central to my theorizing. I do not elaborate on F-PEC, as it is less granular than the '4Cs' framework in differentiating between internal (community) and external (connections) effects. However, I submit that the theoretical propositions presented in this study (Chapter 3) would not be substantially altered if one assumed Carney or Astrachan's description of the attributes of family businesses; rather would they be less rich and less detailed.

2.2.2.1 4Cs-Framework

The '4Cs' framework, as summarized in Table 4, describes four inherent characteristics of family influenced businesses—*continuity*, *command*, *community*, and *connections*.

Table 4: 4Cs—Manifestations of Family Influence

Dimension	Definition	Description	Relevant Literature
Continuity	Degree of long-term as opposed to short-term orientation of strategic activities	Based on the desire to pass on the business to future generations and to keep wealth within the family	Gómez-Mejía et al., 2007; Miller & Le Breton-Miller, 2005; Miller, Le Breton-Miller, & Lester, 2010
Command	Level of decision-making authority and autonomy of the dominant coalition within the company	Enabled by intertwining of ownership and control and high levels of independence from external stakeholders	Carney, 2005; Miller & Le Breton-Miller, 2005
Community	Level of cohesion among the employees of an organization within and across hierarchical levels	Based on the family firm's culture; enhanced by long employee tenures; "thick social walls" (Carney, 2005) separate employees from externals	Astrachan, Klein, & Smyrnios, 2002; Miller & Le Breton-Miller, 2005
Connections	Level of depth of ties to external stakeholders such as suppliers	Personal relationships and interconnectedness as identifying elements in family firms	Gómez-Mejía Nunez-Nickel, & Gutierrez, 2001; Miller & Le Breton-Miller, 2005

Continuity refers to the observation that, compared to their non-family influenced counterparts, family influenced businesses tend to strive for more longevity, as their organizational leaders wish to transfer the business to the next generation (Miller et al., 2010) and to keep wealth within the family (Gómez-Mejía et al., 2007). Besides this forward-looking aspect, striving for continuity allows members of owning families to hearken back to the treasure trove of experience gained throughout previous generations (Astrachan et al., 2002) as well as large stocks of tacit knowledge (e.g., Cabrera-Suarez, Saa-Perez, & Garcia-Almeida, 2001; Sharma, 2004; Sirmon & Hitt, 2003).

Command denotes the link between family influence and greater decision-making authority and autonomy of the dominant coalition (Carney, 2005). Such high levels of command result from the intertwining of ownership and control within family influenced firms as well as the above-average independence of family owners from external stakeholders, especially public shareholders (Miller & Le Breton-Miller, 2005a). It is important to note some interdependence of *continuity* and *command*: Family owners do not only possess high levels of *command* but also strive to *maintain* that level of control for the future (Gómez-Mejía et al., 2007).

Community refers to the number and intensity of relationships among employees, both within and across hierarchical boundaries (Miller & Le Breton-Miller, 2005a). In an archetypal family business, employees constitute a "pseudo-family" (Tan & Fock, 2001: 128). In turn, given certain levels of perceived fairness and trust among the family members (Lubatkin, Ling, & Schulze, 2007), high levels of family influence typically attract employees who value social relationships, which gives rise to steward-like behavior (Davis, Schoorman, & Donaldson, 1997; Karra, Tracey, & Phillips, 2006) and above-average employee tenure in family businesses (Lansberg, 1999).

Finally, *connections* captures the notion that companies strongly influenced by families typically establish profound and stable relationships with their stakeholders, including their suppliers and complementors (Miller & Le Breton-Miller, 2005a). This is because families view interconnectedness and personal relationships as defining elements of their identities (Gómez-Mejía et al., 2001). In terms of network theory, family firms in general prefer a limited set of strong ties over a broad set of weak ties (Granovetter, 1973, 1983). Research on family firms and social capital has identified the relationships of family firm members as resources unique to this type of organization (Pearson, Carr, & Shaw, 2008; Sirmon & Hitt, 2003).

A key premise of my theorizing is that *continuity*, *command*, *community*, and *connections* are reflective, covariant indicators of family influence. In other words, I assume that a marginal increase in family influence entails a marginal increase in all four domains, although this increase is not necessarily equally distributed. In addition, I assume two important boundary conditions. First, my model refers to medium-sized and large companies. This assumption is reasonable because the organizational phenomenon of inertia is typically described for larger, formalized businesses with multi-level resource allocation processes (Bower, 1970). Second, I restrict my hypothesizing to those family influenced businesses in which the individual value systems of the various family members involved are largely consonant and, thus, constitute a coherent family system. As such, I exclude all companies in which dysfunctional conflicts among family members superpose major strategic decisions (Gersick et al., 1997; Schulze, Lubatkin, & Dino, 2003b) and hinder adaptation processes a priori.

2.2.2.2 Socioemotional Wealth

In the next subchapter I will summarize the concept of 'socioemotional wealth' (SEW), which is an alternative approach to describe the particularities of family influenced as opposed to non or less family influenced firms. This concept emphasizes the idiosyncratic goals and intentions of family owners and is of particular avail when discussing framing and interpretation of decision-makers in family influenced firms.

One decade ago, scholars started to highlight the fundamentally different reference points of family as opposed to non-family influenced firms resulting in different framing processes among those organizations' decision makers and variation in strategic behavior (Berrone et al., 2010; Chrisman & Patel, 2012; Gómez-Mejía et al., 2007). Introducing the behavioral agency model (Wiseman & Gómez-Mejía, 1998), researchers argued that while non-family influenced firms assess any strategic activities as a loss or gain in financial performance as compared to an aspiration level (Kahneman & Tversky, 1979), family influenced firms also consider gains and losses of the family's non-economic utility function, such as transgenerational control or reputation (Gómez-Mejía et al., 2007). These non-economic components of the utility function are summarized as socioemotional wealth (SEW) (Gómez-Mejía et al., 2007). Prior research has identified distinct dimensions of SEW (Berrone et al., forthcoming; Gómez-Mejía, Cruz, Berrone, & Castro, 2011a; Zellweger, Kellermanns, Chrisman, & Chua, forthcoming-b):

(1) Preservation of *family ties*

(2) Preservation of *family's power and influence* on the organization

(3) Preservation of *family's status and reputation*

(4) Preservation of *affect* and *emotions*

Preservation of family ties denotes the family owners' tendency to be altruistic towards family members, e.g., by providing them with jobs. Moreover, similar to the concept of *continuity* described in Chapter 2.2.2.1, preservation of family ties refers to the family's desire to pass the firm on to the next generation. Similar to the concept of *command*, preservation of family power and status is linked to the owning family's control over the respective firm, exerted either via ownership share or management positions. As family owners are "inextricably tied to the organization" (Berrone et al., forthcoming: 5), they are particularly concerned about the reputation of the firm as well as the implications on their own status. Lastly, a further consequence of the intertwinement of family and business is the abundance of affect and emotions that family owners associate with their business. Although affect and emotions are also present in non family firms, Berrone and colleagues (forthcoming: 6) argue: "Because the boundaries between family and corporation are rather blurred in family businesses [...], emotions permeate the organization, influencing the family business's decision-making process."

Referring to the umbrella concept of SEW, scholars have gathered abundant empirical evidence on how SEW-considerations influence an organization's strategic behavior. For instance, researchers have investigated the acquisition behavior (Dehlen & Zellweger, 2012), the behavior under institutional pressure (Berrone et al., 2010) and the R&D expenditures of family businesses (Gómez-Mejía, Hoskisson, Makri, Sirmon, & Campbell, 2011b). Similar to all above-mentioned studies is their proposal that family firms are willing to accept business risks and 'greater performance hazards' (Gómez-Mejía et al., 2007) in order to preserve their status quo of SEW. However, whilst there is abundant empirical evidence on the importance of SEW in general, there is still a lack of scale to measure SEW. A relatively new framework to measure SEW—FIBER—still lacks empirical testing (Berrone et al., 2012).

2.3 Literature Review of Family Influence and Organizational Adaptation to Discontinuous Technological Change

2.3.1 Debate on Benevolent vs. Malevolent View

In the early years of family firm research, Gersick and colleagues already pointed out the necessity for family firms to balance change and stability and emphasized that "sooner or later the pressure to change becomes irresistible" (Gersick, Lansberg, Desjardins, & Dunn, 1999: 291). Other scholars alluded to idiosyncratic challenges of family firms during adaptation processes due to the non-rational components in organizational behavior induced by the family system (Davis & Stern, 1988). In recent years, an emergent stream of family business research has begun to describe the ways in which the idiosyncratic characteristics of family firms shape their strategic behavior (Astrachan, 2010; Craig & Moores, 2006; Donckels & Fröhlich, 1991), and, in turn, their innovativeness and their responses to change (Chirico & Salvato, 2008; Chrisman, Chua, & Sharma, 2005; Hatum & Pettigrew, 2004). Most of these studies have adopted the logic of agency theory to examine the particularities and outcomes of principal-agent and principal-principal conflicts in family businesses (Chua, Chrisman, & Bergiel, 2009)[17].

However, despite their shared theoretical lens, the conclusions of studies on family influence and change are highly contradictory (Le Breton-Miller & Miller, 2009), and it is unclear whether the characteristics of family businesses manifest themselves in adaptive or maladaptive strategic responses to change. Two perspectives in particular have emerged, which can be differentiated according to their main assumptions, the presumed key drivers of strategic behavior, and the resulting characteristics and adaptation patterns: I call them the *benevolent*[18] perspective (Davis et al., 1997; Eddleston, Kellermanns, & Sarathy, 2008; Miller, Le Breton-Miller, & Scholnick, 2008) and the *malevolent*[19] perspective (Chandler, 1994; Ward, 2004). Notably, both perspectives are supported by abundant empirical evidence.

Benevolent perspective. Proponents of this perspective argue that family businesses incur lower agency costs than non-family businesses, particularly due to the close relationships in family businesses and the prevalent concentration of ownership and management control in the form of a 'family CEO' resulting in decreased principal-agent

---

[17] Other frequently applied theoretical perspectives that aim to explain the strategic behavior include resource based view (e.g., Cabrera-Suarez et al., 2001; Habbershon & Williams, 1999) and Gómez-Mejía's behavioral agency model (Wiseman & Gómez-Mejía, 1998). Resource based view points to the idiosyncratic, useful, and non-imitable resources available in family business, summarized as familiness (Habbershon et al., 2003) including, for instance, tacit knowledge (Greene & Brown, 1997; Miller & Le Breton-Miller, 2006). Sirmon and Hitt (2003) amend this perspective by identifying five relevant sources of capital in family firms including human and social capital. Building on behavioral theory of the firm (Cyert & March, 1963) and prospect theory (Kahneman & Tversky, 1979), Gómez-Mejía and colleagues propose that family firms have a fundamentally different reference point used for decision-making due to characteristic risk preferences. This model was later further developed into the concept of socioemotional wealth, a construct idiosyncratic to family firms, that influences strategic behavior.

[18] Labeled as stewardship perspective by Miller, Le Breton-Miller, and Scholnick (2008).

[19] Labeled as stagnation perspective by Miller, Le Breton-Miller, and Scholnick (2008) or agency perspective by Le Breton-Miller and Miller (2009).

conflicts (Fama & Jensen, 1983). Furthermore, according to these authors, the strategic behavior of family businesses is predominantly shaped by stewardship and productive manifestations of altruism (Chrisman, Chua, & Litz, 2004). In addition, as a consequence of their *community* culture and the low number of decision makers, family firms develop lean structures, flat hierarchies, and low levels of formalization (Geeraerts, 1984). Together, these factors allow family companies to be more innovative and flexible than other companies (Zahra, Hayton, Neubaum, Dibrell, & Craig, 2008).

Those scholars also suggest that family businesses are willing to take considerable risks (Zahra, 2005). According to Miller and Le Breton-Miller (2005a), this is primarily because family businesses, which are less driven by short-term return targets than other companies, focus on innovative, long-term strategies. Furthermore, proponents of the benevolent perspective argue that, due to their *community* culture, family businesses are able to build up substantial and intangible assets. These assets often encompass high levels of social capital, including large pools of talented, well-trained employees, and close relationships with suppliers and customers (Miller et al., 2008).

Malevolent perspective. In contrast, scholars proposing a malevolent view argue that agency costs are higher in family businesses as a result of increased principal-principal conflicts and various 'moral hazards' (Schulze, Lubatkin, & Dino, 2002). Most importantly, a lack of external control mechanisms in family businesses promotes owner opportunism (Schulze, Lubatkin, Dino, & Buchholtz, 2001). Furthermore, power concentration in family businesses can engender nepotism and 'parental altruism' (Bertrand & Schoar, 2006), resulting, for instance, in the appointment of incapable offspring to executive positions (Lee, Lim, & Lim, 2003). This effect is further exacerbated by resource scarcity, which is typical in highly family influenced companies (Schulze et al., 2002). Overall, proponents of this theoretical view submit that governance mechanisms induced by family influence encompass incentives for family and non-family employees to free-ride and shirk responsibilities (Schulze et al., 2002).

A core tenet of this perspective is that family businesses develop stagnant, inflexible routines as a consequence of these idiosyncratic moral hazards (Basly, 2007; Le Breton-Miller & Miller, 2009) and that, ultimately, they engage in path-dependent, incremental strategies (Lester & Cannella, 2006). According to stagnation theorists, the lack of managerial talent, both within and outside of the family, results in increased levels of perceived uncertainty and, ultimately, in conservatism and risk averseness (Le Breton-Miller & Miller, 2009; Morck,

Wolfenzon, & Bernard, 2005). In addition, owner opportunism results in the extraction of assets from the business to meet the needs of the owner-family. This further enhances resource scarcity (Cabrera-Suarez et al., 2001; Chandler, 1994), which, in turn, constrains the innovativeness of family influenced companies (Le Breton-Miller & Miller, 2009). Figure 9 summarizes the two literature streams.

Figure 9: Benevolent and Malevolent View on Family Businesses

	Benevolent Perspective	Malevolent Perspective
Underlying Theory and Assumptions	• Agency Theory (e.g., Jensen & Meckling, 1983) , Stewardship Theory (Davis, Schoorman, Donaldson, 1997) • Lower agency costs in family firms because of fewer principal-agent conflicts (due to ownership concentration) • Families act as stewards, thus productive manifestations of altruism	• Agency Theory (e.g., Jensen & Meckling, 1983) • Higher agency costs in family firms because of more principal-principal conflicts; lack of monitoring • Families act as agents, resulting in moral hazards such as nepotism and parental altruism • Perspective also labeled as "stagnation perspective" or "agency perspective" by some scholars (Miller & Le Breton-Miller, 2005)
Effect on Firm Structure and Assets	• Structure: flat hierarchies, lean structures (Behr et al., 1997; Geeraerts, 1984) • Managers and employees: Well trained staff (Allouche & Amann, 2002) • General assets: abundance of intangible assets	• Structure: bureaucratic management (Kets de Vries & Carlock 2007) • Managers and employees: less capable staff (Bennedsen et al., 2007) • Assets: resource scarcity (Morck et al., 2005; Schultze et al., 2003)
Effect on General Adaptation Behavior of Firm	• Innovativeness and flexibility (Zahra et al., 2008) • Risk taking (Zahra, 2005) • Superior R&D investment (Miller et al., 2009; James, 2006)	• Conservatism and myopia (Allio, 2004) • Risk aversion (Morck & Yeung, 2003) • Underinvestment in renewal (Bloom & Van Reemen 2007)

56

2.3.2   Gaps in the Literature on Family Influence and Discontinuous Technologies

Why are the findings on the effect of family influence in the context of strategic behavior in general and adaptability in particular so contradictory? To address this question, I conducted a systematic literature review based on explicit selection and evaluation criteria (David & Han, 2004; Tranfield, Denyer, & Smart, 2003). The review focused on 19 leading peer-reviewed management, innovation, and family business journals (see Table 5) published between January 2000 and December 2010 (Chrisman, Chua, Kellermanns, Matherne Iii, & Debicki, 2008). Given my specific interest in the integration of family business and general organization research, I created a search dictionary of synonyms and related terms for 'family business' and '(discontinuous) change'[20]. To complement this analysis and ensure sufficient comprehensiveness, I set up additional search routines based on Google Scholar[21]. In total, I identified 43 papers published in the above-mentioned journals that contained the respective terms in their abstracts[22]. Most of these papers were published in *Family Business Review* (25), *Entrepreneurship Theory and Practice* (7), and the *Journal of Management Studies* (4), as shown in Table 5. Appendix A.A-1 provides a detailed overview of all 43 articles.

---

[20] The search dictionary included the following term: 'family firm,' 'family business,' 'family controlled,' 'family influenced,' 'family led,' and 'family owned' as well as 'adaptability,' 'adaptation,' 'adaptiveness,' 'adoption,' 'change,' 'discontinuities,' 'discontinuity,' 'discontinuous,' 'disruptive,' 'disruptivities,' 'disruptivity,' 'disruption,' 'inertia,' 'innovation,' 'organizational flexibility,' and 'radical change.'

[21] One main purpose of conducting this additional search was to include papers published in *Family Business Review* before 2005 (not included in EBSCO at that time). I searched for papers containing the terms 'family business' (alternatively: 'family firm') and 'change' (alternatively: 'adaptability,' or 'adaptation,' or 'discontinu*') in their title.

[22] Search conducted with EBSCO Business Source Premier database access. I identified papers that included at least one of the family related and one of the change related keywords in their abstract.

Table 5: Literature Review

Journal	Number of Papers[a]		
	Total	Adaptation (to external change)	Adaptation to discontinuous, external change
Academy of Management Journal	0	0	0
Academy of Management Review	0	0	0
Administrative Science Quarterly	0	0	0
British Journal of Management	2	1	0
Entrepreneurship Theory and Practice	7	2	1
Family Business Review[b]	25	12	1
Industrial and Corporate Change	0	0	0
International Journal of Industrial Organization	0	0	0
Journal of Business Venturing	3	0	0
Journal of Economics & Management Strategy	0	0	0
Journal of Management	0	0	0
Journal of Management Studies	4	2	1
Journal of Product Innovation Management	0	0	0
Management Science	0	0	0
Organization Science	1	0	0
Organization Studies	1	0	0
R&D Management	0	0	0
Research Policy	0	0	0
Strategic Management Journal	0	0	0
Total	**43**	**17**	**3**

[a] Articles published in 2000-2010 containing synonyms of 'family business' and 'change' in their abstract. I provide the total number of papers identified (first numbers column), the number of papers thereof referring to adaptability[23] (second numbers column), and the number of papers referring to adapting to discontinuously (as opposed to incrementally) changing environments.

[b] not included in EBSCO database before 2005.

---

[23] I attempted to distinguish between changes initiated organizational-internally (e.g., successions, changes in management practices, ongoing R&D efforts) and changes induced by external triggers (e.g., crisis, discontinuous technologies). Papers referring to the latter are summarized as 'adaptation'-work.

My review reveals significant gaps in the literature and suggests that the inconsistent findings on the implications of family influence in the context of technological change result from three factors. First, despite the increasing amount of attention given to this topic, the implications of family influence in the context of external change appear to be under-investigated. Only 17 of the 43 identified papers dealing with family influence and change address the response of family businesses to a changing environment. The remaining 27 papers describe internal organizational changes, such as a shift from one generation to the next or adaptation of internal management processes.

Second, the majority of research on family business and change fails to differentiate between two paradigmatically different types of technological change: continuous change and discontinuous technological change. Of the 17 papers addressing the response of family businesses to change, only three specifically discuss circumstances of discontinuous technological change (Hatum & Pettigrew, 2004; Zahra, 2010; Zahra et al., 2008).

Third, and most importantly, the studies in the literature on family business that are related to discontinuous technological change lack theoretical or practical implications. This is because this research rarely focuses specifically on technological shifts and because these studies are not integrated into the vast array of research in the organization literature that deals with discontinuous technological change. One of the three articles that mention discontinuous technological change (Hatum & Pettigrew, 2004) explores the adaptation of family businesses to political, rather than technological, turmoil. The other two articles touch upon discontinuous technological change but do not investigate this issue in detail: Zahra et al. (2008) empirically demonstrate that stewardship in family firms is positively associated with overall strategic flexibility. However, they merely mention that stewardship may be beneficial in terms of discontinuous technological change (Nadkarni & Narayanan, 2007). Zahra (2010) empirically studies how family businesses leverage their organizations' social capital in general, and connections to new ventures in particular, to quickly adapt to radical change. Nevertheless, while Zahra identifies relationships with new ventures as a core determinant of successful adaptation to radical change, he does not explore other factors that foster or inhibit the adoption of discontinuous technologies in family firms. His research does not specifically discuss any of the economic-strategic, organizational, or cognitive-emotional barriers to adaptation that are identified in the inertia literature.

The low degree of integration between the family business literature on change and non-family related research on organizational adaptation to discontinuous technologies is further reflected in one more observation. Of the 17 papers on the adaptation of family businesses to change, only one compares the adaptation behaviors of family businesses to those of non-family businesses (Craig & Dibrell, 2006). The others observe family businesses exclusively, however, mostly envisage them as a homogeneous group of organizations. Thus, there is surprisingly little understanding how variance in family influence, ranging from zero for non-family influenced firms to high for substantially family influenced firms effects on organizational behavior in the context of change.

To further corroborate the findings of low integration of these two research streams, I searched for papers published in *Family Business Review* and *Entrepreneurship Theory and Practice*, two of the most important and most cited outlets for family business related work, that cite Clayton Christensen's (1997) groundbreaking work on disruptive technologies[24]. Only two articles published in *Family Business Review* refer to Clayton Christensen's work (Beck, Janssens, Debruyne, & Lommelen, 2011; Craig & Moores, 2006); moreover these authors do not build on disruptive change theory and refer to Christensen only once throughout the paper. Ten articles published in *Entrepreneurship Theory and Practice* contain references to Christensen's work. Most of them, however, refer to new ventures, only two contain the term 'family business' or 'family firm' in the body of their paper (Sharma & Salvato, 2011; Wiklund & Shepherd, 2009), and refer to Christensen's work only once or twice throughout the paper.

In summary, the extant literature on family influence highlights idiosyncratic characteristics that are likely to give rise to differences in the strategic responses of family businesses relative to the responses of other companies. However, this literature neither encompasses theoretical underpinnings from related studies on the behavior of non-family businesses nor is it integrated with that stream of research. I aim to bridge this gap in the following chapter.

---

[24] Search conducted in early 2012 using Google Scholar. No filters regarding time were set. In total, Christensen's book had been cited 7.178 times at that point in time.

3 Theorizing—Towards a Conceptual Framework of Family Influence and Adoption of Discontinuous Technological Change

In this chapter, I provide a conceptual basis to bridge research on family businesses and literature on organizational adoption of discontinuous technologies; two pivotal yet previously disconnected streams of management science. The research question that drives my conceptualization is:

> *How does family influence, manifested in continuity, command, community and connections, affect an organization's adoption of discontinuous technologies regarding speed, intensity, and stamina of adoption, as well as routine flexibility?*

I thus lay the foundation of a theory that links varying levels of family influence in incumbent firms to differences in how those organizations adopt discontinuous technologies. Ultimately, the integration of the two literature streams allows me to suggest that highly family influenced firms respond in a significantly different manner from firms that are less influenced by families or not subject to family influence at all.

This theorizing is rooted in the notion that, by linking family business research with the organizational adaptation literature, I can derive formal propositions regarding the effect of family influence on the adoption of discontinuous technologies by incumbent firms. My theoretical framework is illustrated in Figure 10. As defined in the section on family influence (Chapter 2.2.2), varying levels of family influence are reflected in (co-)varying levels of *continuity, command, community* and *connections* (Chrisman et al., 2009; Miller & Le Breton-Miller, 2005a). Furthermore, as noted in my synthesis of the organizational adaptation research, determinants of organizational adaptation—including sources of inertia and enablers of change—shape the speed, intensity, stamina, and routine flexibility of an incumbent's adoption of a discontinuous innovation. The central tenet of my theorizing, which bridges the two streams of theory, is that varying levels of *continuity, command, community* and *connections* affect the levels of barriers to, and enablers of, technology adoption by organizations.

In the following, I build on this logic and first investigate how each of the 4Cs affects each of the five barriers to and two enablers of organizational adoption discussed in the previous chapters (2.1.2.2.). Moreover, I outline for each of the seven determinants, how family influence would affect organizational adoption of the new technology if no other barriers and enablers existed. In a second step (Chapter 3.8), I integrate the findings relating to the individual determinants and outline a comprehensive picture how family influence affects the speed, intensity, stamina, and flexibility of technology adoption. I propose that firms with higher levels of family influence adopt discontinuous technologies faster, less aggressively, with more stamina, and less flexibly than their less- or non-family influenced counterparts. I then explore the pivotal role of individual and team-level characteristics of the dominant coalition, which are expected to moderate the associations between the manifestations of family influence and the determinants of organizational adaptation, and, ultimately, the link between family influence and technology adoption (Chapter 3.9). Finally, I explore how resource investments into the exploration of the discontinuous technology might alter over time, triggered by alterations in organizations' threat framing and perceived levels of control (Chapter 3.10).

63

Figure 10: Framework of Theorizing

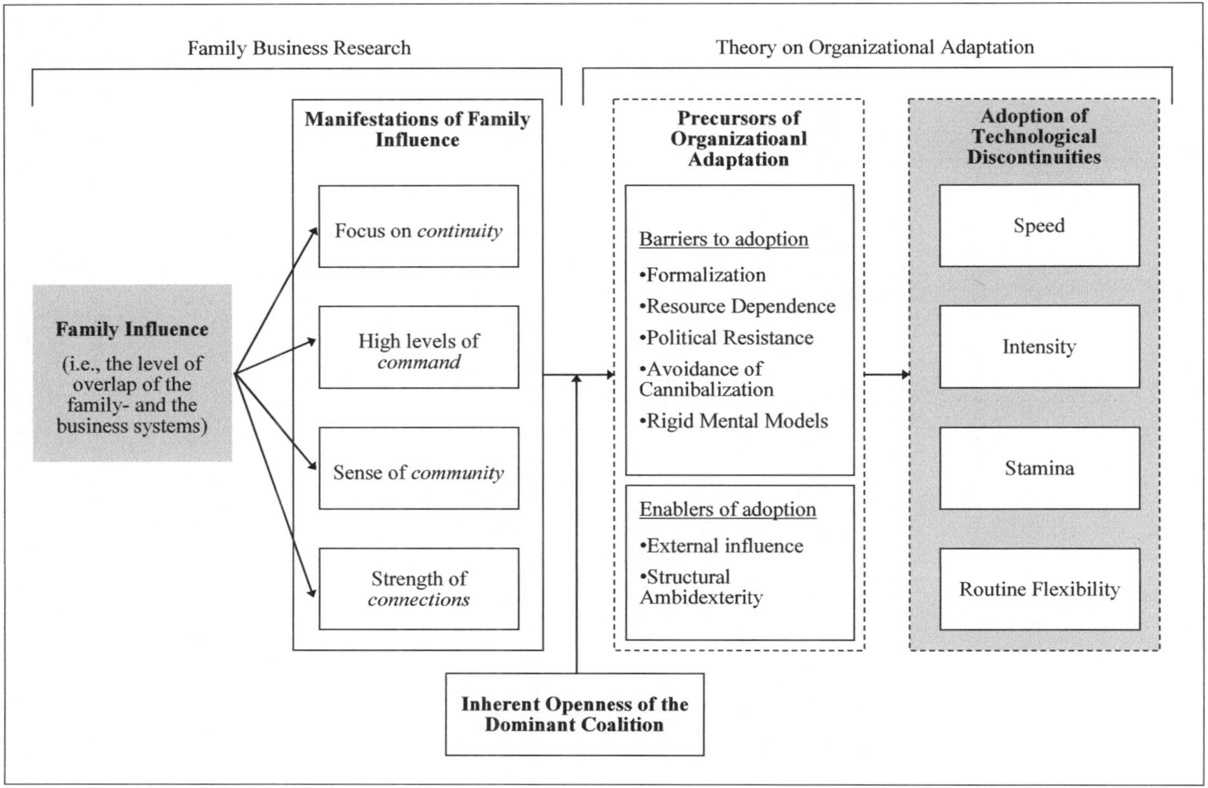

Theory on Organizational Adaptation

Adoption of Technological Discontinuities

- Speed
- Intensity
- Stamina
- Routine Flexibility

Precursors of Organizatioanl Adaptation

Barriers to adoption
- Formalization
- Resource Dependence
- Political Resistance
- Avoidance of Cannibalization
- Rigid Mental Models

Enablers of adoption
- External influence
- Structural Ambidexterity

Inherent Openness of the Dominant Coalition

Family Business Research

Manifestations of Family Influence

- Focus on *continuity*
- High levels of *command*
- Sense of *community*
- Strength of *connections*

Family Influence
(i.e., the level of overlap of the family- and the business systems)

3.1 Family Influence and Formalization

Formalization, as outlined in chapter 2.1.2.2.1, is a critical determinant for the speed, intensity, stamina, and routine flexibility of an organizations' adoption of discontinuous technologies. To investigate the impact of family influence on the level of formalization, I will next discuss how each of the 4Cs implies either high or low levels of formalization.

First, I suggest that *continuity* results in a decreased level of formalization, in particular of routines laying the groundwork for the exploitation of new business opportunities. Striving for continuity depicts the organization's focus on transgenerational wealth creation, manifesting itself in a focus on the long-term, rather than short-term performance targets (Miller & Le Breton-Miller, 2005a). Organization research suggests that long-term, rather than short-term, performance criteria create room for grounded, non-formalized research, open search, and experimentation (Farjoun, 2010). Grounded, open research and experimentation reduce decision makers' myopia to discontinuous technological change (March, 1991), thus leading to earlier awareness of the innovation.

Second, I propose that *command* implies lower levels of formalization. Due to their financial independence of external stakeholders, decision makers in family firms do not need to legitimize their decisions, neither internally nor externally. According to Hannan and Freeman (1984), legitimization is one of the key drivers for accountability, and thus for the firms' motivation to establish formalized routines. In other words, the lack of necessity to create legitimacy towards externals such as shareholders, results in a lack of necessity to establish formalized routines. One immediate consequence of this general trend toward low formalization is that decision-making in family influenced organizations is less bound to "highly calculative or instrumental rationality" (Carney, 2005: 252) as compared to non-family influenced firms. This leads to decision-making based on 'back-of-an-envelope' calculations and gut feeling rather than rational criteria and extensive calculations (Carney, 2005). Thus, high levels of command allow for "opportunistic investment" (Carney, 2005: 259) and ultimately accelerate decision-making processes.

Third, I argue that *community* entails low levels of formalization. Due to their sense of community, family influenced firms are concerned about building up strong social capital throughout their organizations (Arregle et al., 2007; Sirmon & Hitt, 2003). Consequential high levels of trust (e.g., Gómez-Mejía et al., 2001; Jones, 1983) and common language (Sirmon & Hitt, 2003) render formalization of internal processes obsolete. Indeed, high levels of formalization have the potential to harm such a community: Employees of organizations with a strong sense of community are expected to act as 'stewards' rather than 'agents' and as thus might feel demotivated by rigid and formalized structures and processes (Chrisman, Chua, Kellermanns, & Chang, 2007).

Fourth, I suggest that the strong *connections* of family influenced firms to external stakeholders such as suppliers and customers lower certain aspects of formalization. The term 'strong connections' refers to the fact that family influenced firms strive to rely on few and deep, long-term ties rather than various, superficial, and short-term oriented ties. High levels of trust, as inherent to those long-term relationships, render formalization unnecessary and even harmful, following the same line of argumentation as applied above when referring to *community*, i.e. internal relations (Miller, Steier, & Le Breton-Miller, 2003; Pearson et al., 2008; Sirmon & Hitt, 2003).

Summarized, my theorizing proposes that family influence, manifested in the 4Cs, leads to lower levels of formalization in family businesses, entailing a higher speed of adaptation, enhanced stamina, and reduced routine rigidity. Formally,

P1-1a. Family influence lowers the level of formalization.

P1-1b. Ceteris paribus, family influenced firms adopt discontinuous technologies faster, more intensively, more continuously, and more flexibly than less or non-family influenced firms mediated by lower levels of formalization.

Manifold real world examples illuminate the prevalence of low levels of formalization in family influenced firms. For instance, due to their desire to keep the business in the family (Gómez-Mejía et al., 2007), many family influenced firms refrain from applying standardized and rational decision criteria when settling succession processes. This frequently results in appointing non-adequate successors that might be dismissed again shortly after, as recently happened in the German publishing house Neven DuMont. Low formalization is also typical when it comes to strategic planning. In an interview, Henry Mintzberg (Pervin, 1997) referred to the Swedish family furniture store IKEA, whose family CEO was reluctant to engage in thorough strategic planning, ultimately resulting in supply chain issues. The deduced proposition is also in line with Gómez-Mejía and colleagues' (2001) empirical study in the Spanish newspaper sector, showing that the tenure of family CEOs is less tied to performance, that is factual and number-based decision criteria. Whereas this scholarly work as well as the above provided practical examples point to the negative consequences which low formalization can entail, I propose, in line with extant literature (Hannan & Freeman, 1977), that in the specific context of discontinuous technologies emerging, low formalization constitutes a competitive advantage for family firms.

3.2   Family Influence and Resource Dependence

Resource dependence, as outlined in chapter 2.1.2.2.2, is an important inhibitor of quick, intense, and flexible adoption of discontinuous technologies (Christensen & Bower, 1996). For the purpose of this study it is important to distinguish between two different (firm external) groups that can exert power on the incumbent organizations (Pfeffer & Salancik, 1978): external capital providers (such as investors and shareholders) and other external stakeholders (such as customers, suppliers, and complementors). As outlined in the theory section in Chapter 3, these two groups have a characteristic impact on organizational adoption behavior, with capital providers mainly influencing time and intensity of adoption, and other stakeholders predominantly affecting the resource intensity and flexibility of routines applied for implementation.

First, I formally suggest that striving for *continuity* lowers resource dependence on capital providers and increases resource dependence on external stakeholders. Related to capital providers, firm owners who intend to pass their business to the next generation and maintain family influence over time (Gómez-Mejía et al., 2007) are reluctant to dilute ownership by handing out shares to external capital providers (Carney, 2005; Sirmon & Hitt, 2003) and are hence less dependent on such individuals and organizations. Related to the second aspect, striving for continuity involves long-term relationships (Miller & Le Breton-Miller, 2005a), which in turn entails higher levels of dependence on the partners than short-term relationships. This effect will be discussed in more detail when I theorize about the effect of *connections* on resource dependence.

Second, I highlight in line with extant literature (Carney, 2005) that high levels of and striving for *command* lower resource dependence on capital providers and have no effect on resource dependence on other external stakeholders. The underlying rationale is that family owners actively strive to keep their control over their own business by avoiding the accumulation of debt or the need to raise equity on capital markets when investing in strategic initiatives (Gómez-Mejía et al., 2001; Mishra & McConaughy, 1999; Schulze, Lubatkin, & Dino, 2003a). As a consequence of this low dependence, criteria that are important to most external capital providers, particularly quick and predictable returns and growth (Christensen, 1997), are likely to be given less priority in the decision-making processes of family influenced firms. The financial success and the growth of a technological discontinuity are, at the outset, difficult to predict, particularly because the new technology often under-performs with respect to established performance criteria and is, in turn, rejected by existing customers (Christensen & Bower, 1996). As a result, top managers in more family influenced firms are less likely to dismiss such innovations in the early stages of their evolution than managers in non or only weakly family influenced companies, which adds to the relative speed of adaptation.

Third, I assume that *community* is not related to variance in the resource dependence on external capital providers or other external stakeholders. *Community* is defined as a purely *internal* characteristic of family influenced firms (Miller & Le Breton-Miller, 2005a), whereas resource dependence relates to dependencies between the organization and its environment (Pfeffer & Salancik, 1978).

Fourth, I theorize that *connections* have no effect on resource dependence on external capital providers yet increase resource dependence on other external stakeholders. External stakeholders provide rare and valuable resources to organizations, such as revenues (customers), goods (suppliers), labor and technology (alliance partners). As such, these constituents are able to exert power on the respective organization (Pfeffer & Salancik, 1978). Pfeffer and Salancik (1978) theorize that resource dependence increases with the scarcity of the resources. Family influenced firms strive to "maintain close or special relations with a subset of trusted partners" (Carney, 2005: 255) and thus rely rather on a few strong connections than on a multitude of weak ties (Granovetter, 1973; Miller & Le Breton-Miller, 2005b). I thus expect the resource dependence on external stakeholders to be aggravated for those firms, which ultimately results in diminished resource investment and routine flexibility as hypothesized in chapter 2.1.2.2.2. This is particularly true if vendors and suppliers with whom the family influenced businesses have strong and long-established relationships are "viewed as, or […] actually [are], members of the family" (Berrone et al., 2010: 90). The notion of a strong dependence on external stakeholders is also supported by Berrone and colleagues (2010: 84) who emphasize that family owners feel "more vulnerable to negative assessments by outsiders," especially on a local level, and thus "pay greater attention to what others think of them."

To summarize, family influence decreases resource dependence on external capital providers and increases resource dependence on other external stakeholders. Thus, family influenced businesses are less affected by the 'innovator's dilemma' (Christensen, 1997); yet they face an entirely different dilemma based on their heightened dependence on a selected value network (Hill & Rothaermel, 2003). I thus formally propose,

> P1-2. Family influence (a) lowers resource dependence on external capital providers and (b) aggravates resource dependence on other external stakeholders.

The case of Diehl, a German family business internationally active in the defense industry, aero systems and metering segments, exemplifies this striving for financial independence. An entrepreneurial portrait published in an economic magazine (Fasse, 2009) describes: "Despite all entrepreneurial risk, one thing is important: Diehl shall never become dependent of banks or investors." [25] Pablo Dell'Antonio, CEO of the family owned liquor producer Mast-Jägermeister, perceives his organization's independence from the capital market as a major competitive advantage: "We are not publically traded, and thus we can dare to invest in an anticyclic manner" (Thiede, 2009). [26]

I propose that family influenced firms are able to react faster and with more stamina to discontinuous technologies as they are not bound to quick, predictable, short-term returns, and consequently are better prepared for the trap of underrating those innovations. Resource dependence on other external stakeholders endangers family influenced firms to avoid adaptation of non-paradigmatic routines and rather rely on 'cramming' (Christensen & Raynor, 2003). Based on proposition P1-2a, one could assume the adaptation of family influenced firms to be faster and more intensive yet less flexible due to the decreased resource dependence on external capital providers. However, I propose that the aspect regarding intensity does not hold true due to a reinforcing effect of the involved constructs, *command* and resource dependence on external capital providers. Family owners do not only *possess* a

---

[25] Quotation originally in German, translated by author of this thesis.
[26] Quotation originally in German, translated by author of this thesis.

high level of command, they also *strive to maintain* this autonomy for the future (Berrone et al., forthcoming; Carney, 2005). As a consequence, they are somewhat constricted in terms of the height of their investments, a phenomenon, labeled as 'parsimony' that has been investigated by various scholars (Carney, 2005). While dependence on investors and lenders may retard responses to discontinuous technological changes of non-family influenced firms, it also enables firms to raise the financial resources needed to implement those innovations. Therefore, the desire for independence that is typical for family influenced firms and their subsequent preference for internal, rather than external financing restrict the investment decisions of these businesses (Gómez-Mejía et al., 2010). In other words, even though family influenced firms might be able to raise external capital to invest in discontinuous technologies, their efforts to retain a high level of command limits their motivation to do so (Cabrera-Suarez et al., 2001). In turn, these firms tend to invest less aggressively in discontinuous strategic initiatives, which is also in line with the increased dependence on other external stakeholders which also results in decreased resource investment (P1-2b). Thus, I propose,

> P1-2c. Ceteris paribus, family influenced firms adopt discontinuous technologies faster, albeit less intensively, and less flexibly than less or non-family influenced firms mediated by their idiosyncratic resource dependence.

3.3 Family Influence and Political Resistance

Political resistance among middle managers can, as outlined in chapter 2.1.2.2.3, impede an organization's adoption of discontinuous technologies by decreasing its response speed, intensity, and flexibility.

I formally suggest that *continuity* can lower political resistance among middle managers. One reason for middle-managers to hamper the adoption of discontinuous technologies is the fact that their incentive systems are typically tied to short-term performance, whereas investment in discontinuous technologies constitutes a long-term investment, typically with scarce financial success in the early phases (Anderson & Tushman, 1990). In contrast to publically or investor-owned businesses that are driven by short-term earnings expectations, family firms, due to their long-term orientation, have the opportunity to build incentive systems tied to the long-term health of the organization. This, in turn, will lower the motivation of middle managers to engage in political resistance.

Moreover, I formally propose that a high level of *command* lowers political resistance, as such authority forestalls the formation of "turf battles" that decelerate the decision-making and interpretation processes (Hill & Rothaermel, 2003; Pfeffer, 1992). It is crucial to note the differences between the context of this study and the settings of the case based study conducted by Eisenhardt and Bourgeois (1988) who, somewhat contrary to Pfeffer (1992), concluded that central authority is a driver for political behavior. They rationalize this finding as autocratic CEOs "preserve their power through use of politics" (1988: 765)[27]. CEOs of family influenced firms, however, do not need to apply such power retaining measures, as their power is typically set and non-negotiable, for instance because of familial ties with the owners or "relational aspects" in their contracts (Gómez-Mejía et al., 2001). Beyond this formal aspect, high levels of command also have a socio-cognitive impact. As discontinuous technological changes are typically tied to high ambiguity (Tushman & Anderson, 1986), strategic actors engage in what Kaplan terms "framing contests," which are "highly political framing practices to make [one's] frames resonate and to mobilize action in [one's own] favor" (2008b: 729). Prior research has suggested that family influence entails socio-cognitive power in the sense that the frames adopted and communicated by family executives are likely to be adopted more rapidly by their subordinates than the frames communicated by decision makers in other companies (Eddleston, 2008). For instance, Berrone and colleagues (2010: 88) note that "[e]ven in publicly traded firms with greater ownership dispersion, the view of family members as a group are likely to demand a great deal of attention compared with those of nonfamily stakeholders" (see also Chrisman, Chua, & Steier, 2003).

Next, I formally propose that a sense of *community* lowers political resistance. As Christensen and Bower (1996) showed, a frequent issue in the early phases of discontinuous technology emergence is that bottom-up information sharing is disturbed. Based on the high levels of trust and strong ties implied by the concept of *community* (Gómez-Mejía et al., 2001; Miller & Le Breton-Miller, 2005a), I assume members of family influenced businesses to be more likely to propel information from the bottom up, thereby accelerating the decision maker's speed of awareness and interpretation.

---

[27] Eisenhardt and Bourgeois do not provide information on any potential family influence on the organizations they studied.

Finally, I assume that *connections* do not influence the level of political resistance. This is because the term *connections* refer to relationships with externals (Miller & Le Breton-Miller, 2005a), whereas political resistance refers to internal power distribution (Eisenhardt & Bourgeois Iii, 1988). Summarized, I propose that,

---

P1-3a. Family influence decreases the extent of political resistance.

P1-3b. Ceteris paribus, family influenced firms adopt discontinuous technologies faster, more intensively, and more flexibly than less or non-family influenced firms mediated by lower levels of political resistance.

The historic and current culture of the fully family-owned furniture store IKEA illustrates several aspects of lowered political resistance in family firms. First of all, in 1956 an employee generated the idea to sell furniture for self-assembly in handy, flat packaging and propelled his plan directly the CEO; soon this concept became the still enduring and successful business model of the new market leader IKEA.[28] Nowadays, IKEA's culture is characterized by a strong and powerful family CEO, Ingvar Kamprad, top managers that exhibit high levels of loyalty, and a work atmosphere that is described as inclusive by most employees (Reise, 2009).

3.4 Family Influence and Avoidance of Cannibalization

Avoidance of cannibalization, as discussed in chapter 2.1.2.2.4, is a barrier to organizational adoption of discontinuous technologies that affects speed, intensity, and flexibility. Although this construct consists of an economic and a psychological component, my further theorizing will focus exclusively on the latter one. The reason therefore is that economic disincentives to invest in discontinuous technological changes are similar for any profit-seeking incumbent organization, independent of the ownership structure.

First, I suggest that *continuity* does not have a clear, directed impact on the avoidance of cannibalization. Striving for continuity comprises of two distinct components—backward

---

[28] See www.ikea.com, 'about IKEA', 'Our history.'

looking with focus on the family firm's history from the founder's days to present and forward looking with focus on future products and scenarios (Miller & Le Breton-Miller, 2005b). Depending on which of these two perspectives prevails in an industry, striving for continuity might either entail enhanced or alleviated fear of cannibalization.

Furthermore, I theorize that *command* heightens the decision makers' tendency to avoid cannibalization. As discussed in the theory chapter (chapter 2.1.2.2.4), avoidance of cannibalization is closely linked to emotional attachment. Due to high levels of command, family CEOs are inextricably intertwined with the business and identify themselves with the firms, their products and employees, in particular when they have worked for the business for many years. As a consequence, family members involved in the business are emotionally attached not only to the business, but also its products and its employees, and in turn reject any radical product innovations that necessitate shedding of any of those resources. But also non-family top managers, such as heads of business units, are expected to be less willing to cannibalize in family influenced as compared to non-family influenced firms. High levels of command lower the business units' internal autonomy—that is the business managers' extent of authority (Sharma, Chrisman, & Chua, 1997). Internal autonomy, however, is an important antecedent of 'internal markets,' which foster intra-organizational competition, and ultimately enhance business unit managers' willingness to cannibalize existing technologies (Chandy & Tellis, 1998).

Besides, I argue that *community* further strengthens the avoidance of cannibalization due to two factors. First, a sense of community is likely to decrease levels of rivalry among decision makers; rivalry, however, constitutes an enabler for willingness to cannibalize (Chandy & Tellis, 1998). Second, because of family influenced firms' striving to build a community, employee tenure is much longer in family as opposed to non-family influenced firms (Berrone et al., forthcoming), which in turn diminishes willingness to cannibalize as, in those cases, the individuals who are supposed to replace one innovation by another are the same people who previously made a success out of innovations. As Chandy and Tellis put it (1998: 478): "A manager's commitment to a current technology can be high if he or she was involved with the current technology and previously rode its S-curve to success. [...] In that case the manager will be less open to a new or future innovation."

Fourth, I suggest that strong *connections*, particularly close relationships with customers, entail enhanced avoidance of cannibalization. Besides the antecedents of willingness to cannibalize discussed in the above paragraphs, Chandy and Tellis (1998) propose that 'future market focus' helps to overcome fear of cannibalization. Future market focus thereby relates to organizational strategies emphasizing potential future as opposed to current customers. As obtaining a future market focus would require the disturbance or harming of relationships with current customers, it is unlikely that family influenced firms will engage in such behavior. Summarized, I propose,

> P1-4a. Family influence increases the decision maker's tendency to avoid cannibalization.
>
> P1-4b. Ceteris paribus, family influenced firms adopt discontinuous technologies more slowly, less intensively, and less flexibly than less or non-family influenced firms, mediated by higher levels of avoidance of cannibalization.

The recent history of Messer Group, a German firm internationally active in the industrial gas market, is an intriguing example of how family firm owners do not want to 'let their business go.' Even years after sales of the business to externals, members of the previous owning family still felt emotionally attached to the firm and its products and hence tried to restore the former status quo: In 1965, when managed by a second generation CEO, Messer Group merged with the much larger chemical business Hoechst, leaving the owner-family merely some veto rights. When this conglomerate broke down in 2001, Stefan Messer, a third generation family member who had worked for the firm since his youth, perceived this event as an opportunity to bring the company back into the family. As a first step, Stefan Messer redeemed Messer Group with the help of two financial investors in 2001. As a second step, in 2004, the family acquired full control over the business again. In an interview, Stefan Messer rationalized this decision by "continuing, what [his] grandfather had built" and abundance of "lifeblood" (von der Hagen, 2012: 8), hence indicating high levels of emotional attachment to the family business[29].

The claim that family business owners are not only emotionally attached to their firm and its products but also to their employees is supported by a recent study of the *Zentrum für Europäische Wirtschaftsforschung* and *Mannheimer Institut für Mittelstandsforschung* (Gottschalk & Keese, 2011): The results of this study show that in 2009, in the climate of the

---

[29] Quotation originally in German, translated by author of this thesis.

financial crisis, the 500 largest family businesses were more negatively affected by the crisis than the 26 non-family DAX businesses (revenue decline of 10% as compared to revenue increase of 3%), however dismissed less employees (dismissal of 1.2% as compared to 2.3%).

3.5 Family Influence and Rigid Mental Models

Rigidity of mental models, as outlined in chapter 2.1.2.2.5 is a relevant cognitive determinant for the speed, intensity, and routine flexibility of an organization's adoption of discontinuous technologies.

First, I formally suggest that *continuity* enhances the rigidity of decision makers' mental models. One side effect of family influenced firms' striving for continuity is above-average CEO tenures (Berrone et al., 2010; Cruz, Gómez-Mejía, & Becerra, 2010; Schulze et al., 2001). These long tenures ultimately entail rigidity of mental models, manifested in "low willingness to explore new market opportunities (Finkelstein & Hambrick, 1990), [...] rigid decision processes (Katz, 1982), "tunnel vision" (Hambrick & Fukutomi, 1991), and an overly conservative posture" (Gómez-Mejía et al., 2001: 86). As Tripsas and Gavetti (2000: 1158) note "[t]urnover in the topmanagement teams is also an important driver of change."

Moreover, I propose that *command* has no effect on the rigidity of decision makers' and employees' mental models. The underlying rationale is that mental frames are mainly influenced by individuals' beliefs, experience, and training (see chapter 2.1.2.2.5). High levels of command can thus streamline and reinforce existing mental models, however, they do not affect the rigidity of mental models per se.

Next, I argue that *community* increases the rigidity of employees' mental models. The line of argumentation I apply here is similar to the one used for the relationship between *continuity* and rigidity of CEOs' mental models: The above-average length of employee tenure and strong identification with time-honored technologies that are typical in family influenced companies (Haugh & McKee, 2003) can be expected to increasingly freeze cognitive frames and enhance the commitment to path-dependent solutions. Moreover, decision-makers of companies with a strong sense of community are less likely to replace extant employees with new, differently trained staff (Miller & Le Breton-Miller, 2006) with cognitive frames appropriate for the implementation of discontinuous technology, which is often required in order to adopt new, fundamentally different routines (Burgelman, 1994). This outcome is also highlighted by extant literature, for instance Gómez-Mejía and colleagues (2010: 227) observe that family firms are reluctant to create "new routines and modus operandi that stray from the firm's 'true and tried' methods of operation (Eisenmann, 2002)."

Last, I assume that external *connections* do not impact the organizational members' rigidity of mental models. As mentioned above, mental models are affected by individuals' beliefs, experience, and training. Although each individual's mental models can be influenced by externals, there is no a priori association between external organizational ties and organizational members' mental models, as, depending on the specifics of the firm contacts and partnerships, the direction of such influence can be in either direction.

In summary,

> P1-5a. Family influence increases the decision makers' and employees' rigidity of mental models.
>
> P1-5b. Ceteris paribus, family influenced firms adopt discontinuous technologies more slowly, less intensively, and less flexibly than less or non-family influenced firms mediated by more rigid mental models.

Several bankruptcies among traditional German family firms within recent years anecdotally support the hypotheses that family firms' rigid mental models impede organizations' adaptation to changing environments. Model railroad manufacturer Märklin filed for bankruptcy in 2009, in the year of its 150th anniversary. Besides cash flow issues caused by the financial crisis, experts attribute most of Märklin's failure to the organization's narrow focus on existing business models rather than broadening it by non-paradigmatically exploiting new market segments (Petzold, 2009). A backward looking focus resulting in bounded cognitive frames was the reason for Schimmel's insolvency. The German newspaper *Handelsblatt* commented on the failure of the piano manufacturer (Weissenborn, 2010): "For too long, piano manufacturers [in Europe] have been working at a point [in the value chain] where radical innovation was unnecessary. However, in the meantime, innovative low cost competition from Asia—at which [players like Schimmel] had been smirking—have become nearly invincible."[30]

3.6   Family Influence and External Influence

External influence, as described in chapter 2.1.2.3.1 is an effective enabler for quick and intense adoption of discontinuous technologies since it helps to quickly create awareness of the technological discontinuity and interpret the environmental shifts adequately. Moreover, bringing outsiders in during the implementation phase can help to break established routines and to replace them with new, non-paradigmatic ones.

First, I argue that focus on *continuity* and high levels of *command* jointly lower the decision makers' willingness to involve externals. Members of owner-families do not only possess and exert high levels of command; they also strive to maintain that level of command for the future (Miller & Le Breton-Miller, 2005a). In particular, passing on high levels of control over the business to subsequent generations constitutes an important part of non-economic utility function of the owner-families (Gómez-Mejía et al., 2007). Consequently, family members are likely to be reluctant to 'dilute' the family influence by involving outsiders as "[h]iring outside managers may also increase information asymmetries and goal conflict, further eroding family SEW" (Gómez-Mejía et al., 2001: 227). Put differently, any delegation of responsibility and influence to externals is in stark contrast to the family firms' striving for continuity and high levels of command. This is in line with empirical evidence of Gómez-Mejía and colleagues (2007) that show the reluctance of Spanish family firms to engage in benevolent alliances. I expect this relationship to be specifically pronounced for granting outsiders access to the 'inner circle' of the company, i.e. involving them in the implementation phase rather than merely asking for advice.

Moreover, I propose that *community* further decreases the level of external influence. Organizational members of family influenced firms often become a "pseudo-family" (Tan & Fock, 2001: 128) with "thick social walls" (2005: 250) that separate insiders of the organization from outsiders. I propose that this isolation of family firm members has two important consequences: First, it lowers employee motivation to involve externals as they are perceived as 'outsiders' rather than 'pseudo-family members' (Carney, 2005). Second, 'thick

---

[30] Quotation originally in German, translated by author of this thesis.

social walls' mitigate the contingencies of proactive outsiders' abilities to exert influence on the family influenced firm, e.g., by giving relevant advice within the interpretation processes.

Next, I suggest that strong *connections* additionally lower external influence absorbed by the organization. Family firms prefer connections to a few, selected partners. Consequently the base of external individuals potentially providing input throughout the various phases of adoption is much smaller for family influenced businesses than it is for non-family influenced ones, which, ceteris paribus, sustain relationships with a larger number of externals than family influenced firms do. Accordingly, the smaller base of potential providers of external input is likely to result in a reduced amount of relevant and knowledgeable external input for family influenced firms as opposed to non or less family influenced ones. In summary,

---

P1-6a. Family influence decreases the level of external influence.

P1-6b. Ceteris paribus, family influenced firms will adopt discontinuous technologies later, less intensively, and less flexibly than less or non-family influenced firms mediated by lower levels of external influence.

P1-6a is in line with a large body of extant literature on family business behavior and preferences (e.g., Schulze et al., 2003a; Schulze et al., 2001). For instance, Gómez-Mejía and colleagues (2010: 224) note that "family firms are less likely to incorporate outsiders' perspectives and opinions in their decision making." Building on a large data set of Spanish oil mills, these authors empirically show that family firms are more reluctant to join alliances than non-family firms.

Moreover several real world cases exemplify this phenomenon. For instance, Anton Schlecker, former CEO of the meanwhile bankrupt drugstore chain Schlecker and known for striving for maximum command, was long known as 'resistant to advice.' This resistance contributed to a lack of organizational adaptability in the changing retailing market, which ultimately resulted in insolvency (Hirn & Jensen, 2011). A somewhat similar case is the rise and fall of the Swiss Erb-Group. Founded in the 1920s as a garage, the family firm rapidly grew to a diversified group with more than one billion EUR in annual sales. However, while the family owners sealed themselves off from external advice, firm operations became more and more characterized by acquisition of businesses without thorough strategic intentions. The owner CEO was later described as follows (Triebelhorn, 2012): "Instead of competence of managers, he relied on family ties[31]." As a result of the ongoing myopic investment, Erb Group went bankrupt in 2003 (Triebelhorn, 2012).

The effects summarized in P1-6b are not supposed to be of equal strength. Admitting externals access to the intimate core of the business and allowing them to change routines contradicts family firms' preferences for *continuity*, *command*, *community*, and *connections* to a much higher degree than simply taking selected pieces of advice. Thus, referring to P1-6b, I assume the effect of family influence on routine flexibility to be strongest and the effect of family influence on speed (particular speed of awareness) to be weakest.

3.7 Family Influence and Structural Ambidexterity

Structural ambidexterity, as discussed in chapter 2.1.2.3.2 is an effective enabler for intense and flexible adaptation to discontinuous technological change.

First, I suggest that *continuity*, which mainly refers to goals and intentions of family influenced firms, has no effect on structural organizational ambidexterity. The underlying rationale for this assumption is that striving for continuity mainly refers to the time horizon of goals and intentions of organizational members, in particular family decision makers, whereas structural ambidexterity refers to the design of the organizations[32].

Second, I formally suggest that *command* lowers the level of structural ambidexterity as decision makers in family influenced firms are unwilling to delegate part of their responsibility and to decentralize their decision-making (Berrone et al., forthcoming). Although ambidextrous structures are ultimately controlled by top management, the creation of spin-offs to develop and commercialize discontinuous innovations largely independently of the parent organization, requires that top managers compromise on hierarchy and provide such subsystems with substantial decision-making autonomy (O'Reilly & Tushman, 2008;

---

[31] Quotation originally in German; translated by author of this thesis.
[32] *Continuity* might have a non-negative effect on contextual ambidexterity (Gibson & Birkinshaw, 2004), however, the direction of this potential effect is unclear: Several empirical studies on socioemotional wealth (e.g., Gómez-Mejía, Hoskisson et al., forthcoming) as well as conceptual work (e.g., Carney, 2005) indicate that family influence is tied to exploitation rather than exploration. To the contrary, organization theorists argue that exploitation is focused on short-term goals whereas exploration is connected to long- term success (March, 1991), which is a primary characteristic of family influenced firms (e.g., Miller & Le Breton-Miller, 2005). Due to this theoretical equivocality, I focus my theorizing on structural aspects of ambidexterity.

Tushman & O'Reilly, 1996). Therefore, ambidextrous structures are a priori in conflict with the family influence-imposed efforts to maintain high levels of command.

Third, I argue that a family firm's sense of *community* might further decrease the level of structural ambidexterity within an organization. Sense of community, by definition, refers to the intense ties among employees across hierarchies and departments present in family influenced organizations (Miller & Le Breton-Miller, 2005a). Such intimate bonds throughout the entire organizations are most likely to be established when the levels of communication, cooperation, and collocation are high within the firm. These conditions become violated when organizational structures are decoupled, that is when cooperation and communication are largely restricted within business units which might even be geographically separated (Gilbert, 2005). Thus I expect an organization's sense of community to be in contrast to building up ambidextrous structures.

Finally, I suggest that external *connections* do not impact on organizational ambidexterity. This proposition is rooted in the fact that connections, as defined here, relate to external stakeholders (Miller & Le Breton-Miller, 2005a), whereas structural ambidexterity is a purely internal organizational aspect. Summed up, I hence propose,

> P1-7a. Family influence lowers the level of structural ambidexterity.
>
> P1-7b. Ceteris paribus, family influenced firms adopt discontinuous technologies less intensively and less flexibly than less or non-family influenced firms mediated by lower levels of structural ambidexterity.

Also in practice, family firms are often depicted as centralized corporations counteracting the principles of structural ambidexterity. For instance, the German magazine *Wirtschaftswoche* recently entitled the traditional German family business Krupp, a player active in the steel industry, as a 'centralistic' business (Wildhagen, 2011). Owner managers of Dielmann, one of the top 15 shoe-sellers in Germany, perceive themselves as 'fanatic centralists' (Jaeger, 2010).

3.8 Integrated Effects

In summary, propositions P1-1 to P1-7, as derived in the previous sections of this thesis, outline a fundamentally different adoption pattern of highly family influenced firms when confronted with discontinuous technological changes as compared to less or non-family influenced businesses. A summary of the effects of each of the 4Cs on each of the barriers and enablers is provided in Table 6. In the following chapter, I aim to synthesize the findings referring to each individual barrier and enabler and to provide integrated propositions on the impact of family influence on speed, intensity, stamina, and flexibility of adaptation.

Table 6: Summary of Effects of 4Cs on Barriers and Enablers of Adoption

Barriers & Enablers	*Continuity*	*Command*	*Community*	*Connections*	**Total Effect**
Formalization	decreased	decreased	decreased	decreased	**decreased**
Resource Dependence on capital providers	decreased	decreased	unaffected	unaffected	**decreased**
Resource Dependence on other external stakeholders	increased	unaffected	unaffected	increased	**increased**
Political Resistance	decreased	decreased	decreased	unaffected	**decreased**
Avoidance of Cannibalization	ambiguous effect	increased	increased	increased	**increased**
Rigid Mental Models	increased	unaffected	increased	unaffected	**increased**
External Influence	decreased	decreased	decreased	decreased	**decreased**
Structural Ambidexterity	unaffected	decreased	decreased	unaffected	**decreased**

3.8.1 Impact of Family Influence on Speed of Adoption

At a glance, my theorizing provides mixed results regarding the speed of adoption in family influenced firms: Propositions P1-1b, P1-2c, and P1-3b propose a fast adoption in family influenced organizations, whereas P1-4b, P1-5b, and P1-6b argue for a slow adoption. These apparent contradictions can be relaxed when analyzing the speed of adoption along the process outlined in

Figure 4. Rigid mental models (P1-4b) and a lack of external influence (P1-5b) are potential inhibitors of early awareness of emerging technological discontinuities. However, delayed awareness less frequently hampers adoption than deferred interpretation and decision-making (Christensen, 1997). Interpretation of the discontinuous technology, or in other words, answering the questions as to whether the innovation is relevant for the company and whether it should be adopted, is affected by two opposed forces. Cognitive-emotional aspects might hamper a quick interpretation as the relevant innovation path to follow (P1-4b, P1-6b). This effect is further enhanced by the absence of counter steering external influence (P1-5b). The degree of cognitive-emotional blockades, however, is not solely driven by family influence; this relationship is rather moderated by important personality dimensions of the dominant coalition, in particular its 'inherent openness' (see Chapter 3.9). Structural, strategic, and political factors allow family influenced firms to adopt innovations in an accelerated way, because interpretation can take place independent of bureaucratic evaluation procedures (P1-1b), without considering the reluctance of stock owners to sacrifice short-term profit (P1-2c), and with a low level of internal resistance based on politics (P1-3b). Once the innovation is interpreted as relevant, the dominant coalition of family influenced firms can quickly decide how to adopt the innovation due to their high levels of external autonomy and internal control (P1-2c, P1-3b). High levels of internal control also allow for a fast implementation. Summarized, I propose

P1-8a. (i) Family influence has no a priori effect on speed of awareness and speed of interpretation of discontinuous technologies. (ii) Ceteris paribus, family influenced firms make faster decisions to adopt discontinuous technologies and faster implement those technologies.

Extant literature on the strategic behavior of family businesses is inconclusive regarding the speed of those organizations' reactions (Miller & Le Breton-Miller, 2005a; Miller et al., 2008). My detailed investigation provides a more nuanced picture of family firms' adaptation patterns, thereby identifying determinants that either promote or foster adaptation. This lays the ground for further empirical research investigating the speed of adaptation of family firms.

3.8.2 Impact of Family Influence on Intensity of Adoption

Arguments relating to the intensity of adoption need to be considered in a hierarchical manner: Most important, there are barriers to adoption that restrain the total amount of resources available for allocation (resource dependence). Next, factors exist that affect the dominant coalition's decision-making and initial resource allocation process (formalization, resource dependence on external stakeholders, avoidance of cannibalization of the decision makers, rigid mental models of the decision makers, external influence). Lastly, some barriers to adoption further diminish the amount of resources actually allocated to the commercialization of the technology despite initial commitment (Christensen, 1997; Gilbert, 2005) throughout the course of the implementation phase (rigid mental models of employees, political resistance, structural ambidexterity). An illustrative way to visualize these steps is the analogy of a funnel that has a certain width at the top (the amount of resources available for allocation; $R_{stock} + R_{new}$) and subsequently narrows down to its final output ($R_{discontinuous}$). As all aspects referring to the 'top' and the 'middle' (except formalization) of the funnel, such as resource dependence (P1-2c), avoidance of cannibalization (P1-4b), rigid mental models of decision makers (P1-5b), and external influence (P1-6b) point to a less intense adoption in family influenced firms, I propose

> P1-8b. Ceteris paribus, family influenced firms adopt discontinuous technologies with less resource intensity than non or less family influenced firms.

In other words, the fact that less political resistance in family influenced firms leads to less re-shuffling of resources back to the established technology, does not affect the overall negative impact that family influence exerts on intensity of adoption, due to the much smaller advance base of resources. In this context, it is specifically important to note the difference between motivation and capabilities (Chen, 1996). Given market efficiency, family influenced firms are to a large degree *able* to raise external capital which can subsequently be invested in the commercialization of the discontinuous technology (Hustedde & Pulver, 1992; Romano, Tanewski, & Smyrnios, 2000). However, I propose that despite this general ability, they lack motivation to do so, meaning they are *unwilling* to take out loans or issue shares because these activities would be in stark contrast to their striving for independence (Mishra & McConaughy, 1999).

3.8.3 Impact of Family Influence on Stamina of Adoption

Regarding the stamina of adoption, family influence results in more persistent investments. This pattern can be traced back to two main reasons. First, due to less formalization and a focus on long-term goals, it is less likely that an investment will be seen as a 'failure' in early phases of technology commercialization in which no dominant design has yet emerged (Anderson & Tushman, 1990) and consequently be stopped (P1-1b). Second, even if the strategic investments are perceived as (temporary) failures, the high levels of independence present in family influenced firms allow for further activities (P1-2c). Summarized,

> P1-8c. Ceteris paribus, family influenced firms adopt discontinuous technologies with more stamina than non or less family influenced firms.

Extant literature on family business has already investigated this phenomenon in other contexts and labeled it 'patient capital' (e.g., Arregle et al., 2007; Habbershon & Williams, 1999; Ward, 1997), however, it has not yet been linked to the emergence of discontinuous changes. Due to the specific technological cycle of evolution including long periods of lack of financial success before a dominant design emerges (Tushman & Anderson, 1986), 'patient capital' is specifically important under those discontinuous circumstances.

Gómez-Mejía et al. (2001) provided empirical evidence on the owners' ways to deal with failures (absent discontinuous innovations). Those authors emphasize the "managerial entrenchment, which refers to executives' holding their jobs past the point where their stewardship is beneficial to owners" (2001: 84). As a consequence of owners' unwillingness to assess the CEOs' performance based on current (short-term) performance and penalize apparent failure, these authors conclude that "assessment of agent performance under family contracting may shift negative performance attributions from the agent to exogenous forces" (Gómez-Mejía et al., 2001: 84). My theorizing, however, shows that this idiosyncratic behavior, perceived as a core disadvantage by Gómez-Mejía et al. (2001), turns into a core advantage in times of discontinuous change.

3.8.4   Impact of Family Influence on Routine Flexibility

Propositions referring to the flexibility of routines in family influenced firms adopting a discontinuous technology are ambiguous: Aspects relating to the environment (P1-2c, P1-6b) and cognitive-emotional aspects (P1-4b, P1-5b) point to a low level of flexibility, whereas internal, political factors (P1-3b) propose the opposite. The effect of structural factors is equivocal with low levels of formalization indicating high (P1-1b) and low levels of structural ambidexterity (P1-7b) suggesting low flexibility. To solve these contradictions, the analogy of the 'funnel,' as introduced in Chapter 3.8.2 is helpful again: Fewer defaults of processes, less bureaucracy, and internal politics hamper flexibility in family firms less than they hamper flexibility in non or less family influenced firms. However, the basic level of flexibility in such firms is lower, that is the 'width of funnel' is narrower in family influenced firms, due to cognitive-emotional restrictions and external pressure induced by valued stakeholders. I thus propose

> P1-8d. Ceteris paribus, family influenced firms adopt discontinuous technologies with less flexible and less non-paradigmatic routines than non or less family influenced firms.

This is in line with recent work of Gómez-Mejía and colleagues (2010: 227): "[T]he family firm (especially with a family CEO) will be more likely to stay closer to the ,core' because it is an option that carries less anxiety and one that given prior experience, feels more comfortable to the dominant coalition." More recent work (Gómez-Mejía et al., 2011b) builds on these assumptions suggesting that family firms are less active in technological diversification. However, my theorizing provides a more nuanced picture by showing that family firms do not *per se* refrain from non-core activities related to discontinuous technologies, but rather engage in them in a less intense and less flexible way. Moreover I propose that the specific risk preference of family influenced firms is not the only factor driving such inflexible behavior. Rather, I provide alternative explanations such as 'perceived resource dependence on external stakeholders' thereby broadening the range of theories applicable to explain the strategic behavior of family influenced firms. In addition I argue that despite the overall reluctance of family firms to adopt new technologies flexibly, there are indeed some family firm characteristic factors such as low formalization and diminished political resistance that counteract the dominant drivers of incumbent inertia and thus foster flexible adoption of new technologies.

3.9 Moderating Effects of the Inherent Openness of the Dominant Coalition

A fundamental premise in my theorizing is that increases in family influence imply increasing levels of command of the dominant coalition—"the powerful actors in an organization who control the overall organizational agenda." As highlighted in upper echelons theory (Hambrick & Mason, 1984), the more influence key decision makers wield in an organization, the more their experiences, values, and personalities shape organizational outcomes (Finkelstein, Hambrick, & Cannella, 2009). In addition, personality-outcome associations are amplified in ambiguous and uncertain situations (Mischel, 1977). Therefore, the characteristics of the members of the dominant coalition are likely to affect organizational responses to technological discontinuities, particularly in family influenced firms (Minichilli, Corbetta, & MacMillan, 2010).

Adopting the lens of upper echelons theory, I maintain that when building a theory on the effect of family influence on firms' adoption of a discontinuous technology, one characteristic of the dominant coalition is pivotal: inherent openness. By 'inherent openness of the dominant coalition' I refer to the level to which a company's dominant coalition, due to generic individual level and group level attributes, thinks and acts creatively and open-mindedly while stimulating path-divergence in its organization (Bantel & Jackson, 1989).

The inherent openness of the dominant coalition is, first, determined by the characteristics of individual members (Wiersema & Bantel, 1992). Certain demographic characteristics of key decision makers, such as high levels of formal education and functional backgrounds in R&D or marketing, have been found to engender a higher propensity for innovation (Barker Iii & Mueller, 2002). An even more fundamental ingredient to the openness of the dominant coalition are high levels of individual 'openness to experience' (Lepine, Colquitt, & Erez, 2000)—the coalition members' generic "receptivity to learning, new experiences, novelty, and change" (Lounsbury, Smith, Levy, Leong, & Gibson, 2009: 201).

Second, heterogeneity among the various members of a dominant coalition seems to be conducive to the group's overall openness (Tushman & Rosenkopf, 1996; Wiersema & Bantel, 1992), at least as long as any interpersonal differences stemming from heterogeneity can be managed (Williams & O'Reilly, 1998). Heterogeneity drives openness because more diverse knowledge bases, skills, and perspectives within a top-management team increase the group's propensity for path-divergent thinking (Cannella, Park, & Lee, 2008). In family influenced firms, heterogeneity in the dominant coalition is often fostered through participative generational involvement, which refers to the number of family generations that simultaneously have a fundamental stake in a family influenced firm's strategic decision making (Chirico, Sirmon, Sciascia, & Mazzola, forthcoming; Kellermanns, Eddleston, Barnett, & Pearson, 2008).

The inherent openness of the dominant coalition is fundamental to my model. In a highly family influenced company, a dominant coalition that is generally open to innovation is likely to use its influence to speed up and persistently pursue the implementation of discontinuous technology. In contrast, a 'closed' coalition is likely to use its authority to protract, reduce, or even stall such initiatives. Furthermore, 'open' individuals actively seek pleasure and risk (Judge, Erez, Bono, & Thoresen, 2003). Therefore, an 'open' dominant coalition is more likely to trade off the risk of losing control for the opportunity to invest in discontinuous, risky innovations than a 'closed' dominant coalition. In addition, decision makers' openness is also positively associated with routine flexibility (Nadkarni & Herrmann, 2010). Thus, the openness of the dominant coalition is likely to reinforce the positive effects of *command* on adoption speed and stamina, and to alleviate the negative effects of *command* on resource and routine rigidity.

Moreover, the openness of the dominant coalition is likely to influence effects of the family-induced *continuity*- and *community* focus on technology adoption. As DeYoung, Peterson, and Higgins state, "Open people are permeable to new ideas and experiences; they are motivated to enlarge their experience into novel territory" (2005: 830). Thus, an 'open' dominant coalition increases the probability that a company chooses path-divergent initiatives to ensure *continuity*, and thus strengthens the positive effect of *continuity* on adoption speed and stamina while attenuating the effect of *continuity* on adoption intensity. Moreover, as 'open' individuals show a "breadth of interest [...] and permeable boundaries" (Dollinger, Urban, & James, 2004: 37), openness of the dominant coalition relaxes the routine rigidity that stems from the entrenched identities of family influenced companies. In comparison with 'closed' key decision makers, 'open' decision makers are also more likely to leverage the opportunities to generate knowledge regarding discontinuous market developments that result from the more intense cross-level communication within the family influenced firm *community*. This tendency, in turn, strengthens the positive association between *community* and decision speed, while it reduces the propelling effect of *community* on resource and routine rigidity.

Finally, the openness of the dominant coalition is influential because it determines whether the associations between family influence-induced *connections* and the four dimensions of adoption of discontinuous technological change are positive or negative. All else aside, a coalition of 'open,' but in other respects heterogeneous decision makers is likely to establish a range of broad, externally-oriented networks, especially with players from other technological and industrial domains (Athanassiou & Nigh, 1999). The broadness of firm ties is important because social capital can become a social liability if the network of an incumbent is dominated by proximate actors within the established industrial 'macroculture' (Abrahamson & Fombrun, 1994) who are constrained by the same economic rationales and traditional cognitive schemata as the focal firm. Broader connections facilitate earlier awareness of discontinuous technological change because they extend a firm's screening radius to outsiders at the peripheries of industries (Gilbert, 2005; McDonald et al., 2008), which is where discontinuous technological changes typically emerge. In addition, a narrow network of relationships with established customers, suppliers, competitors, and complementors reinforces resource and routine rigidity as these contacts are typically unwilling or unable to offer new, radically different inputs or to cooperate in the commercialization of non-paradigmatic innovations (Christensen, 1997). Along the same vein, a broad network decreases the likelihood that companies will stall adoption initiatives on the basis of negative feedback from partners who oppose adopting the new technology. In other words, strong, family-induced connections with other actors can serve as enablers of change, but only if they are broad enough to include relationships with actors outside the established innovation ecosystem.

To sum up, incumbent organizations led by 'open' dominant coalitions are more likely than those led by 'closed' coalitions to overcome barriers to adoption and to install structures that enable adaptation. The strength of this effect is reinforced by increases in family influence, because the characteristics of the upper echelons are more important in family influenced organizations and because various effects induced by the openness of the dominant coalition interact with the manifestations of family influence.

P1-9a/b/c/d. The higher the level of the inherent openness of a firm's dominant coalition, (a) the stronger the positive association between family influence and the speed of adoption; (b) the weaker the negative association between family influence and the intensity of adoption; (c) the stronger the positive association between family influence and the stamina of adoption, and (d) the weaker the negative association between family influence and the routine flexibility

3.10 Phase Dependency of Adoption Patterns

So far, I have assumed a constant or step-like commitment of resources, as well as a stable level of routine rigidity. However, this simplification is likely to not hold true in practice as real life examples show more complex investment patterns that alter with time. For instance it is thinkable that a company could apply a 'wait and see' strategy, starting with a low resource commitment that is enhanced over time as the technology develops further. Vice versa, it is also conceivable that an organization could start with a significant investment, be it out of enthusiasm or threat, which abates over time due to initial failures and/or decreasing motivation.

I suggest that, over time, changes in resource commitment to the new technology are mostly caused by changes in the decision makers' framing of the discontinuous technology. Prior research (Gilbert, 2005) has identified 'threat-framing' as a potential relaxant of resource rigidity while at the same time aggravating routine rigidity. Building on this work, and extending it, König (2009) identified two dimensions of framing (gain vs. loss; high vs. low perceived control) as determinants of adaptation (Dutton & Jackson, 1987; Lazarus & Launier, 1978): Given threat or loss framing, perceived levels of control are associated with resource intensity in a reverse U-shaped curve, whereas this relationship is positive and linear under gain framing conditions (König, 2009). Although abundant empirical evidence (Gilbert, 2005; König, 2009) indicates that framing changes over time, to the best of my knowledge, no study has thus far systematically investigated the evolutionary development of different variants of 'framing[33].'

---

[33] I refer to framing in the same way as Gilbert (2005) and König (2009) do. Recent conceptual work of Kaplan and Tripsas (2008) links framing to the technological cycle proposed by Tushman and Anderson (1986). Kaplan and Tripsas propose a mutual interdependence of framing and technological outcome and suggest that frames alter throughout the evolutionary life cycle. These authors build heavily on the theory of bounded

Nevertheless, alterations to the decision makers' framing of the discontinuity over time are important, particularly when investigating behavioral differences between highly family influenced and non or less family influenced firms. As explained in Chapter 2.2.2.2, family firms possess a different reference point for strategic decision-making, induced by socioemotional wealth considerations (Gómez-Mejía et al., 2007). Consequently, family firms do not only assess alternative options as a gain or loss in terms of financial performance, but also as a gain or loss in their non-economic utility function (Gómez-Mejía et al., 2007).

Figure 11: Phase Dependent Resource Investment Patterns

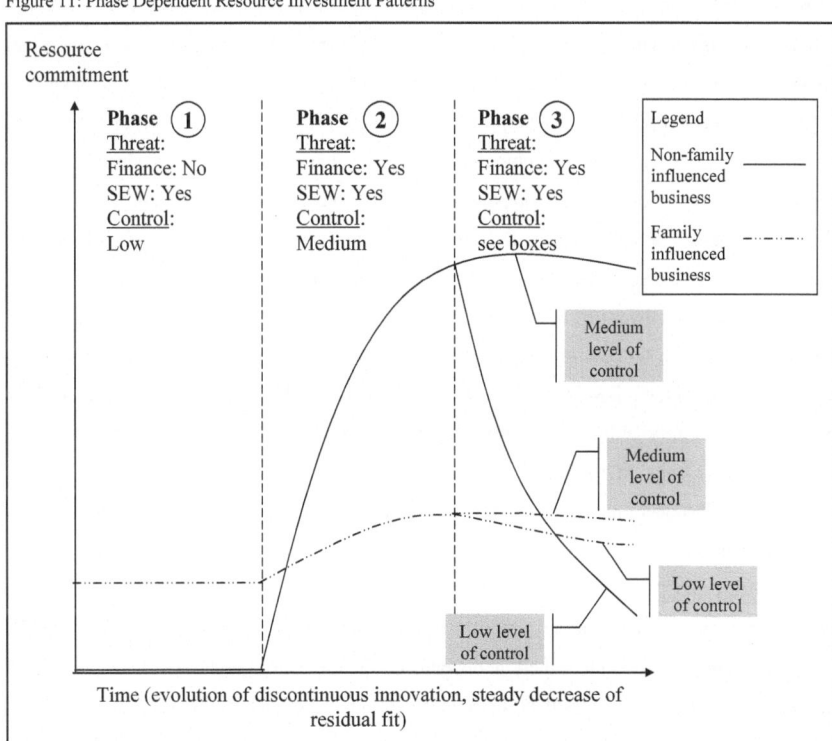

---

rationality (March & Simon, 1958) linking frames to an organization's history and previous industry affiliation. For the effects identified by Gilbert (2005) and König (2009), however, a prospect view (Tversky & Kahneman, 1974) differentiating between gain/loss-framing is more adequate.

In the following chapter, I present a first model (see Figure 11) referring to the phases of framing when a discontinuous technology emerges, based on literature on SEW and theory on discontinuous technological changes[34]: When a technological discontinuity initially emerges (phase 1 in the figure), it frequently attracts only customer groups at the low end of a market or in market niches (Christensen & Bower, 1996). Thus, the new technology has only a scarce direct impact on the short-term financial performance of incumbent players (Anderson & Tushman, 1990) and ultimately entails only minimal levels of threat for non-family influenced firms. Consequently, numerous decision makers ignore discontinuous technologies in these early phases of perceived low levels of threat and control[35] (Kaplan & Tripsas, 2008). Despite lacking the immediate impact on financial performance, family influenced firms might perceive the discontinuity as a threat to their future socioemotional wealth. As transgenerational control and reputation are important to decision makers in family influenced firms (Gómez-Mejía et al., 2007), I propose that those organizations are more sensitive to emerging technologies with the potential to disrupt the market. This supports the notion of P1-8a, proposing a fast technology adoption by family influenced firms. However, though present, the threat to SEW is rather vague in the early stages. Indeed, it is not even possible to foresee the future market potential of the discontinuity ex ante (e.g., Hill & Rothaermel, 2003); hence I expect the level of the threat to SEW to be rather low. Together with the low levels of perceived control that accompany the high uncertainty and lack of dominant design in the emerging market and the general resource scarcity prevalent in family influenced firms (Carney, 2005), this results in timid resource commitments (König, 2009), in line with P1-8b.

In a subsequent phase (phase 2 in the figure), established customers begin to switch to the new technology (Christensen, 1997). Consequently, profit pools of established players start to shrink and thus induce rising threat perception (Gilbert, 2005). Threat perception, in turn, triggers aggressive resource commitment in non-family influenced firms (Gilbert, 2005). The exact height of investment thereby depends on the degree of perceived control as well as the intensity of threat—factors that are determined, among others, by the pace of technology development and intra-organizational factors such as stock of available knowledge and competences and prior experience with changing environments (Hill & Rothaermel, 2003).

---

[34] As routine and resource rigidity are somewhat intertwined (König, 2009) I will concentrate my theorizing in this chapter on resource rigidity, leaving the phase-dependent effects on routine rigidity for further research.

[35] As the commercialization of a discontinuous technology requires fundamentally different competences and the future dominant design is still largely unclear in those early phases (Anderson & Tushman, 1990), the level of perceived control is likely to be low among incumbent organizations. Hence, an investment of incumbent players based on opportunity framing is unlikely (König, 2009).

Response to threat perception is somewhat different in family influenced firms. As the market potential and future prevalence of the discontinuity become clearer in this phase, family- influenced firms' threat to SEW remains high and might even become more pressing[36]. However, due to the above-mentioned 'parsimony' (Carney, 2005) rooted in those organizations' striving for control (e.g., Carney, 2005; Gómez-Mejía et al., 2007; Miller & Le Breton-Miller, 2005a), there is typically little space for family companies to further augment resource commitments in the new technology without contravening their inherent desire to maintain control. Summarized, it is likely that the resource investment of non-family influenced firms will exceed the investments of their family influenced counterparts in that phase of the technology department.

Divergent pathways are conceivable for the further stages (phase 3 in the figure) in the technological development cycle. Two potential developments are possible: The first one assumes that, due to routine rigidity, commercialization attempts of incumbents fail. A logical consequence of such failures is a decrease in perceived control that gives a sense of inability to shape the future development of the market, often combined with a feeling of resignation. According to König (2009), organizations that perceive a technology as threat, while at the same time perceive low levels of control, suffer from enhanced resource rigidity. As decision makers in organizations engage in continuous, iterative interpretation and decision-making processes on resource allocation in general, and resource investments in discontinuous technologies in particular (see Chapter 2.1.2.1.3), such a decrease in perceived control is likely to reduce resource investments of incumbents (König, 2009). This line of argumentation in principle holds true for organizations, regardless of their ownership structure and level of family influence. However, the potential loss at stake is much higher for family businesses compared to non-family ones. This is because the financial wealth of the owner-family is tied to the success of the family business. Moreover, unlike external managers, family CEOs cannot leave the company but are forced to become the 'last man standing,' an effect that researchers term as 'caught in family handcuff[s]' (Berrone et al., 2010; Gómez-Mejía et al., 2007). Business failure and particularly bankruptcy of the firm endanger the family members' reputation, in particular when the firm name is identical to the family name (Berrone et al., 2010; Sirmon & Hitt, 2003; Zellweger, Eddleston, & Kellermanns, 2010).

---

[36] Family influenced firms also perceive threats to their financial performance; however, prior research has shown that threats to SEW are more relevant for decision makers in family influenced firms. This holds true regardless of the family status of the CEO. Berrone et al (2010: 105) state: "socioemotional motives are so strong among family owners that it does not matter if the CEO is a family member or if the CEO serves as a board when it comes to institutional responses."

Hence, I assume that family owners, despite decreasing levels of perceived control, will strive to maintain their investments in the new technology, even in times of (temporary) failure (compare P1-8c), and thus the investment of family influenced firms will decrease less compared to non-family firms.

The second conceivable path is deduced from the premise that (some) incumbents could be apt to successfully commercialize the new technology. In this case, the level of perceived control among organizational members of established organizations is likely to be higher than assumed in the pathway of failure[37] described above. As argued by König (2009), the combination of threat perception and medium levels of control leads to an intensified resource commitment, particularly in non-family firms because due to their inherent resource scarcity (Carney, 2005) family influenced firms will commit less financial resources than companies with low or no family influence. Figure 11 summarizes the proposed temporal investment patterns described in this chapter.

3.11 Concluding Remarks

3.11.1 Summary

My initial goal was to enhance the understanding of how varying levels of family influence lead to different firm responses to discontinuous technologies. To address this question, I adopted the perspective of family business research, which emphasizes *continuity*, *command*, *community*, and *connections* as the main manifestations of family influence. I subsequently explored how these four characteristics affect the fundamental barriers to, and promoters of, new technology adoption that are described in discontinuous technological change literature.

The combination of family business literature with research on organizational adaptation allowed me to propose a model of a specific response pattern that is induced by family influence in situations characterized by discontinuous technological change. This response pattern is summarized in Figure 12. I argue that companies with high levels of family influence will adopt emerging discontinuous technologies quicker than companies with

---

[37]According to König's approach (2009), I expect a moderate level of perceived control for those companies. The level of control is not low due to the organizations' initial successes with commercialization. The level of control, however, is not extensively high either, due to the characteristics and uncertainty inherent in any discontinuous technology. Any company that has already invested resources and management attention in the commercialization of the respective technology—and I assume any company successfully commercializing the innovation to be one of those—will know about the complex nature of the technology and lack of precise predictability, and will therefore not perceive extensively high levels of control.

low levels of family influence (P1-8a). However, the speed of adoption will not necessarily co-vary on par with the intensity or flexibility of the family influenced businesses' responses. Rather, high levels of family influence engender a 'silent' introduction approach, as family owners aim to protect their families' fortunes and independence for future generations (P1-8b). Furthermore, I extend the period of time under consideration by proposing that businesses with higher levels of family influence are more likely to continuously invest in discontinuous technologies, even if initial designs of the innovation fail (P1-8c). Finally, I predict that the adoption initiatives in companies with higher levels of family influence will be more closely tied to the established technological trajectory than the adoption initiatives seen in less family influenced companies (P1-8d).

Guided by upper echelons theory and by family business research emphasizing the heightened importance of key decision-makers in family influenced businesses (Minichilli et al., 2010), I incorporate one additional pivotal determinant into my theorizing: the inherent openness of the dominant coalition. Openness not only amplifies the positive effect of family influence on adoption speed and stamina but also relaxes the tendencies of family influenced firms to under-invest in discontinuous technological change and to adhere to path-congruent routines (P1-9a/b/c/d).

Figure 12: Family Influence and Adoption of Discontinuous Technologies (Theoretical View)

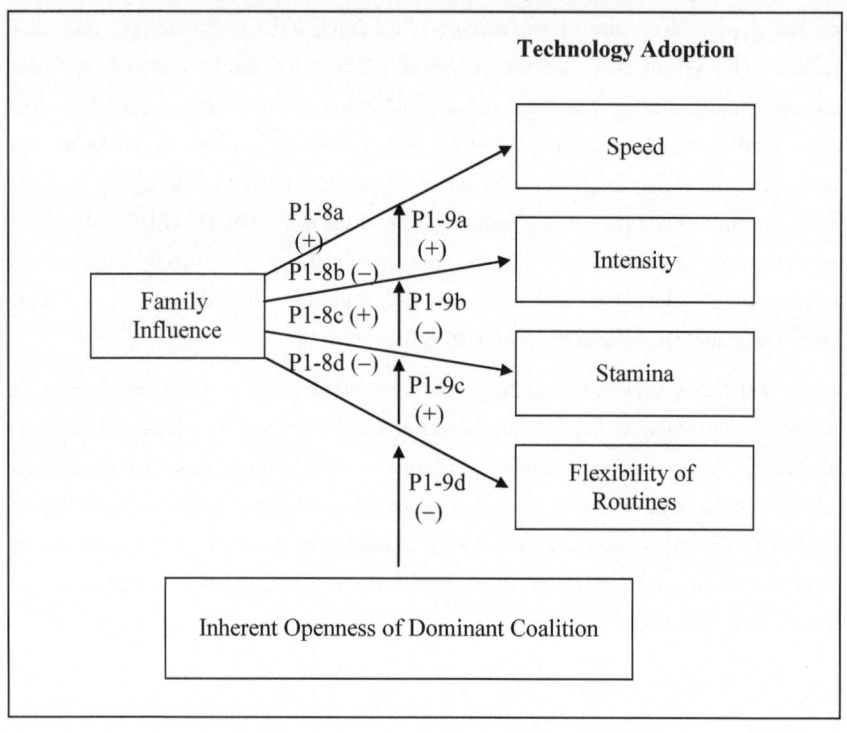

3.11.2 Contributions

My findings add to the research on organizational adaptation by providing a new and more nuanced explanation for the varying responses of incumbent firms to discontinuous technologies. The relevance of the social context of owners—particularly family owners— has been largely neglected in prior studies on incumbent inertia. This is a notable research gap given that family owners think and act differently than other owners, and as family influenced businesses contribute significantly to our economies worldwide (Anderson & Reeb, 2003).

A key contribution of my hypothesizing is rooted in the cross-pollination of discontinuous technological change literature with family business theory, which is a research domain characterized by high momentum (Debicki, Matherne Iii, Kellermanns, & Chrisman, 2009). A multitude of investigations have sought to identify the mechanisms underlying organizational inertia on various levels (Christensen & Bower, 1996; Henderson & Clark, 1990; Hill & Rothaermel, 2003). I add to this research by going further back within the causal chain to explore those owner qualities that define the extent to which companies develop the previously studied determinants of technology adoption and thus vary in their adaptation behaviors.

My analysis echoes recent advances in literature (Gilbert, 2005; Tripsas, 2009) by envisioning incumbent response as a much more fine-grained phenomenon than traditionally portrayed. Incumbents are not necessarily sluggish in their responses to discontinuities. In fact, it is not even possible to dichotomously categorize organizational adaptation into sluggish responses on the one hand and flexible responses on the other. Instead, as first highlighted by Gilbert (2005), established organizations vary with respect to different dimensions of responses to radical change and an antecedent of adaptation can spur flexibility in one domain (e.g. speed) while, at the same time, exacerbating rigidity in another domain (e.g., intensity). I show that the variance in the different adoption domains depends on firms' ownership structures. High levels of family influence trigger a more adaptive response with regard to speed and stamina. However, more family influenced businesses are generally more reluctant to invest aggressively in technological discontinuities than businesses less influenced by family systems.

As a result, my theory not only improves the predictability of the responses of incumbents that are influenced by families but also provides theoretical insights into all organizations by explaining various anomalies to established inertia theory. Specifically, I show that several dimensions of inertia can be explained by effects other than those proposed in extant literature. For example, the 'resource rigidity' (Gilbert, 2005) of family influenced firms is not rooted in resource dependence on short-term-focused providers of capital, as described in the disruptive innovation theory (Christensen, 1997), but is instead driven by family-owners' efforts to maintain independence from external capital providers and by long-term focused family systems. My analysis also runs counter to standard theory by stressing that *command*, which has traditionally been associated with lower levels of innovativeness and flexibility (Staw et al., 1981), can spur radical changes under certain circumstances.

My theorizing also has critical implications for family business research. To date, research on family influence and responses to change has been largely self-referential and has culminated in the development of two opposing views (Le Breton-Miller & Miller, 2009): the 'benevolent' perspective, which argues that family influence renders firms more adaptive to change; and the 'malevolent' perspective, which holds that family influence exacerbates organizational inflexibility. I suggest that one key to the reconciliation of these two views lies in the disentanglement of the various dimensions of adoption (speed, intensity, stamina, and routine flexibility). Another key lies in the differentiation between discontinuous and continuous change because predictions of family businesses' responses to change are likely to differ fundamentally depending on the type of change. For instance, the effect of family influence on the relative speed of adoption of new, but continuous, technologies might be insignificant or even negative, as less efficient routines are likely to reduce the relative speed of interpretation in continuous circumstances (Levitt & March, 1988). In contrast, family influence is likely to have a significant positive impact on response speed in the context of discontinuous technological change. Moreover, applying a benevolent view, one could argue that resource scarcity is compensated or even overcompensated for by loyal and well trained staff (Allouche & Amann, 1997; Miller et al., 2008). However, this line of argumentation does not hold true under discontinuous circumstances. When radical technological innovations emerge, loyalty as well as experience and training in old, established routines might entail emotional attachment, narrow cognitive frames and rigid routines (Kaplan, 2004; Kaplan & Tripsas, 2008; March, 1991), ultimately resulting in inflexible adaptation. My theorizing also shows, that patience and ongoing commitment even when lacking initial success, an organizational behavior often envisaged as disadvantageous (Brockner, 1992; Gómez-Mejía et al., 2001; Staw, 1981), can turn into something positive when the market is disrupted by a discontinuous technology. In such cases, the patience and long-term focus allow family firms to continuously invest in the new technology.

Finally, this research contributes to managerial practice. As I will demonstrate, managers face the challenge of balancing various mechanisms that represent 'double-edged swords.' For instance, executives in businesses who are not at all or only weakly influenced by family systems are likely to benefit from a stronger awareness of the benefits and drawbacks of dependence on external financing. Simultaneously, managers in family influenced firms are likely to be more successful in navigating their companies through times of radical turmoil if they are conscious of the behavioral idiosyncrasies engendered by family influence. For example, although well-trained, loyal employees might be a viable asset during periods of continuous technology development, they can hamper adoption of discontinuous technologies because of their focus on path-dependant solutions and their rigid mental schemata. Managers and owners can adopt numerous measures identified in previous research to mitigate this source of routine rigidity. In particular, they can seek external support for implementation decisions (McDonald et al., 2008) and establish ambidextrous structures (O'Reilly & Tushman, 2008). Another recommendation for family influenced firms that emerges from my theorizing, is that these firms should be particularly eager to choose 'open' executives and build heterogeneous mixes within the dominant coalition.

3.11.3 Limitations and Research Agenda

As with any theory, my model builds on a number of assumptions that serve as important boundary conditions, but also provide avenues for future research. Most importantly, I have defined family influence as the overlap between a coherent family system and the business system that arises via formal mechanisms, such as ownership or management, and behavioral mechanisms, through which family owners actively shape the firm's vision, strategic decisions, or firm culture (Chua et al., 1999). This definition has two consequences. First, in contrast to research building on more formal definitions of family influence (see Chua et al., 1999 for a summary), I consider family influence to be low for those (partly) family-owned firms in which family members are not involved in business and control activities, or merely aim to exploit the business (Le Breton-Miller & Miller, 2009). Second, deviating from other studies (Gersick et al., 1997), I exclude those firms in which behavior is shaped by systematic conflicts among family members. Such conflicts violate the condition of a comprehensive family system because systematically conflicting owners constitute a set of disjunctive, individual 'family' systems. Therefore, future research is advised to study the implications of various mixes of formal and informal mechanisms of family influence in the context of discontinuous technological change and to refine my theory by studying cases of systematic conflicts among family owners.

A related limitation of my model is grounded in my assumption that family influence has similar effects on organizational adaptation regardless of the specific, case-idiosyncratic weight with which each of the 4Cs 'loads' on family influence. In this regard, one might expect that a specific source of family influence enhances one indicator of family influence more than another source. For instance, power from formal ownership could reinforce the *continuity* dimension more than the other three dimensions, whereas family influence exerted through management could have a disproportional effect on the *command* dimension. For the sake of parsimony, I also refrained from taking into account the possibility that different forms of owner influence might affect each of the 4Cs differently. For example, variance in the mix of voting rights and cash flow rights, which owners use to exercise control, could entail variance in capillary aspects of *command* (deAngelo & deAngelo, 1985). I maintain that considering the effects of variations in the weighting of the '4Cs' and in elements of firm governance, would exceed the scope of my current theorizing. Nevertheless, future research could extend and refine my model by exploring the relevant sub-mechanisms.

I have, to a great extent, chosen to discuss how the characteristics of the dominant coalition moderate the effect of family influence on technology adoption because understanding those characteristics is necessary for determining the precise effect of family influence on new technology adoption. However, I have not discussed the impact of other contingencies that could affect the mechanisms underlying my model. For example, I disregard several factors often highlighted in family business research, such as the exact number of the family members involved (Astrachan et al., 2002), the stages of the family firm (Gómez-Mejía et al., 2007), and firm performance hazard (Gómez-Mejía et al., 2010). I also ignore differences in organizations' cultural backgrounds, which have been found to impact family influence and its effects (Sharma et al., 1997) as well as decision-making under uncertainty (Barr & Glynn, 2004). Future studies might examine the role of such variables in the context of my model.

Another promising area for subsequent studies is the interaction among different types of owners and how such relationships affect organizational adaptation. In this regard, it is noteworthy that few companies are completely non-family influenced or fully family influenced. Most established companies are owned by a highly diverse mix of differently acting owners. However, how this mix affects decision-making in the context of discontinuities remains unclear.

Finally, future research should extend my theory to the performance implications of family influence in the context of discontinuous technological change. For instance, the early, aggressive adoption of discontinuous technological change has often been envisaged as necessary for long-term firm prosperity. However, depending on the respective isolating mechanisms (Lieberman & Montgomery, 1998), late followers and careful step-by-step investors can also derive competitive advantages (Suarez & Lanzolla, 2007). In fact, under certain conditions, non-adoption may be the most promising alternative (Adner & Snow, 2010). Therefore, even though I maintain that the four dimensions of technology adoption investigated in this study are critical to adaptation performance, further research is necessary to derive definitive predictions on how family influence plays out in this respect. Somewhat similar, it might be interesting to further explore variance in resource investment patterns. In large parts of my theory, I assume a constant or step-like resource pattern; only in Chapter

3.10 I relax this assumption and take phase-dependent resource investments into account. Further research can follow this path, investigating more complex resource investment patterns and their effect on adoption success.

I encourage researchers to empirically test the rich theory provided in this paper in an integrated longitudinal approach that combines multiple methods. Qualitative case-based fieldwork is likely to provide profound insights into the various mechanisms that link family influence and technology adoption. I suggest quantitative panel studies in single industries affected by technological discontinuities. These industries should comprise a larger number of at least mid-sized players and encompass a sufficient degree of variance in family influence, such as the publishing industry (affected by electronic publishing), the energy industry (affected by smart grid/smart metering technologies), and the automotive industry (affected by electric mobility). Unobtrusive textual vehicles to capture family influence and adoption initiatives, such as websites, conference calls, and company publications seem particularly suitable. To measure family influence, I suggest to other scholars to extend the F-PEC scale of family influence by adding a category for the *connections* dimension of Miller and Le Breton Miller's 4Cs (Astrachan et al., 2002; Miller & Le Breton-Miller, 2005b). Any empirical study that scrutinizes my model must control for variables such as size, prior performance, and the influence of other owners, e.g. state-owners. All in all, regardless of the methodology applied to scrutinizing this research, I hope that my theorizing may stimulate manifold conversations and empirical studies, which could lead to a deeper understanding of established organizations' responses to discontinuous technological change.

4 Empirical Evidence on how Family Influence Affects Organizational Adaptation to Discontinuous Technological Change

4.1 Introduction

In the following chapter of my thesis, I aim to illuminate the relationship between family influence and organizational adaptation to discontinuous change from a different epistemological perspective, thereby drawing on empirical evidence from multiple case studies, which aim to answer the following research questions:

> *(1) How do family influenced firms adapt to discontinuous technological changes and why?* and
>
> *(2) Does variance exist within the response patterns of family influenced firms to discontinuous technological changes, and if so, why?*

I believe that this qualitative empirical evidence will add to our understanding of how and why family influenced firms do or do not adapt to discontinuous technologies. My theoretical deduction of hypotheses (Chapter 3) identified a large number of factors potentially influencing family firms' adoption patterns such as those organizations' supposed lower levels of formalization, their idiosyncratic resource dependence, and their reluctance to cede responsibility to outsiders. With the empirical findings based on ten case studies within three different industries, presented in this chapter, I strive to enhance our knowledge on which of these theoretically identified determinants are, in practice, the main drivers of (non-) adoption. Moreover, this empirical approach allows me to relax the assumption that family influence 'loads' equally on each of the 4Cs (see Chapter 2.2.2). Findings from my case studies will show that this is not true; although all of my sampled firms were substantially influenced by families, their adaptation to discontinuous change varied to some degree. I will show that different non-economic goals and intentions of the family owners affect their respective organizations' adaptation to discontinuous change, and thus cause variation among the group of family influenced firms. Although family owners might have a variety of goals and intentions at the same time, decision-makers of the sampled companies always had one dominant motive that drove their strategic activities. In line with recent literature on the non-economic goals and intentions of family influenced businesses (Gómez-Mejía et al., 2007; Zellweger et al., forthcoming-b), I will categorize these dominant motives along four dimensions of socioemotional wealth, and show how variation in those dominant motives can explain variance in family firms' adaptation behavior.

I will proceed as follows: First, I will outline my research setting and methodological approach (Chapter 4.2) before providing a detailed description of each case this study builds on (Chapter 4.3). Next I will show and discuss the results of my study (Chapter 4.4) before concluding with some remarks on this study's implications and limitations (Chapter 4.5).

4.2 Methodology

4.2.1 Research Design and Setting

Given the limited knowledge about the causal connections between discontinuous technological change and family businesses' responses, as well as the inconclusive findings in extant literature, I applied an inductive, case-based approach for my study (Yin, 1994). Such a research design is particularly appropriate since my research questions aim to shed light on interpretation processes and answer questions on 'how?' and 'why?' rather than 'what?' and 'how many' (Yin, 1994).

My research setting is the German retail, toys and gaming, and energy sector—industries that have recently been challenged by discontinuous technological changes. The retail industry has been affected by online business models and dis-intermediation—changes that have been categorized as discontinuous by various authors (e.g., Christensen & Raynor, 2003; Jelassi & Enders, 2008). The German retail market, a sector with annual revenues of more than 400 billion EUR (HDE, 2012), was disrupted by the launch of 'online shops' at the end of the 1990s, particularly, when the US based online pioneer, Amazon, opened its subsidiary for the German speaking market in 1998. While at the beginning, online shopping attracted only few customers—those being generally affine to technological novelties and possessing a computer as well as broad band internet connection—it has become more and more accepted and utilized by mainstream customers during the last decade (HDE, 2012).

Online retailing can be classified as a discontinuous technology, because necessary core competencies of retailers shifted, for instance, from deep knowledge on physical store set up or friendliness and helpfulness of the checkout clerk to knowledge and experience related to electronic technologies: Retailers had to learn how to set up secure and stable online shops, promote their shops and products online, adopt prices dynamically, and react to online customer queries instantly. The core performance dimensions as perceived by customers changed from 'shop close by,' 'good atmosphere when shopping,' and 'friendly staff' to 'availability anywhere and around the clock,' 'optimized search functions,' and 'possibility to purchase goods without security concerns.' Delivery from the retailer to the customer was traditionally the responsibility of the customer (except home delivery services offered by some retailers to customers living in the respective shop's neighborhood), but shifted to the retailer entirely when online shops emerged. Integrating this part of the value chain into the business also required new competencies. The possibility to compare prices of numerous providers with little effort required many retailers to rethink their unique selling points. Moreover, the entrance of product manufacturers, who leveraged the low entry barriers of online shopping to engage in direct sales of goods to customers, into the retailing market further threatened retailers' positions. To summarize, online shopping constituted a radical, disruptive (Christensen, 1997), competence destroying (Henderson & Clark, 1990) technological innovation, affecting the internal processes as well as the 'product' offered (König, 2009).

The emergence and increasing prevalence of computers and electronics did not only affect the retailing sector, but also disrupted the traditional toys and gaming market whose incumbent players found themselves in a situation competing against vendors of video games and electronically enhanced toys. In the 1970s, the first gaming consoles were brought to the market and successful commercialization began in the mid-1980s, when Nintendo launched its Nintendo Entertainment System NES. In the following years the mostly Asian- and US-based companies Nintendo, Atari, Sega, Microsoft, and Sony introduced a series of new product innovations, such as handheld devices for electronic games, and thereby substantially increased the market size for electronic games and toys. In Germany, the relative contribution of electronic based products such as video games to the toys and gaming market increased from 12% in 1994 to more than 30% in 2009, partly at the expense of traditional toys and games, thereby being the main driver of market growth in that sector (BVS & eurotoys, 2010). The emergence of electronic toys and games challenged the incumbent producers of toys and games by demanding that they process new materials (software and hardware supplement replace textiles, plastic, and wood), launch new product innovations at a substantially different pace (life cycles of electronic products being much shorter than those

104

of traditional toys and games), and react to new dimensions of customer requirements regarding the product features. Moreover, this trend forced incumbent producers that aimed to adopt the new technology to build up new IT competencies. Summarized, the emergence of electronic toys and games involved non-paradigmatic changes of the internal processes as well as the nature of the final products. Moreover, the discontinuity was disruptive (Christensen, 1997), in that it first attracted only fringe customer groups—technology-affine teenagers—before spilling over to established customer groups.

In contrast to the retailing sector and toys and games industry, where new dominant designs of the discontinuous technology (Anderson & Tushman, 1990) had already developed over the last decades, the German energy metering sector was still in an 'era of ferment' (Anderson & Tushman, 1990) at the time of the study, being increasingly affected by new smart energy grid technologies. Energy metering had constituted a relatively stable market segment throughout the last decades and the mostly small and medium-sized incumbent players had been specialized on developing, assembling, and selling mechanical meters to measure the amount of power a household consumes. Driven by regulatory forces[38], energy meters were becoming more 'intelligent' at the time of this study, i.e. the measuring and counting systems of the devices were supplemented by data processing and communication functionalities that allowed for a bilateral data transfer between the meter and energy consuming devices on the one hand and the meter and the energy grid on the other hand. This change entailed a switch from analog to digital products that offered new, previously unavailable features to customers such as data storage, tracking of energy consumption, communication, load control, and remote control. As a consequence, incumbents adopting the new technology had to build up new IT-related competencies that substantially deviated from their previous processing experience. Moreover, the value chain of the entire metering sector was heavily affected by the discontinuous technological change including threats to supplant existing business models such as reading off meters (Nuttal, Zhang, Hamilton, & Roques, 2009). Furthermore, adopting the new technology fundamentally changed the cost structure of the players and created new categories of revenue such as IT-based after sales services. Smart metering constitutes a discontinuous technology that substantially affects the product features, the internal processes, and the value capturing models (König, 2009). However, the innovation was competence-enhancing rather than competence-destroying (Henderson & Clark, 1990), as established knowledge still remained valuable to the players.

---

[38]Since 2010, it has been mandatory by law in Germany to install smart energy metering devices in newly constructed or completely refurbished buildings.

Thus, all three industry sectors had one thing in common—after decades of relative stability, a discontinuous technological change had disrupted the industry; however the point in time when the discontinuity emerged varied, ranging from the mid-1980s for toy producers, the end of the 1990s for retailers and the end of the 2000s for metering producers. In each of the sectors, family businesses with varying sizes and degrees of family influence play an important role, thus rendering this setting specifically adequate for my analysis. Subsequently studying the reaction of five retailers, three toys and game producers, and two producers of metering devices allowed me to replicate my study and thus enhance the generalizability of my findings (Eisenhardt & Graebner, 2007).

As suggested by literature (Eisenhardt, 1989; Glaser & Strauss, 1967), I built on a theoretical sample of businesses, including firms of the above-mentioned industries with variation in size, family influence, and also reaction to the discontinuity. Appropriate firms were identified through information available on the internet and data provided by industry associations. All ten family businesses discussed here had been led by a family CEO of the second or later generation throughout the emergence of the respective discontinuity. The companies were all majority owned by one or two families who significantly influenced the respective organizations' cultures and values. All interview partners confirmed that they perceived their respective businesses as 'family firms.' In nine out the ten firms, the founder's family name is identical with or part of the firm name, and in seven firms this relationship still holds true for name of the current CEO. Conformity of names is an important predictor for family members' identification with the business (Zellweger, Nason, Nordqvist, & Brush, forthcoming-a).

My final sample consisted of ten family influenced firms as presented in Table 7. In addition to readily available factual data on the respective industries, sizes (indicated by number of employees), and ownership structures (percentage of shares held by the owner-families), Table 7 also provides information on family influence, the company's history (previous major changes), the focus of the organization, and the image of the firm.

Table 7: Family Influence and Adaptation—Overview of Sample

Dis-guised Cases[a]	In-dustry	No. Emp-loyees[b]	Family Owner-ship[c]	Family Influ-ence[d]	Previous Changes	Focus of Business[e]	Customer Image[e]
Alpha Star	Retail	~2.000	>50%	++++	Extension of product portfolio	Family first	Conser-vative
White & Blue	Retail	~2.000	100%	++++	Extension of product portfolio	Family first	Innovative
Anything & More	Retail	>5.000	>50%	++	Extension of customer services	Business first	Innovative
King's Goods	Retail	>5.000	100%	+++	Globalization, change of product portfolio	Business first	Conser-vative
Retail 2000	Retail	>5.000	100%	++	Globalization	Business first	Conser-vative
Kiddies	Toys & Games	<100	100%	+++	Change of product portfolio and customer groups	Family first	Conser-vative
Play & More	Toys & Games	~200	100%	+++	Extension of product portfolio	Family first	Conser-vative
Toys 2000	Toys & Games	~200	100%	+++	Extension of product portfolio	Business first	Innovative
Walter & Coll.	Energy	~200	100%	++++	None	Family first	Conser-vative
Power-house	Energy	>5.000	100%	+++	Globalization, acquisitions	Business first	Innovative

[a] cases ordered along industry, size, (small to large), family ownership (low to high), and alphabetically
[b] in 2009
[c] when discontinuous technology emerged
[d] when discontinuous technology emerged; measured based upon family ownership, family members in top management positions, generation active in management, family influence on decision-making, family influence on shaping firm culture. (o = not available/very low, + = low, ++ = medium, +++ = high, ++++ = very high)
[e] based on consolidated interview statements and secondary data

Based on these authors' considerations, I assessed the family influence of all sampled firms on a 5-point scale. Moreover, I used information from interviews and secondary data to extract whether the businesses are inclined to a 'family first' or 'business first' focus as well as whether they are perceived by customers and themselves rather as 'innovative' or 'conservative,' because these attributes might influence the organizations' strategic activities (Chirico & Salvato, 2008; Distelberg & Sorenson, 2009), and ultimately also their willingness to adapt to changing environments.

4.2.2 Data Collection

For each case, I developed historical, top-level accounts of the sense and decision-making as well as the implementation processes. To do so, I collected data from two main sources. The first is a set of 11 open-ended, in-person interviews conducted in mid-2010 based on semi-structured templates, with managers and owners of the organizations (including 8 family CEOs). Each interview was 60-150 minutes in length. As suggested by extant literature (Bechhofer, Elliott, & McCrone, 1984), interviews were conducted by two interviewers[39], were audio-recorded, and transcribed shortly after the interviews. The interviews were structured in four parts: The first part focused mostly on the organization as a subject of analysis. I asked about the firm's culture, most important goals, strategy, and historical events. A focal topic of this section was the structure of typical decision-making processes within the firm. In the second part of the interview, I switched my focus to the owner-family, aiming to understand the family members' influence on the company, specifically on decision-making processes. The third and most extensive part of the interview dealt with the organization's adaptation to the discontinuous change. First I posed general questions on how the family-owners inform themselves on ongoing changes and trends. Next, I aimed to understand when and how the CEOs became aware of the discontinuity, how they interpreted it, and whether and how this attitude has changed over time. Moreover, I requested the informants to provide detailed information on the decision-making processes and to describe specific activities within the organization that had been implemented to adapt to the change (only in cases, in which an active reaction took place). Ultimately, I asked my informants about their opinion of their companies' adaptation performance. In the fourth and final

---

[39] The typical setting for such interviews included me as the main interviewer accompanied by a master student, who took additional notes and asked further questions.

section, I questioned the informants about what, in their opinion, would have been different regarding the reaction to discontinuous change, if they had not been a family business. Moreover, I engaged in a discussion on my preliminary results (when available) with the informants, thus getting immediate feedback and allowing me to fine-tune my findings in rare cases. The full interview guide can be found in Appendix A.i.

In addition, I conducted 14 interviews with industry and family business experts, which I used to get a better general understanding of the industry, in order to evaluate informants' statements in the correct context, but also to triangulate my findings. The second source of information encompassed several hundred pages of primary and secondary data (including company brochures and catalogs, company websites, and press articles from a 15-year timeframe) and observations (mostly from store visits), which I used to validate my findings. One specifically important data source for longitudinal data was historic websites of the respective companies retrieved via the 'Waybackmachine,' a free service offered by Internet Archive (http://www.archive.org). Internet Archive is a non-profit library based in California, USA, which has stored 'snapshots' of websites starting from 1996 up to today. At frequent intervals, mostly a matter of weeks, data from websites is accessed and permanently saved, resulting in a total database size of 150 billion websites equivalent to 3 petabyte data volume[40]: Although not all historic data is available, for instance due to incongruent data format or deliberate exclusion from archiving as decided by the owners, data quality could be considered sufficiently good in this study. Only one company (*Anything & More*) made use of the 'opt-out' option and prohibited access to their historical websites. I systematically accessed websites of each company in at least two-year steps (1996, 1998, 2000, 2002, 2004, 2006, 2008, 2010, as well as the current page in 2012). I browsed through all website contents (including sub-sites) with special focus on the 'about us' section and the section on 'our services,' as well as 'news' and 'press releases.' Table 8 summarizes the data sources.

---

[40] Data volume refers to 2009; http://www.archive.org.

Table 8: Family Influence and Adaptation—Overview of Data Sources

Company	Interviews	Years of Analyzed Websites	Further Primary Data & Observations	Number of Secondary Data
Alpha Star	Family CEO	1998, 2000, 2002, 2004, 2006, 2008, 2010, 2012	Store visit	17
White & Blue	Family CEO	1998, 2000, 2002, 2004, 2006, 200, 2010, 2012	Book on *White & Blue*'s organizational history, store visit	9
Anything & More	Manager	2012	Press releases 2002-2012	75
King's Goods	Family CEO	2000, 2002, 2004, 2006, 2008, 2010,	Annual reports 2002-2012, press releases 2002-2012, store visit	18
Retail 2000	Family CEO	1999, 2000, 2002, 2004, 2006, 2008, 2010, 2012	Store visit	14
Kiddies	Family CEO	1997, 1998, 2000, 2002, 2004, 2006, 2008, 2010, 2012	None	8
Play & More	Family CEO	2000, 2002, 2004, 2006, 2008, 2010, 2012	Product brochure	3
Toys 2000	Family CEO	1999, 2000, 2002, 2004, 2006, 2008, 2010, 2012	Annual report 2010	12
Walter & Colleagues	Family CEO, manager	2000, 2002, 20004, 2005, 2007, 2012	Factory visit, annual report 2010	6
Powerhouse	Manager (family-external, regional CEO)	1998, 2000, 2004, 2007, 2012	Press releases 2007-2012, annual report 2010, several brochures	7
Experts	14	n/a	Several industry overviews and market outlooks	4
Sum	**25**	**67**	**n/a**	**173**

4.2.3 Data Analysis

In line with Eisenhardt's (1989) eight-step process to build theory on cases, I iteratively engaged in a data analysis and data gathering process. First, I summarized the findings of the data collection process to single case studies. In a subsequent step, I conducted a cross-case analysis searching for common patterns among the individual cases. To do so, I created an initial coding scheme based on extant literature (see Appendix A.A-3), which I iteratively refined during the data analysis process. Supported by software (NVivo, Excel), I used axial coding (Charmaz, 2006; Strauss & Corbin, 1998) to detect patterns both cross-case/within industry and cross-case/cross-industry (Eisenhardt, 1989). Following best practice in case study research (Bechhofer et al., 1984; van Maanen, 1979a), this process included multiple coders[41], subsequent discussion of the results, and triangulation with secondary data rather than purely relying on the quantitative outcome of the coding analysis (van Maanen, 1979a).

4.3 Detailed Case Description

In their intriguing article illustrating the factors that discriminate between strong and weak theory building, Sutton and Staw (1995) criticize the established practice in well-established social science journals of shortening the description of the qualitative characteristics to a minimum length due to tradeoffs between data description and theory development. The authors state: "The resulting description may end up as little more than a small sequence of vignettes or a summary table of quotations, illustrating those concepts or hypotheses formulated in a paper. Such paring can deplete a manuscript of much of its value" (1995: 383).

To avoid those pitfalls, I include a description of each case in the following subsections before I dwell on the results of my study. However, as anonymity was granted to each interviewee, not only regarding their individual identity but also the organization's identity, general information about the organizations' key data and main products need to remain scarce. Thus I will not reveal precise information about the age of the organizations,

---

[41] Typically, master students engaged in the interviewing process were also involved in data analysis. Prior to coding, I trained them in the use of the software and the coding rules in a one-day workshop.

the gender of the respective CEOs, or the exact product portfolios[42]. In rare cases, anonymity concerns require that I remain vague about exact dates[43].

I will structure the case descriptions as follows: First, I will provide general facts on the organization's market segments, size[44], and family ownership. Moreover, I will outline whether the product portfolio has significantly changed over the last decades as well as the company's degree of internationalization (Becker & Ulrich, 2011). Furthermore, I will describe how family influence manifests itself in each of the companies. Second, I will provide retrospective information on the sense and decision-making processes of the CEO (and other key stakeholders, where appropriate) when faced with the discontinuous technological change. Third and last, I will describe the company's adaptation over time to the discontinuous technology. I will supplement this information by factual data such as revenue shares wherever available and ready to share without violating anonymity. Data marked with quotation marks refer to statements from the personal interviews. These statements were originally expressed in German and translated to English by myself for the purpose of this thesis. Back-translation by students was applied to ensure the correctness of the translation. Data obtained from other sources such as annual reports, press releases or website information are indicated as such.

4.3.1 Alpha Star

4.3.1.1 Background Information and Family Influence

Alpha Star is a large family-business active in the retail industry. Several family members, who own the business, fill top management positions within the company. Besides retail, *Alpha Star* is also active in wholesale business and markets its own products. The product range had been selectively expanded in the years before the study. *Alpha Star* has been internationally active since the early 1990s (according to the website).

---

[42] Information on these aspects is stored in the project database and can be revealed to readers upon request provided that anonymity will remain unviolated.

[43] Again, information is stored and available upon request given that anonymity is preserved.

[44] Definition according to IfM Bonn (2012; http://www.ifm-bonn.org); small companies employ up to 9 employees with annual revenues of less than 1 million EUR. Medium sized refers to those companies with 10—499 employees and annual revenues higher than 1 and lower than 50 million EUR. Companies with more employees and higher annual revenues than medium-sized firms are labeled as 'large.'

Most important to the family is maintaining the company's reputation that has grown over the last decades. Hence, the family is highly concerned about the company's image and strong brand building belongs to the major strategic goals of the family business.

The level of formalization is very low in this company, as the junior CEO stated: "Many things are not codified, not written down here, and yet it works anyway—maybe it even works because of that." Two written 'codes' exist in the company, one relating to sustainability and ethics, the other referring to the owner influence on the business. The level of power, the family, and particularly the CEO possesses, is rather high. Due to the intermingling of ownership and control, the family can make decisions independently. The 'family influence code' allows the CEO to take most decisions without asking other family members.

The junior CEO emphasized the importance of *community* effects in the company: "The employees back the company as one man. Everyone knows the key points—that is part of the culture. [...] Culture also means that people should take pleasure in working here." New employees that join the company often perceive the culture as "liberating, providing more air to breath as there are less rules [...], more impact for each employee, [...] and fewer meetings."

Connections to outsiders play a crucial role for the junior CEO who believes that "it is important to have an ear to the ground," and to talk to stakeholders along the entire value chain of *Alpha Star*'s industry. He/she believes that these strong connections, at least partly, compensate for the fact, that his/her company does not run its own R&D department due to resource constraints. Connections on a 'semi-private basis' are perceived as most useful by the junior CEO, while he/she thinks that idea exchange and discussions at industry association meetings are not useful for innovativeness.

Somewhat special to this case is, that the junior CEO I spoke to had not yet been involved in the sense and decision-making process related to the emergence of online business models. Thus, the following notes of this subchapter will refer to *Alpha Star*'s interpretation of and reaction to another recent, product-related discontinuous technological change that affected one of *Alpha Star*'s main segments. The junior CEO was deeply involved in the adaptation of the product substitution that threatened one of *Alpha Star*'s own core products.

4.3.1.2 Interpretation and Decision-making Processes

The product discontinuity was first commercialized around the turn of the century and two of *Alpha Star*'s competitors immediately reacted by imitating the product. As a consequence they were accused of patent infringement and had to temporarily stop their product delivery. At this time, *Alpha Star*'s owner-family began to discuss whether to react. The junior CEO described this process as follows: "In that situation we considered entering the market. [...] It was almost sure that we would get sued. It was uncertain, how large the market segment would become. And, honestly, we did not like [the product]. We believe, it is not a good product—technically speaking, it is not a good product. It's neither fish nor fowl [...] Moreover, we had no experience in this category." However, despite all these concerns, the family felt that they "must not miss the boat" (junior CEO, personal interview) to ensure long-term business success and engaged in "day and night-long discussions" about the topic. Ultimately, to deal with the uncertainty, the family asked (potential) consumers for feedback. They selected 10 to 20 decision-makers in their industry segment as well as private consumers whom they "personally knew and of whom [they] thought they would be able to provide somewhat sophisticated feedback" (junior CEO, personal interview). Instead of engaging a market research institute, the family members themselves engaged in the discussion processes that ultimately delivered the result that "consumers were enthusiastic" (junior CEO, personal interview). Due to these discussions, the senior CEO, "initially, the most skeptical of all [family members]" (junior CEO), changed his opinion about the new technology from highly negative to highly positive. According to the junior CEO, this change of mind "did not stem from any calculation, any presentation or any market research study. It stems from a gut feeling, the feeling that one has to keep options alive. And if you are not active from the start, it will be difficult to enter the market later on." The junior CEO described the further processes as follows: "Every day [the senior CEO] asked me: 'When will we have the product?' In the beginning I thought, 'wait, yesterday you told me, it was crap.' But he had changed his mind and everything had to be implemented as fast as possible." The junior CEO followed the senior CEO's guidelines: "In a 'chop chop' manner, we designed a product in cooperation with a partner as we didn't have our own production lines for such products yet. Next, we ran a series of tests. We engaged in discussions with business customers without having the product yet. We did not exactly know what the production costs would be, but we started to list these products, without having them and without knowing the delivery date. Everything happened very quickly." Two or three months later, *Alpha Star*'s first products based on the new technology were available for sale.

This interpretation and decision-making phase occurred at a time, when *Alpha Star*'s revenues in its core segments were declining. As early as in 1998, the senior CEO admitted in an interview that young (non-)customers would have associations such as "conservative and outdated" when thinking about *Alpha Star*'s products (press article, 1998).

4.3.1.3 Adaptation to Discontinuous Technological Change

Twelve to fifteen months after the first commercialization of the discontinuous technology, *Alpha Star* entered the market as the third follower (personal interview, press articles 2004, 2005). Due to several issues the first two followers faced (mostly because of the patent war), *Alpha Star* was able to quickly win market shares. The market has significantly grown since then.

Internally, *Alpha Star* did not build up any new business unit or hire new employees. The CEO's rationale for the decision for an integrated organizational setup was the "lack of time to build up a new unit." The investment required to commercialize the new technology was rather low: for instance, the company did not invest in advertising for the new products; *Alpha Star* preferred a "silent introduction" (junior CEO, personal interview) of its new products.

For the sake of completeness of this study I will next provide information drawn on press articles and historical websites illuminating *Alpha Star*'s adaptation to online business models: According to its historical websites, *Alpha Star* launched its first internet store in 1999, one year after starting a phone and fax-based home delivery service (press article, 1998). After piloting e-commerce in one segment until early 2000, the store comprising four of the five core product segments of *Alpha Star* went live between September and October 2000. The online shop was integrated in the company's main website (launched no later than January 1998).

4.3.2 White & Blue

4.3.2.1 Background Information and Family Influence

White & Blue is a large retailer, one hundred percent owned by the founding family. The business is managed by a family CEO and runs numerous specialized stores in Germany. The children of the family CEO were also employed by *White & Blue* at the time of the interview, and several members of the next generation were preparing to succeed the current CEO in the near future.

Transgenerational conflicts (across three generations), caused by "high levels of lifeblood [for the family business]" are frequent at *White & Blue*, however, the family has found constructive ways to deal with them. Continuity of management, leadership, and also rituals are crucial for the CEO, whereas growth plays a subordinate role.

Employees and customers are seen as the company's key stakeholders (CEO, personal interview; historic websites). As the website states, "providing 'Heimat[45]'" to employees and customers is the vision of *White & Blue*. "Enthusiastic employees entail enthusiastic customers," is one of the slogans of *White & Blue* according to the CEO and information provided by the company's website. Maintaining this sense of community over time is one of the CEO's main concerns. *White & Blue* organizes festivities and jubilees to bring the employees together. Besides this general focus on *community*, "social and ecological activities" and "training for employees" are two core pillars of *White & Blue*. In turn, *White & Blue*'s employees are "proud" of their employer and show high levels of "loyalty" and above-average tenures (press article, 2010). During the interview, the CEO also mentioned disadvantages that come along with the high level of *community* at *White & Blue*: Long-term employees "regret bygone times" and hence have become reluctant to change.

The CEO describes his/her past decision-making as "hitting my head against a wall— whenever I was convinced, I decided for it and we subsequently implemented it. I listened to all contributions to discussions, but ultimately, I went my own way." The CEO continued: "Power is the key, [...] as well as conviction and trust in decisions." "Allowing for mistakes," is another important aspect of *White & Blue*'s culture.

Despite his/her general retentiveness to granting externals access to the company, the CEO has occasionally drawn on consultancy advice in the past. The CEO and members of *White & Blue* are rarely active in associations—the CEO "avoid[ed] that kind of intermingling due to a lack of time." When building alliances, the CEO is concerned about a "shared image of humanity."

---

[45] 'Heimat' denotes an untranslatable German term, referring to a specific geographical region or social group that offers individuals a feeling of belonging.

4.3.2.2 Interpretation and Decision-making Processes

The CEO perceived the changes to the retail sector during the last years as "continuous and evolutionary developments." Due to this framing, it was not the CEO but *White & Blue*'s employees that interpreted the discontinuous technology as relevant at the end of the 1990s and propelled their ideas for adaptation to the CEO. The CEO, relying on and trusting *White & Blue*'s employees, quickly decided for adaptation but, as indicated by interview statements and secondary documents, without much passion for the new technology[46].

The CEO further believes that "price, quality, and service" are the three determinants of success in the retailing market and that those three will also remain important in the future. E-commerce "will change a lot, but many things will remain the same," the CEO explains his negative attitude towards e-commerce, "I think, maybe I hope, that human contiguousness will not be replaced by computers, iPhones, or iPads." The CEO admits: "In the beginning, I was very diffident. However, now I see that we have to go this way."

4.3.2.3 Adaptation to Discontinuous Technological Change

In 1999, *White & Blue* first entered the e-commerce market by setting up a central online ordering system for its customers in a selected region (historical website, 2000). After putting in their order via internet, phone or fax, the goods were delivered to the customers by the nearest store. Customers could select from a number of 'catalogue products' (roughly equivalent to the minimum number of products offered in a store). Delivery was free as long as a certain threshold value of sales was reached and delivery time approximated 2 to 3 days. This business model can be traced back to independent home delivery activities driven by individual stores that were subsequently consolidated by the management. However, these early attempts were not profitable for *White & Blue* due to the high cost caused by delivery (CEO, personal interview). The online shop closed again at the end of 2002 (historical websites, 2002, 2003). Based on the company's chronicle, a new nation-wide internet store was launched in cooperation with a partner. Again, this effort was not successful and ceased business operations shortly after. Instead of investing in online businesses, *White & Blue* invested heavily in offline and online customer retention programs for several targeted customer groups. Moreover, *White & Blue* has increased the number of physical stores during the last decade and put substantial effort into building its brand as a sustainable retailer. In

---

[46] The 200-page chronicle of *White & Blue* only contains the term "internet" (or related terms such as "online") twice.

2007, *White & Blue* launched internet based ordering and home delivery of flowers, but abandoned this service after a short time. One small, but still existing online-based revenue channel of *White & Blue* refers to online photography services (available since 2002 according to chronicle).

At the time of the interview, *White & Blue* was working on a new concept to re-enter the new market, which ultimately went live in mid-2011 according to press articles (2011). This web shop, offering core products of *White & Blue*, combined 'classical' online shopping with more advanced mobile purchasing techniques (promotion video, website). Interestingly, the main website of *White & Blue* neither refers to nor mentions this newly launched online shop.

4.3.3   Anything & More

4.3.3.1 Background Information and Family Influence

Anything & More is a large retailer, which is majority-owned by the founding family and runs numerous specialized stores in Germany. In the early 2000s, when the discontinuity 'online retailing' emerged, a family CEO managed *Anything & More*. Moreover, the family exerts influence through positions on the supervisory board.

My informant, a manager of *Anything & More*, described his motivation to join the retailer more than 10 years prior to the interview as due to his "initial high level of identification, not with the products, but with the family business." The owner and thus also the company stands for "social engagement, environment, and ecology." In particular, sustainability is a 'pet project' of the family CEO (e.g., press releases 2005; 2010; 2011)—at least as long as it does not harm the business' continuity (press article, 1995).

External influence, for instance via consultancy firms, has gained increasing importance during the last years. At the time of the interview, *Anything & More* was in a major portfolio restructuring process. Reputation, in particular related to social responsibility, and longevity are important yet not dominant concerns of the owner-family. Keeping the control over the organization within the family is one of *Anything & More*'s most important goals and the CEO frequently "categorically eliminated any possibility [of the core business] to go public" (press article, 2001).

A mid and long-term view is important for the family owners. According to several press articles (2001, 2002), any important strategic project gets a grace period of five years before first successes need to be reported.

4.3.3.2 Interpretation and Decision-making Processes

E-commerce is seen as the "key innovation" of the last decades by members of *Anything & More* and it has been framed as an opportunity from the beginning onwards: For instance, in the mid-1990s, the family CEO emphasized e-commerce's "growth potential" (press article, 1997) and talked about a "completely new shopping experience" realized by "interactive TV" as the CEO paraphrased online ordering at that time. In 1998, the CEO called internet based retailing and its current development "exciting" (press article, 1998).

The manager described in the personal interview: "We had a pragmatic view and said: 'ok, it's a new way for the customers to get their goods.' [...] 'Okay, internet is something new and we have to offer an online shop now.' [...] Pure internet [as a substitute for traditional retailing] was far away. Thus, no one perceived it as a threat; it was more like a completion, a 'must have'." Besides this pragmatic view, there was also "a substantial amount of fantasy among the decision makers: 'Hum, there could be more in it. We have to pursue that topic. Maybe there are successful concepts that differ from the core business nowadays." As a consequence of this belief, a venture capitalist was founded (see Chapter 4.3.3.3). Perceiving e-commerce as complementary rather than a substitute was a crucial determinant for the company's decision-making and adaptation process as a different strategic decision illustrates: Strong emotional attachment of the family CEO had substantially protracted an economically required portfolio restructuring process in the early 2000s. In 1995, the CEO believed: "E-commerce will not repress stationary retail."

The manager described the decision-making processes during the family CEO's incumbency as follows: "Before a proposal was discussed in the board meeting, one went to the CEO and said: 'So, [name of the CEO], in the next board meeting we'll discuss the following topic.' The CEO then had a look and responded: 'Hum, I don't understand that point.' Or 'That's good, I like it.' Or 'No, I don't like it, we should do it differently'. [...] When the last was the case, one could remove the respective agenda point again. One had to rework the concept until the CEO liked it. But once the CEO said 'yes, I like it,' the board meeting took place as follows: One introduced the concept in the board meeting, the CEO invited the executive committee members to the discussion by saying: 'Thank you for the presentation. I wait for your feedback.' Then it became really silent in the room. One could hear the rustling of the air conditioner. Few seconds later, the CEO said: 'I believe, it's a great idea, we should it do it that way.' Subsequently, the first voices rose, saying, 'Yes, [name of the CEO], I think so, too.' Ultimately, the decision of the executive board was unanimous."

Decisions were typically taken instantly—given that the factual basis was sufficient—without any respite required by the CEO. The manager described it like this: "When [name of the CEO] sits at the table and is asked a concrete question, then he/she will take a decision, given that there is enough information. [...] Frequently those decisions are based on an amount of data, of which I would assume that more details would have been beneficial." The motto of the CEO thereby is: "I know the business. My father built the business; I took it over from him. Buck up! That's the direction we'll go now."

In 1999, the family CEO called e-commerce a "more and more important strategic determinant of success" (press article, 1999). Two years later, at a time when the internet hype had abruptly declined, the CEO stated in an interview: "At that time [comment: mid-/end 1990s] there was an overblown euphoria, today exaggerated skepticism" (press article, 2001). In early 2002, the family CEO expressed his/her "joy in the investments in the strongly growing segments of the future, such as e-commerce" (historical website).

In 2002, *Anything & More* worked on its first mobile ordering services (press article, 2002). In 2006, *Anything & More* complemented its online shops with interactive elements allowing customers to rate products (press article, 2006).

According to a press note released in early 2009, the CEO perceived "e-commerce as the only remaining growth market." In the same year, one of *Anything & More*'s external directors stated in an interview (press release, 2009): "[Our core business] will remain [besides e-commerce] as long as it is useful for us."

4.3.3.3 Adaptation to Discontinuous Technological Change

Anything & More started its first internet based activities in the mid-1990s by following a two-fold strategy. First, the entire product range was offered online, thereby envisaging internet solely as a new sales channel for *Anything & More*'s extant products. Second, *Anything & More* founded a venture capitalist to invest in external internet startups as well as spin offs. The manager described the ex-post assessment of those firms supported by the venture capitalist as follows: "Unfortunately, it did not work. [...] There was a series of projects in which *Anything & More* was involved—to cut a long story short: It all failed. That was at the time of the first internet bubble. [...] Most projects ended up in the online chamber of horrors." For example, one of the early investments was stopped after six months due to lack of customer demand (press article, 2001). Any of these early investments were seen as "further steps to gaining experience in e-commerce" (press article, 1995). Since 1997/1998, *Anything & More* offered all of its products online (press article, 1997). At that time, the sales manager announced: "the first phase of experiments to have ended" (press article, 1999).

Around the turn of the century, *Anything & More* transcended its internet presence from pure offering of goods to native internet-based activities (such as customizing products and previewing them online; press article, 2001). At the same time, the company started to invest in one e-commerce segment known as particularly complex (press article, 2000) but failed. Up to the time of the study, *Anything & More* had not managed to be successful in that field and had, after another failure in 2010—at least temporarily—ceased to invest due to low profit ranges: "We have to accept this" (external manager, press release, 2010). Customer retention was perceived as one of the most pressing challenges at the turn of the century (press article, 2002). At the same time, besides its e-commerce activities, *Anything & More* still invested heavily in expanding its physical retail stores.

At the time of this study, *Anything & More* was running a successful, and award-winning online business in its core segment, accounting for more than 50% of the total revenues (according to several press releases 2009—2012). Furthermore, the company joined cooperations, owned a venture capitalist investment in internet startups, and invested heavily in e-commerce activities involving mobile devices. According to the website, one of *Anything & More*'s guiding principles is "[...] to be the leading player with new media."

Internally, *Anything & More* set up a project group called 'new media' that reported directly to the executive board. An external manager, experienced with new media, was hired as director of this department. To stay informed about emerging trends, *Anything & More* closely cooperates with academic research institutes.

4.3.4 King's Goods

4.3.4.1 Background Information and Family Influence

King's Goods is a large retailer which is owned one hundred percent by the founding family. The company runs numerous specialized stores and also markets its own products. Besides its main focus on Germany, *King's Goods* is also internationally active. The company is managed by a group of family members. The portfolio of offered goods had been changed over the last years due to several acquisitions and divestitures.

Continuity of family involvement in the firm plays an important role for *King's Goods'* owner-family and is also included in the company's family constitution. The previous generation had "hoped" for a family internal successor as a CEO and had fostered their children's emotional attachment to the family firm from early on. Still, familial ties are important for the CEO, as presence of 'family' in *King's Goods* provides "identification" and "hold[s] the values." To ensure the long-term goals of the company and to foster future prosperity, incentive systems of all top managers are closely tied to long-term rather than short-term goals.

In an interview in 2008 (press article), the CEO stated: "A loss of control [over the firm] is out of the question. We would rather do less, but do it more thoroughly." In another interview (press article, 2010), the CEO rephrased his/her motivation and explained this striving to maintain high levels of command as follows: "I once experienced being dependent on banks. I do not want to experience such a situation ever again." The owner-family is reluctant to cede responsibility and decision-making power to outsiders. Yet the CEO is indeed willing to take external advice for interpretation and decision-making and relies on an informal network of weak ties that are not particularly strong.

An important element of *Kings Goods'* culture is "to give freedom [to employees], particularly financially." [New innovations] don't have to be a giant success within the first or second year. Of course, we do not like total failure either. But providing free space is important."

Emotional attachment to the business plays an important, yet ambiguous role in the family firm. Due to this attachment, the CEO of the previous generation did not divest any business units although there were good economic reasons to do so (according to the current CEO, press article, 2010). One of the current CEO's first official acts was to sell those unprofitable business units. However, the CEO is not free of all emotional attachment, and in this vein, stated in an interview (press article, 2008): "We will not secede [one of *King's Goods*' business units]. It is the traditional core of our business."

4.3.4.2 Interpretation and Decision-making Processes

Early on, *King's Goods*' CEO was aware of the emergence of e-commerce and owned several Amazon shares privately until the first market breakdown at the beginning of the 21st century. The CEO described his/her own interpretation process from initial awareness through skepticism to perceiving market entry into e-commerce as inevitable as follows: "I was always interested in following Amazon's activities. However, later I made the mistake of saying at the beginning of the 21st century: 'They [Amazon] have not earned money for nine years now. I doubt whether this is really working right now.' Then the market broke down, the value of the shares fell and I sold them... From a business perspective, we were busy with some internal transformations at that time [and therefore had no capacity to invest in new economy]. The reason to enter the market was trivial, yet somehow logical: I have kids, and of course, they possess all the gadgets that have come onto the market in the last years: mobile phones, computers, apps, and so on, and they have always been used to them. At first I used to think: This is a temporary trend only. But then they grew older and began to buy that stuff themselves. And then I realized: 'Gosh, this is a generation that uses technology to consciously differentiate themselves from previous generations by saying: 'Listen, you still purchase and communicate the old way—we do it differently.' I realized that [e-commerce] would not go away again and that it would win an important role as a sales channel in the future." After becoming aware of the importance of e-commerce in general, the CEO pondered about its relevance for *King's Goods*: "It was obvious: I could have continued for another ten years without e-commerce and afterwards hand over the business to the next generation. But then, my children or the nephews and nieces from [my siblings'] families would say: 'Didn't you realize that something was happening here? Why didn't you enter the market?' And in principle, this was the motive [to become active in e-commerce]. Before you ask: No, this was not tied to any great strategic plans, scenarios, or discussions and so on. […] It was gut feeling. A feeling that we had to become active, that we had to enter the market." In an interview with the media (press article, 2010), the CEO described *King's Goods*' e-commerce activities as "laying the foundation for the future business of [our] children." The time horizon the CEO considered in this planning is "approximately 30 years" (press article, 2010).

The CEO commented on the decision-making process to enter the e-commerce market and the implementation as follows: "One of the advantages of an owner-managed business is you can get an idea into your head and just say—given a certain congeniality and assertiveness—'we'll do it that way.' Some people that sit at my table still do not want to accept that this is part of the future. They say: I don't want to do that. In the next meeting, I'll show them an iPad." The CEO shortly discussed the decision to heavily invest in e-commerce with the supervisory board (low number of members; equal representation of family internals and family externals) and subsequently "just did it" without lengthy discussions with top managers.

At the time of the interview, the CEO perceived *King's Goods'* situation to be as follows: "We are about to enter a fascinating world which is not easy at all; a world that is very transparent, incredibly fast, and direct. [...] When this world arrives, one has to take part." The CEO also admits, that e-commerce is a "totally different way of thinking" (press article, 2010).

4.3.4.3 Adaptation to Discontinuous Technological Change

In 1997, the company, in cooperation with regional partners, started its first home delivery services in selected geographic areas based on telephone and fax-based ordering, and including minimum order quantities of 30 Deutsche Mark[47], delivery costs of roughly 10 Deutsche Mark per order, and delivery times of two to three days (press article, 1997). At the same time, *King's Goods* temporarily experimented with online-based retailing in a few selected geographic areas (press article, 1999). Since 2001, one of *King's Goods'* subsidiaries ran its own online shop for products outside its core business. This shop had constituted a "shadowy existence" for a long time and had mediocre success over a long time span (less than 1% of revenue share in 2007, according to the annual report of 2008). *King's Goods*, however, continued the online shop as the CEO "had the dumb feeling that something is going on and we have to stay tuned to it" (press article, 2010). Asked about the reasons for the initial lack of success of this shop, the CEO responded: "We were too fainthearted, not revolutionary enough. Furthermore, our perspective was still that of a stationary retailer, the old world. [We ran the online shop] as an add-on and were not sufficiently embedded into the online world." Despite this retrospective skepticism about *King's Goods'* early online performance, an online shop that *King's Goods* launched with a partner around 2003 was

---

[47] Former German currency.

bestowed with an honorable online award (annual report). In the following year, the company's online shop, which was among the most frequently visited German online shops at that time, was seen as an "important unique selling feature" (annual report).

When the CEO started to concern him/herself with that online shop at the end of the 2000s and when the online shop was separated from its stationary equipollent, e-commerce got new momentum within *King's Goods*. The number of employees of that online shop doubled from 2006 to 2009 according to annual reports. In order to break established routines, an external manager with experience in e-commerce was hired to lead the new online department. Moreover, the product portfolio was broadened. In 2009, *King's Goods*' annual report put emphasis on the company's online activities, stressing the importance of that medium as a marketing and sales channel. At the same time, *King's Goods* launched further internet shops. To enable fast growth in e-commerce, which became a "strategic business unit" of *King's Goods*, the company started investing in internet shops. The CEO assessed the risk of these investments as relatively moderate, as the amount of money invested is "reasonable" (press article, 2011). These online shops are seen as an important strategic activity of *King's Goods*, however, the CEO stated in a 2010 interview (press articles): "Our physical stores will continue to exist for a long time."

4.3.5 Retail 2000

4.3.5.1 Background Information and Family Influence

Retail 2000 is a large retailer, fully owned by descendants of the founder. The family business has steadily grown throughout its history and runs numerous specialized and non-specialized stores in Germany and other selected European countries (since the mid-1990s). A family CEO manages the company in a decentralized manner. Other family members are involved in several top management positions. On its historic websites (2002), *Retail 2000* puts emphasis on growing organically without any acquisitions. The company's main goal is "to be number one in specific regions and markets" (press article, 2004).

At the time of the interview, one member of the next generation was preparing for taking on management responsibility in the medium-term, whereas his/her siblings were not involved in the business at all. The current family CEO is actively engaged in associations that aim to foster the regional economy.

According to the CEO, and confirmed by historical website information (e.g., 2002, 2004), *Retail 2000*'s culture rests on two main pillars—"customer orientation" and "qualified employees." Particularly compared to industry average, employees are granted "much space for own decisions," thus indicating a strong sense of community within the organization. Numerous employees are "silent partners" of *Retail 2000* with information rights and profit shares, but, with no voting rights. Employees also influence *Retail 2000*'s strategy. For instance, *Retail 2000*'s decision to expand internationally was originated on impulses of one employee who propelled that idea to the CEO. The CEO in turn quickly assessed the potential of such an expansion and decided to enter the new geographical markets despite initial reluctance within the family. The underlying rationale for this expansion was to ensure the long-term goals of *Retail 2000* could be met. Due to its focus on continuity, the CEO was willing to accept high cost in the short-term, as he/she hoped to be able to harvest profits in the long-term. On first sight, this international expansion constitutes a paradox, because, within Germany, *Retail 2000* is very focused on certain regions. The CEO emphasized in interviews that "[r]etail is a local business" (press article, 2004). However, when taking a closer look, one notices that *Retail 2000* solely expanded to geographical regions to which either family members (in particular of the new generation) or employees had personal connections.

The CEO of *Retail 2000* is deeply involved in all decision-making processes, also regarding the decisions of the European subsidiaries; however, he/she strives for a "cooperative manner" of decision-making, resulting in mostly "joint decisions." All decisions are based on "factual, objective arguments" rather than on gut feeling. This is seen as a measure to attract and keep talented employees. The CEO self-assesses *Retail 2000* as follows: "Maybe, we are not a typical family business, we do not first contrive decisions within the family." However, members of the previous generation, more than 90 years in age, are still, on a regular basis, involved in discussions about the firm's development, although this happens in an informal way. The CEO is reluctant to involve any firm-externals in decision-making processes as he/she sees "tackling difficult tasks as a core task of managers [rather than externals]."

Connections to externals play a subordinate role for *Retail 2000*. The CEO prefers to spend time internally, e.g., with managers and employees, rather than attending external events. Several decades ago, *Retail 2000* experienced a radical change in its environment that resulted in the discussion among family members whether to "be loyal to customers and hazard the potential decline" or to exasperate the existing customers and enter new market

126

segments. *Retail 2000* decided to enter the new market, offering some compromises for the existing customers.

4.3.5.2 Interpretation and Decision-making Processes

At the turn of the century the CEO of *Retail 2000* became aware of the discontinuity and stated in an interview (press article, 2000): "We perceive pure e-commerce, by that we mean selling things online and delivering them to the customers, as secondary; it requires too many resources." Rather, the CEO assessed the internet to be important for supplier networks and as a consequence, managers discussed how to utilize the new technology.

Over the years, with the rise of Amazon, *Retail 2000* started to recognize that e-commerce is an "important topic" (CEO, personal interview). However, "[*Retail 2000*] has not seen the way to get there yet" (CEO, personal interview). The CEO admitted that "new ideas" were required to be successful in the new field, and, as a consequence, market entry into e-commerce regarding products of the company's core was perceived as non-attractive, since a pure transfer of the old business model to an online business model was not possible. The CEO believed that profit margins would become negative when selling their products online. Although he/she was aware of international competitors successfully offering those goods online, he/she did not have any promising concepts in mind.

Interestingly, when the CEO was asked about "major changes in the company and/or industry" and "major turmoil" in the first half of the interview, he/she did not refer to online businesses, but instead talked about internationalization and increased competition. Only when asked explicitly about online businesses, did he/she make the above-mentioned statements.

4.3.5.3 Adaptation to Discontinuous Technological Change

Instead of entering the new domain of online retailing, the company long focused on expanding its existing business to other selected geographical regions. First online market entries comprised digital photography services (early 2004 according to historical website, still available in 2012) and 'order online—collect from store' (available for international markets no later than early 2006, according to historical websites). However, all these initiatives failed according to the CEO. At the end of the 2000s, *Retail 2000* set up a project team dealing with potential online strategies. Similar to *Retail 2000*'s internationalization process, this online team was initiated by some employees. Just before the interview, *Retail*

2000 had acquired the majority of shares of a small online retailer, active in a segment not related to *Retail 2000*'s core business segments. Sixteen months after the interview, by the end of 2011, *Retail 2000* re-entered the online business segment by combining online ordering and physical collection, a model similar to those of several competitors. According to press articles, this was an investment of several million EUR.

4.3.6 Kiddies

4.3.6.1 Background Information and Family Influence

Kiddies is a medium-sized, one hundred percent family-owned, toy manufacturer, managed by a group of family owners. The company exports more than half of its goods. In 2008, two years before this study began, *Kiddies* had to dismiss one third of its staff due to economic issues.

The product portfolio has barely changed since the company's foundation, as the current CEO describes: "My grandfather manufactured [a special kind of] toys, my father manufactured [a special kind of] toys, and I also manufacture [a special kind of] toys, because I have not learnt to do anything different." In the 1970s, *Kiddies* was the market leader in its specific product segment in the German speaking part of Europe. At the same time, the company broadened its product portfolio for several years before it was consolidated again. At the time of this study, the number of newly developed products (incremental innovations) and product variations was high (approximately 50 per year) and new products were mostly closely tied to current social trends. The owners themselves, employees, and also customers, generate ideas for new products. One of *Kiddies*' core competencies is its high product development speed resulting in short time-to-market. *Kiddies*' self-image can be described as follows (press article, 1999): "We see ourselves as preservers of the past [...] but also as renewers."

The level of formalization within *Kiddies* is low, "nothing is written down, everything is in the heads of the individuals" (CEO, personal interview). The company's main goal is "survival" and to constitute a "company with integrity" where employees "enjoy working." Thus, the company is highly focused on *community*. Much competency is delegated to the employees, according to the CEO and staff turnover is very low at *Kiddies*. At the time of the interview, the 'newest' employee had worked for *Kiddies* for eight years, and the most senior employee for 35 years.

Connections play a subordinate role for this company. Although the CEO is active in various industry associations, he/she does not value the information obtained from such exchanges. Moreover, no family externals (except employees and customers) are involved in sense or decision-making processes.

Decisions are taken based on intuition rather than on calculations, but are transparently communicated to the employees. Additionally, the CEO stated: "When some [activity or project] fails, it is not the end of the world—it is, what it is."

4.3.6.2 Interpretation and Decision-making Processes

The previous CEO, parent of my informant, was one of the first to import components and products from Asian manufacturers in the 1970s, however "burnt his/her fingers" by receiving low quality goods only. This CEO assessed that successful cooperation with Asian partners would require presence of a *Kiddies* representative in Asia. He/she rejected that idea, and as a consequence, renounced any activities in that geographic region. The current CEO admits, that this 'exclusion' lead to several problems in the following years, however, he/she had never tried to revoke his/her parent's decision—"it is what it is," the CEO interposed frequently during the interview. The previous CEO plays a crucial role in *Kiddies*' history. From 1997 to 2010, a separate subsection of the website was dedicated to describing the role this CEO who was described as a "straight and earnest" person who "cut [his/her] own path" and "whose handshake counted."

The CEO aims to be attentive to "weak signals" of change such as altering customer demands observed at industry fairs. When Asian competitors entered the German market with cheap, electronic products, the CEO assessed the situation as follows: "There was one segment [where we] had difficulties [due to Asian competitors] and another segment with much demand [*comment: the niche, this company later retreated to*]. Of course, in such a situation one strengthens the easier segments." The CEO believes that *Kiddies*' niche is 'safe' due to a lack of material, technology, tradition, and flexibility among the potential competitors.

4.3.6.3 Adaptation to Discontinuous Technological Change

To "avoid competition, particularly [by] Asian competitors entering our market" (CEO, personal interview), *Kiddies* ultimately retrenched into a market niche in 1989, focusing on products for a specialized target group. The CEO described this move as "fleeing" into a niche and "not serving the markets [threatened by electronic toys and games, mainly produced by Asian competitors] anymore." The CEO concluded that part of the interview by stating "It is the only [possible] way."

The CEO summarized the strategy as follows: "We have not changed our technology, we have changed our product range, we have changed our products, we have changed our markets, and we have changed our customers." Internally, *Kiddies* strongly involves employees in the product development and innovation processes, e.g., one of the family members explains product issues to the employees in detail, ensuring that they understand the product as a whole.

In mid-2007, *Kiddies* began to refer to online shops selling the companies' products and in 2012, *Kiddies*' own online shop went live.

4.3.7 Play & More

4.3.7.1 Background Information and Family Influence

Play & More is a medium-sized game producer, led by two members of one family. The business is one hundred percent owned by the founding family, however, ownership is dispersed among roughly 20 persons. *Play & More* owns branch offices in several European and international cities. Besides development and manufacturing, *Play & More* is known as an importer of games and toys. *Play & More*'s mid and long-term goal is growth by further expansion in the European market as well as an increase in market shares through further development of sales channels. The company positions itself among the high quality providers of games and toys. Information given on the company's websites show that the number of products sold had declined by 50% within one decade, for two of *Play & More*'s core segments (historical websites 2000, 2010).

The CEO described his/her relationship to the firm as follows: "Unfortunately I was born with the name [of the family and the family business...] and from early on, my parents pushed me in the direction of becoming their successor. After an apprenticeship, I joined the family firm at the lowest hierarchical level and had to climb up the greasy pole." While the CEO appreciates the atmosphere within the firm and the good manner of all stakeholders, he/she does not identify with the products at all.

According to the self-description in the 'about us' section on the company's website, *Play & More* has changed significantly in the past. However, the last major changes occurred in the 1950s (broadening of the product range to today's three core segments). The last minor change happened in 2003, when the company expanded one of its factories.

Play & More's culture is shaped by "tradition as passing on of fire, not the worship of ashes" as well as "social responsibility" of the CEO. This perceived responsibility is also the reason that prohibits *Play & More* from shifting production to non-European countries. Strategic decisions are fact based, whereas decisions on topics around human relations are based on gut feeling. Once a decision has been made, the CEO sticks to it, even if it turns out wrong.

Membership in associations of various related industries is important to the CEO in order to get "information, [ideas for] innovation and to have the finger on the pulse of time." The CEO states, "one has to make use of any information available [to get informed about new trends]," and thus the CEO, for instance, relies extensively on discussions and reading of relevant material. Also listening to his/her children's opinion in order "to remain fresh" is important. External impulses play an important role for *Play & More*'s strategic moves, for instance, many of its successful product developments, the market opportunities for those new products, as well as its current focus on individualized products (see Chapter 4.3.7.2) were triggered by externals approaching *Play & More* with respective ideas.

On average, employees work for *Play & More* for more than 25 years, before they leave the company or retire. The CEO assumes that this leads to a "conservative attitude" among employees and reluctance to think differently: "'We have always done it like that,' is the philosophy of most employees," the CEO summarized.

4.3.7.2 Interpretation and Decision-making Processes

The CEO attributes much of *Play & More*'s past successes to "a large portion of luck." Most previous innovations are seen as incremental and ideas for new products were delivered to *Play & More* by game and toy inventors who either actively presented their ideas on industry fairs or contacted *Play & More* directly. Before committing to new products, *Play & More* invests in thorough psychological and pedagogical testing and calculations as a basis for decision-making. Due to the low ratio of tops versus flops, the CEO rejects having an internal group of product developers.

The CEO expresses his/her belief like this: "I reckon, the 'homo ludens' will not die off and thus I believe that there is also room for [*Play & More*] in the future at least as long as 'computers do not start laughing.' Electronics is not communicative. [...] And even online games are not very communicative. [By playing traditional games and using traditional toys] one can talk, one can have fun—I can't do that when playing electronically. And that is our right to exist." A statement on the company's historic website (2000, first available website of *Play & More*) supports this view: "[Our traditional games and toys] bring relief in a digitized world."

This is in line with the attitude of the second family CEO who stated in an interview in 2004: "Computers are successful, but they suffer from lack of immediate reaction of the teammate. Furthermore, cordial laughter is not possible with computers" (press article).

My informant perceives electronic games and toys to be "a hype that could not have been foreseen." Ex-post, he/she justifies the decision not to enter the market (as described in Chapter 4.3.7.3): "I have to tell you anyway, that I am very happy that we did not join in. We would have lost our shirts. Only the really big players such as [...] can do that [i.e. can be successful...]. We did not have the capacity, the financial resources, to take part. [Compared to the competitors], *Play & More* is flyspeck." The CEO believes: "I think we should continue working on [our core segments]. If we are not able to do so, we have to close our business."

4.3.7.3 Adaptation to Discontinuous Technological Change

Due to the above (Chapter 4.3.7.2) mentioned reasons, mainly resignation caused by capacity constraints and denial of electronic games and toys' right to exist, *Play & More* rejected entering the electronic market. For long years, the company behaved passively, before starting to invest in domain offense activities in the 2000s. These activities largely relied on broadening *Play & More*'s portfolio of 'traditional' products, and for instance, individualizing its products as much as possible (since the end of 2009, according to historical website). In earlier years, *Play & More* relied on sustainable product innovations (website 2002, 2008) and luxury editions of its products (website 2006). Moreover, *Play & More* has steadily increased its emphasis on "quality" and "sustainability" (word counts, historical websites 2000—2010).

Now, in 2012, it is still not possible, to purchase *Play & More*'s products via an online shop—despite the fact that all products are described online, including pictures of each product.

4.3.8 Toys 2000

4.3.8.1 Background Information and Family Influence

Toys 2000 is a medium-sized specialized toy manufacturer that is fully owned by the founding family and managed by a team of two CEOs, one from the family, and one outsider. A member of the next generation was preparing for succession at the time of the interview. The company undertook major changes in its portfolio during the early years after its founding, ultimately entering the toy segment approximately 60 years ago. In the last decades, the company has substantially invested in the reputation of its toy brand. During the last decades, *Toys 2000* has substantially broadened its product portfolio, including addressing new customer groups. To remain competitive, the company possesses its own factory in China.

The CEO describes *Toys 2000* as a company with "personal leadership, [with processes that are] not fully formalized, [... characterized by] personalized [as opposed to fact-based] decision-making and fair treatment of employees." Compared to companies of similar size within the industry, *Toys 2000* has a "large development and construction department." Employees of *Toys 2000* strongly identify themselves not only with the company itself but also with its products. *Community* plays an important role for the CEO who, several times during the interview, spoke about *Toys 2000*'s "responsible and prudent [...] really good employees" as a key factor determining the firm's success.

Due to economic reasons the CEO has shifted the production to several Eastern countries, first by cooperation with a supplier, later by their own factory, although this "was a risky step for a company of [*Toys 2000*'s] size." The CEO aimed to compensate for that risk with meticulous preparation of that strategic move. When cooperating with a supplier, it was very important to the CEO to find a partner that "*Toys 2000* was able to influence, to tell them exactly what we want and in what quality."

Connections and exchange with externals play a crucial role for *Toys 2000*. The CEO is active in many, mostly local, associations and unions, where he/she is "engaged in many discussions and an intense exchange," since he/she believes that "discussion partners, sparring partners are absolutely necessary." Besides this continuous involvement, the CEO perceives visits at national and international industry fairs as an important source of information to "create an overview of what is happening in the industry [as] there is important contact not only with other manufacturers but also with customers, not only nationally but also internationally." To account for external influence on any major strategic decisions and "to avoid myopia," *Toys 2000* has installed an advisory board whose composition and rights are frequently adapted to best serve the company. At the time of this interview, the advisory board consisted of several external members—mostly 'friends' of the CEO who are successful entrepreneurs of firms within different industries—as well as the family owners. Several strategic decisions could only be taken if at least one non-family member of the advisory board agreed[48].

4.3.8.2 Interpretation and Decision-making Processes

Due to *Toys 2000*'s close connections to the Asian market, particularly fostered through partnership with a high quality component supplier in China, the CEO of the company was aware very early of the trend toward electronic games and toys. An internal project group with representatives of product development, sales and marketing discussed the development and made a proposal to the CEO. In 1998, the CEO described Asian competitors as "murderous" and being responsible for the decline of German incumbents in the toy and gaming market (press article). In the 2010 annual report, these competitors are listed as one the main risks of *Toys 2000*'s business. The opportunity of *Toys 2000* to successfully master this challenge, lies in the company's "swiftness and flexibility," according to the same annual report.

Two things were important to the CEO when deciding about *Toys 2000*'s 'electronic toys strategy': First, they aimed to develop and market products "that have not been on the market yet." And second, *Toys 2000* is very concerned about "biting the hand that feeds [them]." Thus, entry into the electronic toys market was only acceptable as a "parallel market segment." This concern is also the reason why products of *Toys 2000* still cannot be ordered online in 2012.

Electronic toys and games were framed as "something completely new to us, something fairly complex." To come to a decision about the adaptation strategy, the CEO asked the respective employees to first create an optical draft of the new products in order to be able to "thoroughly assess the concepts." The CEO describes *Toys 2000*'s employees, specifically the engineers, as drivers of innovation who "work new concepts internally"[49].

---

[48] Of course, in principle, the family members could just dismiss the external board members if they did not agree with a decision or proposal from the family members. However, as the CEO stated: "That was not the object of this exercise."

[49] This is in line with a 2003 press article that describes a situation in which employees convince the initially reluctant CEO of a new product idea.

Once there was the idea to enter the electronic market, a "jolt went through the company and the employees enjoyed working [on the new developments]."

4.3.8.3 Adaptation to Discontinuous Technological Change

Several decades ago, in the early 1980s, *Toys 2000* started to react to the discontinuity by engaging in strategic activities with Asian partners. However, development of products based on the new technology did not begin before 2001. Since the knowledge and competencies required for this type of development significantly deviates from that of traditional toy development, the company hired several new employees with specific capabilities. The first electronic products were brought to market at the end of 2004 (according to historical website) and nowadays constitute an important part of the product catalogue, together with classical products. *Toys 2000* has broadened its product portfolio within the last decade with focus on 'classical' products on the one hand and 'electronic' products on the other hand. For all product segments, "high quality" is crucial to *Toys 2000*, as many certifications show. The company's website includes many customer retention elements. An important industry association acknowledged *Toys 2000*'s above-average financial development over the past years (annual report, 2010).

4.3.9 Walter & Colleagues

4.3.9.1 Background Information and Family Influence

Walter & Colleagues is a medium-sized, established development and manufacturing company in the power-metering segment. It is one hundred percent owned by the founding family and several family members manage the company jointly. With a focus on metering components and distribution techniques, the company is internationally active. During previous years, *Walter & Colleagues* has invested in a few selected acquisitions, *Powerhouse* (Chapter 4.3.10) is perceived to be the main competitor of *Walter & Colleagues*. Continued growth is one of *Walter & Colleagues*' major goals, and, at the time of the interview, the company was expanding its office and logistics buildings (manager, CEO). The CEO described this strategy as follows: "We invest in new products and new markets [...] but we refrain from risky investments. This has not changed since my grandfather's times." In the concluding remarks of the interviews, the CEO summarized: "Being successful means that we are able to finance our investments. We want to be a leading player in our industry. We do not need to become the number one; that is not our pretension. We can also be number two, three or even four in some segments. But we want to have a relevant role."

Continuity is most important to the CEO: "We want to remain a family business. We do not want to give the firm into the hands of strangers." The manager described, "the company and the family are inextricable. There are no decisions without the agreement of all three family managers." Particular focus is put on hiring decisions. The family managers "are careful about who comes into 'the family,' who matches [the business culture]" (manager). This also leads to a specific sense of *community* and "a special kind of employee" working for the company (manager).

When describing the company's culture, the CEO heavily emphasized the external connections and the internal link to the family as "the family represents the company externally." This is consistent with historical and current websites, where relevant messages to customers are accompanied by a photograph of the senior CEO. The CEO continued regarding the importance that "customers, employees—both groups are our most important partners—and of course also the suppliers have direct access to the owner-family." This too is in line with the company's self-presentation on the internet. Remarkably, the family firm offers the acquisition of (non-voting) shares in their company to business partners and customers—and only to those.

The manager believes that "early awareness of market development," is what distinguishes *Walter & Colleagues* from its competitors. More generally, "speed of reaction" is seen as the company's unique selling point. The manager describes: "When [the CEO] averred, 'We'll do it that way,' then we did it that way. There was no time-consuming questioning. The CEO said it, therefore we did it."

4.3.9.2 Interpretation and Decision-making Processes

Early awareness of trends is created by the "networking activities of the owner-family" (manager). The family managers are active in industry associations and involve customers as well as potential customers (i.e. companies from different positions of the energy value chain) in discussions on new technology and product trends.

So it is the "gut feeling of the seniors" that determines strategic directions. The manager describes his/her own experiences: "The senior CEO has a gut feeling and says: 'Yes, this direction could be promising.' And concerning the CEO, the directors ask: 'How can the CEO take this decision? It is doomed to failure. Particularly the numbers guys, like me, calculate the cases and conclude: 'no.' But then, a year later, everyone is happy about the CEO's decisions." The CEO confirmed the important role of "gut feeling" for any strategic decisions and justified it by a lack of resources available to assess opportunities and risks on a more factual basis.

Walter & Colleagues' CEO became aware of smart meters around 2003, when the first industry cooperation was set up to develop an electronic meter. This product fundamentally affected the design of one of Walter & Colleagues' main products, a component part. The CEO instantly decided to adapt the metering component, because: "If we didn't adapt to technical developments or even shape the topic, then we would have become a has-been and we would have had to tell our customers that we were unable to deliver due to a lack of technology. Therefore we instantly decided on adaptation. [...] It was a 'must-decision.'" From that time onwards, the CEO of the company was persuaded of the prevalence of the new technology and perceived the development as a "strategic chance" that metering products would ultimately "get more attention of end customers." However, the CEO somewhat curtailed the growth expectations: "This [increased interest of customers] will not be measurable within two or three years. I believe one has to look thirty years ahead. "

Besides adapting its core product, the metering components, the company ultimately perceived the market turmoil as an opportunity allowing Walter & Colleagues to successfully enter the metering market itself. In 2008, after one year of interpretation, Walter & Colleagues decided to produce electronic meters, although this was "uncharted technological territory" for the hardware producer because of substantial software requirements.

Smart meters are perceived as something that will definitely come: "It will come, it has to come. If you watch the trend towards refrigerators placing orders autonomously, a smart meter constitutes a small component to that direction" (manager).

4.3.9.3 Adaptation to Discontinuous Technological Change

Smart meters were first mentioned on Walter & Colleagues' website in 2005. In 2008, Walter & Colleagues acquired a competitor that possessed the software competencies required for smart metering development and production. Moreover, the company engaged in a cooperation with the engineering department of a university to improve the smart metering development. The new product is designed by a development team and its advancement is supervised by the family managers who give large portions of their managerial attention to the new meter and are even involved in decision-making on a detailed, technical level. Roughly

20% of *Walter & Colleagues'* employees are engaged in smart meter development to some level. No additional staff was hired for the new market entry. The CEO wanted to wait for first (financial) success before committing further resources. At the time of the interview, the smart components had been successfully launched and the smart meter was about to be introduced to the market. According to the company's website, the product was launched in mid-2011. At the same time, the company invested in electronic mobility features, which constitute other important aspects of smart energy besides smart meters (press article, 2011).

Customers reacted positively, yet with astonishment, when realizing that *Walter & Colleagues* had entered a new market. In principle, *Walter & Colleagues* assessed the smart metering projects as successful at the time of the interview.

4.3.10 Powerhouse

4.3.10.1   Background Information on the Company

Powerhouse is a large family business active in the development and manufacturing of meters, as well as related product segments. The company is one hundred percent owned by several family members and managed by a family CEO. My interview partner was the regional CEO of Germany, a non-family member, who, at the time of the interview, had been working for the company for 45 years; approximately 25 thereof in his/her current position. The CEO spends approximately one third of the working time on information exchange and discussion with members of the owner-family. *Powerhouse* is internationally active and holds a leading position in several market segments. In previous years, *Powerhouse* has acquired several national and international companies.

For the (non-family) CEO, focus on continuity is important: "We would like to remain a family business. [...] It is really important for us to keep the owners interested in the business, because interest in the company reduces anonymity and reduces the hazard of selling shares." The CEO names "strong emotional attachment" and "independence from the capital market" (note: the company has no bank loans at all) as two relevant advantages of *Powerhouse*. "Enduring continuance" is *Powerhouse*'s most important goal.

The organizational culture is characterized by "sustainability, endurance, [...] teamwork, [...] loyalty and trust [...]. People, not numbers are in the center of *Powerhouse*." The company's values "loyalty, trust, enthusiasm" have remained constant during the manager's tenure, whereas "team work" was added several years ago. Due to flat hierarchical structures, decisions can be taken quickly.

The founders, aged over 80, remain actively involved in the company by "maintaining contacts, not solely to employees but also to industry associations and to selected customers." "Closeness to customers" is *Powerhouse*'s unique selling feature according to the CEO. The company realized these close connections through seminars, trainings, a strong sales department, and prominent presence at industry fairs. Altering its customer groups is not an option for *Powerhouse*: "We remain in this track; we do not want to bypass [our customers], we want to remain predictable."

Powerhouse voluntarily installed a supervisory board, consisting of two family members and one external person, more than 20 years ago. At the time of the interview, *Powerhouse* was restructuring its board to a distribution of power that leaves family members only one third of the votes. Decision makers at *Powerhouse* aim for consensus-based decisions. For instance, the regional CEOs discussed the closure of *Powerhouse*'s first factory extensively with the owners. The owners felt emotionally attached to the location and initially voted for "leaving everything as it is." Ultimately, however, they were convinced and agreed to closure.

Powerhouse shows high levels of stamina: After the failed acquisition of a company in the past, *Powerhouse* re-engaged in the activities and ultimately purchased the company almost 10 years after the first trial.

4.3.10.2 Interpretation and Decision-making Processes

In general, the owner-family is rather enthusiastic about investing in new technologies. For instance, in the past the owner family's members jointly invested private money in solar energy generation despite the risks involved. The main motivation to invest a 7-digit amount in that field was the family's striving for "sustainability."

Smart metering first attracted attention within *Powerhouse* at the beginning of the 2000s, when media commenced discussions on the topic and regional power distributors fostered the 'smart development.' The new technology was seen as something that "positively influences [*Powerhouse*'s] business model." To evaluate its strategic options and decide on one, the company did an initial calculation of the predicted market size for smart meters.

For *Powerhouse*, it has always been an important "basic philosophy [...] to shape the standards, [...] to participate in development from early on. When you feel that something is happening, then you need to investigate it and try to find out whether there is a market opportunity. And if there is a market opportunity, then join in or don't join in and bear the consequences. But you have to show stamina. [...] We showed stamina when we first started developing [a certain product component] in 1983. It took 10 years until the product was ready to sell. But today, it is an evergreen." The manager remarked that this shaping of standards means that employees are "happy about their potential contribution" and become flexible.

Regarding smart meters, members of the organization felt that they "must participate [in the market development] and that [they] must be able to handle the technology." Thus the family CEO decided to enter the field. In an interview (annual report, 2010) the family CEO remarked: "Before us there is a completely new, electronic world [...] that offers manifold market opportunities to [*Powerhouse*]. This pie we are talking about—and how it will be sliced—is not yet here—but it will come." Somewhat similar, in a media interview, the CEO described *Powerhouse*'s strategy for a future world based on intelligent grids and mobility (press article, 2010): "We have to participate. We have to position ourselves. And we have to watch the development."

4.3.10.3   Adaptation to Discontinuous Technological Change

After the family CEO's decision to enter the market, *Powerhouse* first engaged in a market study to evaluate the potential market segments and market participants. During that time, *Powerhouse* built up several partnerships with other players in the energy sector and drew on many of its contacts in order to collect as much information as possible. As a next step, *Powerhouse* started its development phase, first with a very "limited capacity." The regional CEO explained: "Smart metering is a slow growing topic, but a topic that will generate huge dynamics." Over the last years, *Powerhouse* hired some new employees who were organizationally allocated to one of the existing business units. Most members of *Powerhouse*'s development team are long-tenured employees who first had to learn the new competencies. A long-tenured employee of *Powerhouse* managed the department. As a first step, the company developed and produced a smart meter and participated in a local field trial. As a consequence of this experiment, *Powerhouse* developed a second generation of smart meters. While the production for these meters is outsourced to a partner, sales are still carried out by *Powerhouse*. First products were commercialized in 2008. At the time of the interview, the partner firm was developing a third generation of the smart meter. Moreover, the company had developed several component parts for smart meters.

Although *Powerhouse* assesses its activities hitherto in the smart metering sector as quite successful, two major difficulties remain according to the regional CEO: First, to foresee the time when market growth will take off and thus to ramp up the capacities at the right point in time. And second, to assess the performance of its products hitherto, as conventional measures such as market shares, revenues and profit margins are inapplicable. To overcome this challenge, *Powerhouse* uses established technological and customer based criteria such as usability, security, quality, and features as a measure for success. Other than those challenges described, 'waiting for takeoff' is not perceived as a problem since *Powerhouse*'s production is fully loaded with manufacturing the traditional goods. *Powerhouse* has ceased planning for the capacity ramp up for smart meters. Moreover it is important for *Powerhouse* to get in contact with potential customers early on. *Powerhouse* is very concerned about not disturbing the market too much, that is, *Powerhouse* tries to avoid threatening players active in other parts of the value chain; the image of being "credent" is crucial to the firm.

4.4 Results

In the following chapter, I will outline and discuss the results of my empirical investigation. I will thereby show that eight out of ten companies adopted the technology, most of them quickly and with high levels of stamina, yet with low resource commitment and somewhat rigidly. Moreover, I will show that despite these general trends, the organizations varied in their response patterns and I will identify socioemotional wealth considerations as the determinants of such variance.

At the time of writing this thesis (early 2012), only two out of the ten companies in this sample had not adopted the discontinuous technology. Of the two non-adopters, one company (*Kiddies*) had deliberately retrenched into a niche (Adner & Snow, 2010), a behavior labeled as 'domain consolidation' by prior research (Ford & Baucus, 1987). The CEO of *Kiddies* actively decided to focus on a selected group of customers (those who value traditional toys) and give up all other parts of the company's previous customer base as a reaction to the emergence of the discontinuous technology. This strategic move entailed various, albeit incremental, product innovations. Another company (*Play & More*) reacted mostly passively by continuing business as usual (Ford & Baucus, 1987). This behavior was rooted in the CEO's resignation ("*Play & More* is flyspeck"[50]) and denial ("I can't [talk and have fun] when playing electronically. And that is our right to exist"). During the last years, *Play & More* has started to engage in some level of domain offense activities (Ford & Baucus, 1987) by offering high quality and individualized products.

All other companies had engaged in domain creation (Ford & Baucus, 1987), meaning they commercialized the discontinuous technology, albeit with different speed, resource intensity, stamina, and routine flexibility as shown in Table 9.

At most of those companies, however, strategic switches towards the new segment were accompanied by offense activities (Ford & Baucus, 1987) (*King's Goods*, *Powerhouse*, *Retail 2000*, *Toys 2000*, *Walter & Colleagues*, and *White & Blue*). Domain offense included, for instance, opening new physical stores, launching customer retention programs, and extending the classical product portfolio. As several interview statements show, the new technology was (at least initially) rather perceived as a supplement than a substitute to the established technology. Most companies aimed at a co-existence of the old and new technology, as reflected in the CEO statement of one adopter company:

> *Our physical stores will continue to exist for a long time. (King's Goods)*

Anything & More was the only company indicating its general willingness to give up the old business model, that is its physical stores, entirely in the future, and instead switch over to business models relying on the new technology:

> *[Our core business] will remain [...] as long as it is useful for us. (Anything & More)*

[50] As described in chapter 4.3, all quotations henceforth were originally provided in German, translated to English by myself, and back-translated by students to ensure correctness.

Table 9: Family Influence and Adaptation—Empirical Findings

Company	Type of Reaction[a]	Speed[b]	Intensity[b]	Stamina[bc]	Flexibility of Routines[b]
Alpha Star	Domain creation	+++	+	(+++)	++
White & Blue	Domain offense/ creation	++++	+/++	o/Re-start after early failure	+
Anything & More	Domain creation	++++	++++	++++	+++
King's Goods	Domain offense/ creation	++++	++/+++	++++	+/+++
Retail 2000	Domain offense/ creation	++	+/+++	o/Re-start after early failure	+
Kiddies	Domain consoli- dation	++++	n/a	n/a	n/a
Play & More	Mostly passive	n/a	n/a	n/a	n/a
Toys 2000	Domain offense/ creation	++	+++	(+++)	++
Walter & Colleagues	Domain offense/ creation	++	++	(++++)	++
Powerhouse	Domain offense/ creation	+++	++	(++++)	++

[a] categorization based on Ford & Baucus, 1987; Zammuto & Cameron, 1985

[b] o = not available/very low (substantially below industry average), + = low (below industry average), ++ = medium (industry average), +++ = high (above industry average), ++++ = very high (substantially above industry average). Speed denotes the time from emergence of the innovation until the first product building on the new technology was brought to market. Intensity refers to the relative amount of resources invested in commercialization of the new as opposed to exploitation of the old technology. Stamina refers to the time span a company maintains its investments even under conditions of (temporary) failure. Routine flexibility refers to a) how radically internal structures were changed (e.g., creation of new business unit/department) and b) how radically the new products differed from products based on the new technology. Data on industry average was obtained by interview statements, expert interviews, and related press articles.

[c] Numbers in brackets refer to observations in a different context, e.g., historic behavior (not referring to the adoption of the focal discontinuous technology discussed in this chapter due to a lack of observation)

4.4.1 Speed

Speed of adaptation refers to the inverse of the time that has elapsed from the emergence of the discontinuous technology until the point in time when the company's reaction to it becomes active, i.e. in case of domain creation the launch date of the new product (see also Figure 4).

I assessed the speed of adaptation by extracting relevant data from historical websites and documents and compared them to industry averages[51]. To triangulate this data and to acquire a more nuanced understanding of the speed of the different phases—awareness, interpretation, decision-making, implementation—I drew on data from personal semi-structured interviews in which I asked my informants to provide a most detailed and precise reconstruction of the entire adaptation process. I coded the interview transcripts in order to extract the reasons for low or high adaptation speed.

4.4.1.1 Speed of Awareness

In the following chapter, I will discuss how early or late organizational members, particularly CEOs, of the sampled firms became attentive of the discontinuous change. Moreover, I will provide evidence how intensive 'screening' of family CEOs that are committed to their businesses accelerates the speed of awareness. As Table 9 shows, five of the eight adopters reacted to the discontinuous technology either fast or very fast , meaning that their first adaptation attempts occurred substantially earlier than those of competitors within the same industry. All of the adopters except *Retail 2000* became aware of the discontinuous technology early on, although most companies (all but *Anything & More*) were reluctant to ask externals for support—a family idiosyncratic behavior (e.g., Berrone et al., 2010; Gómez-Mejía et al., 2010) that is proposed to hamper speed of awareness (Gilbert, 2005)[52].

However, this does not mean that family influence results in isolation of the respective organization and its decision makers; indeed the CEOs of more than half of the family businesses[53] were active members in industry associations and participated in industry fairs. As stated by many respondents, such activities gave executives the opportunity to receive information on ongoing trends and to participate in 'peer' discussions 'at eye level'.

---

[51] Data on industry averages extracted from press articles, expert and CEO interviews.
[52] Besides mere power-related reasons, this unwillingness to ask for advice was also rooted in reputational concerns, such as a quotation of *Retail 2000*'s CEO illustrates: "Tackling difficult tasks [is] a core task of managers [not of external advisors]."
[53] *Alpha Star, Anything & More, Kiddies, King's Goods, Play & More, Toys 2000, Walter & Colleagues.*

Moreover, CEOs of all organizations that adopted the technology fast or very fast, (i.e. all except *Kiddies*, *Play & More*, and *Retail 2000*[54]) emphasized the effort they put into 'scanning their environment' regularly on a daily basis. Such screening also included, but was not limited to, talking to a variety of stakeholders involved in the industry and reading relevant industry-specific and unspecific news. As the CEO of *Alpha Star* noted,

It is important to have an ear to the ground. (Alpha Star)

Most of the contacts utilized by the CEOs of these organizations to remain informed about market trends, were "semi-private" (*Alpha Star*). Although, the family firms investigated in my study might have had *few* weak ties, they draw on ongoing market developments for catching signals, and *used them extensively*. This can be traced back to the high commitment of family CEOs. Other than non-family CEOs, there is no clear distinction between 'family' and 'professional' life for members of the owner-family (Habbershon & Williams, 1999). Allio (2004: 3) refers to this behavior as the "family business leaders' inability to leave it at the office; it follows them to the dinner table, the golf course, the piano recital, and Sunday brunch." As a consequence, family CEOs feel committed to the business and engage in searches for relevant information in their private time, too.

All three discontinuities investigated in this study were discussed early on, for instance, by media, in industry associations and at industry fairs. Although the tenor of these discussions was not necessarily positive and many reports depicted the developments like online shopping as 'temporary hypes' and 'relevant for niche customer groups only,' they resulted in a general awareness of the discontinuity among the CEOs.

These findings are in line with results of previous studies (e.g., Christensen & Bower, 1996) illustrating that 'becoming aware' is less of a problem than interpreting, deciding, and implementing the new technology. However, my findings transcend prior research by differentiating between the number of ties used and the time my informants spent on building on those relationships in order to become early aware of discontinuous technological changes. Based on the findings described above, I conclude

---

[54]*Retail 2000* and *Play & More* became aware of the discontinuity at a substantially later point in time than their comparable competitors. *Kiddies* was aware of the ongoing environmental shifts early due to feedback and observations at trade fairs.

> P2-1a. The more CEOs of family businesses engage in regular 'scanning of the environment' by making intense use of their weak ties, the earlier they become aware of an emerging technological discontinuity.

Some of the sampled companies (e.g., *Alpha Star* and *Play & More*) abstained from installing internal R&D departments because of resource scarcity. Such 'parsimony' has frequently been observed in both small and larger family businesses (Carney, 2005). Having no dedicated R&D department of their own might constitute a severe disadvantage regarding the speed of continuous change (Allen, 1977; Cohen & Levinthal, 1990; Mowery, 1983; Tilton, 1971). However, in times of discontinuous technologies, family influenced businesses can compensate for this downside and avoid delayed awareness, if they engage heavily in external information exchange, thereby mostly draw on semi-private contacts, and 'keep eyes open day and night.' In particular, when family CEOs show high levels of commitment to the business (Koiranen, 2002; Lansberg & Astrachan, 1994), they might be better able and more willing to engage in such a persistent scanning process.

4.4.1.2 Speed of Interpretation

In the following subchapter I will discuss the speed of interpretation; that refers to the time between first awareness and the general agreement among decision makers which adaptation strategy fits best for the respective organization. To explain the observed response patterns, I will draw on the well-established concept of socioemotional wealth (SEW) (Gómez-Mejía et al., 2007), referring to four specific, non-economic parts of the owner-family's utility function, as outlined in Chapter 2.2.2.2. In particular, I will highlight that depending upon which of the four SEW dimensions is most relevant for the owner-family, the speed of interpretation is either short or long.

After the organizational members of the companies I observed, specifically their respective CEOs, had become aware of the radical changes in their environment, these actors interpreted the technological discontinuity as either relevant for their own business or non-relevant. I split the process between awareness and the decision on how to react into two parts—interpretation and decision-making[55]. The reason is that the underlying drivers for the

---

[55] Extant literature is inconsistent on the use and concise interplay of both constructs. Daft and Weick propose

146

speed of each of those two phases are fundamentally different, as I will show in the following, although, there might be some interdependencies between the two phases[56]. Questions answered within the first phase, interpretation, include, for example, 'do we believe that this technology has a future?' and 'if we believe so, should we adopt the innovation?' As such it mainly refers to cognitive framing of organizations' decision makers as well as their perceptions of their respective organization's identity and capabilities. In the more politically and structurally driven decision-making processes, decision makers are concerned about questions such as 'how do we do it?', 'who is responsible?', 'how much do we invest?', and 'how do we justify our decision?'.

Six of my sampled organizations (*Anything & More, Kiddies, King's Goods, Toys 2000, Powerhouse*, and *Walter & Colleagues*) quickly interpreted the technological discontinuity as relevant, whereas three[57] businesses (*Alpha Star, Retail 2000*, and *White & Blue*) experienced delays in this phase, which in some cases were significant. For instance, *Alpha Star*'s CEO spoke about "day and night-long discussions" in this context.

The rationale for the variance in the length of interpretation phases was rooted in the variance of the owner-family's desire to preserve socioemotional wealth (Berrone et al., 2010; Chrisman & Patel, 2012; Gómez-Mejía et al., 2007; Zellweger et al., forthcoming-b). As outlined in Chapter 2.2.2.2, researchers identified four relevant dimensions of the SEW utility function (Gómez-Mejía et al., 2011a; Zellweger et al., forthcoming-b)—preservation of (1) family ties, (2) power and influence, (3) status and reputation, and (4) emotions and affect—, however, not all dimensions are equally important for each family influenced business[58]. I assessed the sampled organizations' striving to preserve socioemotional wealth along these four dimensions by abstracting and analyzing interview statements as well as triangulating the findings with secondary data. Although most organizations pursued socioemotional wealth of several dimensions (in particular, family influence and power was, at least to some level,

---

that "[d]ecision making generally is part of the information" (1984: 292) whereas March and Heath, to the contrary, see interpretation as a means to handle complex decision-making: "Decision making systems deal with [inconsistencies through]... interpretation" (1994: 134).

[56]For instance when quantitative calculations conducted in the decision-making phase result in a re-interpretation of the relevance of the technology.

[57] I neglect *Play & More* in several of the following considerations due to its mostly passive behavior, which does not allow me to assess the interpretation time.

[58] This somewhat relaxes the assumption in my theorizing chapter, that family influence manifests itself in equal loads of the 4Cs. For the purpose of this empirical study, it is more reasonable to draw on family influenced firms' motives (SEW) than on the manifestation of family influence (4Cs). However, the two concepts are closely intertwined, for instance, high levels of command frequently entail striving to preserve power and influence, high levels of community provide a positive affect; and deep connections as well as a sense for community might both imply concern for reputation; striving for continuity is closely intertwined with aiming to provide positions to family members and, as such, preservation of family ties.

important to most firms), it was possible to identify one dominant dimension that was pivotal for the respective organization's interpretation of the discontinuous technology. In the following, I categorize the ten cases according to their dominant dimension of SEW. Furthermore I discuss how their specific SEW focus influenced the interpretation phase, and finally how this affected the respective organizations' adaptation. Table 10 summarizes the dominant SEW dimensions of the sampled firms and their respective interpretations and responses.

Table 10: Family Influence and Adaptation—Socioemotional wealth

Company	Dominant SEW Dimension	Description	Time for Interpretation	Reaction
Alpha Star	Identity & Reputation (2nd: Power & Influence)	Known as traditional family business	Medium/ Long	Domain creation but no promotion / advertising of new product
White & Blue	Identity & Reputation (2nd Power & Influence)	Strong focus on welfare of staff and customers	1st move: short 2nd move: long	1st move: domain creation initiated by staff; focus on domain offense after failure; 2nd move: domain creation in selected segment
Anything & More	Power & Influence (2nd: Identity & Reputation)	Concerned about keeping the business profitable and securing family influence	Short	Domain creation: Early felt that discontinuity might be a future harm to the business and thus also to family influence
King's Goods	Family Ties (2nd: Power & Influence)	Concerned about the shape of the company when passing it over to the next generation	Short	Domain creation; felt that no opportunity must be missed; wants to avoid question of offspring 'why didn't you react?'

Company	Dominant SEW Dimension	Description	Time for Interpretation	Reaction
Retail 2000	Identity & Reputation	Rooted in local *community*	Long	Domain offense for extended time; end of 2000 domain creation by combination of virtuality and locality
Kiddies	Affects & Emotions	Focus is on good working atmosphere and on products (emotional attachment)	Short	Domain consolidation; Sticking to decisions of forefathers; Incremental innovations of products that CEO is emotionally attached to
Play & More	n/a (Affects & Emotions)	Business mainly seen as income source	n/a	Passive reaction; *Play & More* does not feel urged to react
Toys 2000	Identity & Reputation	Identity & reputation as innovative, high quality provider	Medium	Domain creation; medium interpretation and decision-making; however long implementation time to meet own standards
Walter & Coll.	Power & Influence	Concerned about keeping the business profitable and securing family influence	Short	Domain creation: Early felt that discontinuity might be a future harm to the business and thus to family influence
Power-house	Reputation & Identity	Concerned about being perceived as innovative and trustful partner	Short	Domain creation: Early felt that discontinuity might be an interesting opportunity to follow

Familial Ties. One firm, *King's Goods*, emphasized the importance of familial ties. The CEO of this retailer was most concerned about the future shape of the firm at the time when ownership and management will be passed to the next generation. In this vein, the CEO is specifically worried about how members of the next generation will evaluate his/her work. To avoid offspring asking 'Why didn't you react?', the CEO swiftly assessed the discontinuity as a potential future threat to SEW, and thus an important strategic field to enter for the company. The CEO described the motivation to act as follows:

> *It was obvious: I could have continued for a further ten years without e-commerce and afterwards hand over the business to the next generation. But then, my children or the nephews and nieces from [my siblings'] families would say: 'Didn't you realize that something was happening there? Why didn't you enter the market?' And in principle, this was the motive [to become active in e-commerce].*

As a consequence of these considerations, the CEO engaged in domain creation activities, building up e-commerce as a second, future-compliant pillar of the company.

Power and Influence. Two companies (*Anything & More* and *Walter & Colleagues*) were mostly focused on maintaining power and influence within the family. *Anything & More* suffered from strong competitive pressure when the discontinuity emerged. Moreover, at this time experts already expected future, long-term stagnation within the established retailing segments[59]. In this context, the CEO perceived early market entry into the e-commerce as an opportunity to escape the ongoing contention in established markets and to ensure long-term power over the company. This power and influence, in turn, allow the CEO to pursue his/her personal 'pet projects' that are fostering sustainability and ecology. In the energy metering market, *Walter & Colleagues* is a rather small player compared to its competitors. The owner-family perceived the step from analogous metering components to smart metering components as a must to successfully stay in the market. After the implementation of those products, they felt that an early entry in the market of developing and metering smart meters (instead of only its components) would allow them to expand their market position and ultimately their family influence. Consequently, *Walter & Colleagues* engaged in domain creation activities[60].

---

[59] For an overview of development of net revenues in the retail market since 1994, see Statistisches Bundesamt: www-genesis.destatis.de .

[60] *Walter & Colleagues'* first smart meters were launched substantially later than those of *Powerhouse*. However, as the technology is still emerging, I counted both as fast adaptation (expert interviews). *Walter & Colleagues'* delay compared with *Powerhouse* can be traced back to the fact that the former is originally a component provider who had to decide to enter a completely new market whereas the latter is a meter provider per se who had to decide to build on a new technology.

Reputation and Status. Five companies (*Alpha Star*, *Powerhouse*, *Retail 2000*, *Toys 2000*, and *White & Blue*) were most concerned about their status and reputation. However, as the 'images' of their respective organizations varied, interpretation, and thus duration of interpretation, also varied.

Retail 2000 is deeply rooted in its local community (Berrone et al., 2010). This embeddedness goes back to the early days of the founder and still substantially contributes to the firm's and the family's present identity. Entering the online market would jeopardize this part of the socioemotional utility function, because 'virtuality' is a priority in contrast to 'locality.' Consequently, *Retail 2000*'s interpretation phase lasted until the late 2000's. In the meantime *Retail 2000* engaged heavily in domain offense activities by expanding the business to selected geographical areas—those countries that either family members or long-time employees had close ties to. At the end of the 2000's, *Retail 2000* identified a way to combine locality and virtuality by offering services that combine online ordering and 'drive-in' delivery. As this concept allows for preservation of SEW, the organization quickly decided and implemented the new technology.

Alpha Star is a high quality producer and vendor with a long history in its business segment. The discontinuity was initially perceived as being of inferior quality, thus contradicting the family firm's image and identity. The junior CEO described:

We believe, it is not a good product.

As a consequence, the CEO was first reluctant to adopt the new technology and the organization engaged in a long-lasting interpretation process. However, as customer feedback to prototypes of the new technology was unambiguously positive, whereas at the same time sales of the traditional goods were declining, the CEO began to feel that pursuing established paths would, in the long-term, endanger family power and influence, a second important source of SEW. Consequently, *Alpha Star* entered the new market. Although *Alpha Star*'s marketing budget is, in general, relatively high compared to competitors, the company did not launch any activities to promote the new product. A thorough analysis of historical websites and press releases revealed "a silent introduction" (CEO, personal interview) approach of introducing the product. This approach ultimately allowed *Alpha Star* to adapt to the new technology without forsaking socioemotional wealth, more specifically its well-established image as a conservative and traditional player.

White & Blue is most concerned about its reputation amongst employees and customers on a local basis. The decision makers quickly assessed the discontinuity as something rather positive to be followed because the idea to explore the business opportunity was generated by employees who proposed first strategic activities in the end of the 1990s. However, this first attempt to enter the market failed due a lack of profitability. As a consequence, *White & Blue* abandoned its online activities and invested in domain offense activities (see also Chapter 4.4.3), in particular customer retention programs, which fit better with the company's socioemotional priorities than 'anonymous' internet shops. However, when competition intensified, and competitors targeted to absorb *White & Blue*, the CEO felt a threat to *White & Blue*'s second relevant SEW dimension, power and influence, and interpreted e-commerce as a relevant investment alternative (press article, 2011). *White & Blue*'s intention is to be a "unique and not easily to copy" business; the new online shop tries to transfer this concept to the virtual world.

Organizational members, and specifically the owner-CEO, of *Toys 2000* perceive their company as a high-quality, innovative toy and game manufacturer. The main SEW goal is to preserve this image. As adopting the new technology is—in principle—congruent with this SEW goal, the company assessed the innovation as a path to be followed without major delay. SEW concerns started to emerge during the implementation time (see also Chapter 4.4.1.3). In order to preserve its image as a 'high quality' provider, *Toys 2000* had to build up various new competencies related to IT and to meticulously develop new products meeting its own quality standards.

Somewhat similar to *Toys 2000*, *Powerhouse*'s main SEW goal is to maintain its reputation as a "credent" partner in the industry that "shapes standards." As such, the emerging technology was congruent with the organization's image resulting in a quick interpretation as relevant, and ultimately domain creation.

Affect and Emotional Attachment. Affect and emotional attachment refer to the positive (and negative) feelings and emotions family owners associate with their business and the daily work there (Berrone et al., forthcoming). The owner-CEO of *Kiddies* is driven by emotional attachment to the firm and its products. "[E]njoy[ing] work" belongs to the CEO's leitmotifs. The CEO's emotional attachment does not only refer to the current state of the company but also to its past. More specifically, the CEO honors and values decisions made by the previous CEO, now deceased, who decided against entering the new markets after the initial failure. The CEO does not question or reconsider this decision and states, "it is, what it is." Thus, to preserve the emotional attachment to the firm and its history, the CEO quickly came to the conclusion that *Kiddies* should not follow discontinuous innovation.

Play & More constitutes a special case. Despite high family ownership (100%), high family presence in management positions (two family CEOs), some transgenerational intentions (offspring of CEO preparing to enter the business), and high family influence in decision-making (family CEOs as sole decision makers), the level of *Play & More*'s striving to preserve SEW is astonishingly low:

> *I think we should continue working on [our core segments]. If we are*
> *not able to do so, we have to close our business.*

The CEO entered the family business decades ago due to pressure from his/her parents and a lack of alternatives. Moreover, the CEO does not identify him/herself with products offered by the company. Although, the next generation is about to join the company, the CEO has mixed feelings about the offspring's abilities to lead the business. *Play & More* is a renowned company; however, the CEO perceives the name as a 'burden,' as stated in the interview: "Unfortunately I was born with the name [...]." The only SEW component detectable is some level of positive affect due to the good firm atmosphere. As this component of the SEW is not threatened by the discontinuous technology in the short or mid-term, the CEO did not perceive the necessity to change.

In summary, drawing from empirical evidence of the cases just described, I assume

---

P2-2a. The more a family firm's predominant socioemotional wealth dimension is congruent with the discontinuous technology, the faster the family firm interprets the technology as relevant.

P2-2b. Family firms with focus on "Power and Influence" and "Family Ties" are most likely to quickly interpret discontinuous technologies as relevant for their own organizations as compared to family firms with focus on other predominant socioemotional wealth dimensions.

These propositions are in line with SEW literature predicting that the preservation of socioemotional wealth affects organizations' reference points and ultimately strategic behavior (Berrone et al., forthcoming; Berrone et al., 2010; Chrisman & Patel, 2012; Gómez-Mejía et al., 2007). However, my findings transcend prior literature in two ways. First, to the best of my knowledge, this is one of first studies to examine the effects of the single SEW dimensions and compare them with one another, assuming that family firms do not equally strive to preserve all four SEW dimensions. Second, this is the first study investigating the effect of SEW on strategic behavior in times of discontinuous change.

Moreover, these findings also contribute to literature on discontinuous change. Recent research in that field has made great advances in identifying cognitive and emotional antecedents of adaptation to discontinuous change (Barr et al., 1992; Kaplan, 2011; Tripsas, 2009). However, so far, no study has been concerned with the effect of various dimensions of SEW in the context of responses to such non-paradigmatic breakthroughs.

4.4.1.3 Speed of Decision-making and Implementation

In the following subchapter I will show that the sampled family firms quickly decided on the adoption of the discontinuous technology and implemented the associated changes quickly. Moreover, I will provide evidence, that intuitive decision-making and high levels of authority and perceived legitimacy drive these high levels of adoption speed.

Once the discontinuous technology was assessed as 'relevant' by the decision makers[61], all adopting companies except *Retail 2000*, *Toys 2000*, and *Walter & Colleagues* quickly decided to adopt the technology and ultimately implemented the associated changes in their business routines in a fast way. Interestingly, in contrast to the predictions of standard literature on discontinuous changes, in these companies decision-making and implementation were not protracted by forces such as formalized processes (Levitt & March, 1988) or political resistance (Eisenhardt & Bourgeois Iii, 1988; Pfeffer, 1992). As the case evidence collected shows, this promptness of response was facilitated by structural aspects stemming from the high levels of command (Berrone et al., 2010; Carney, 2005; Miller & Le Breton-Miller, 2005a) present in all family firms. Roughly 90% of the interview statement codes as referring to standardization of structures reflected a low level of formalization in the sampled companies. The following quotations from my interviews illustrate this aspect:

> *Many things are not codified, not written down here, and yet it works anyway—maybe it even works because of that. (Alpha Star)*

---

[61] This was the case for all companies except *Kiddies* and *Play & More*.

> *Nothing is written down, everything is in the heads of the individuals. (Kiddies[62])*

> *[Toys 2000 is characterized by processes that are] not fully formalized. (Toys 2000)*

High levels of command in combination with non-standardized processes implied that CEOs of family influenced firms made their decisions based on gut feeling rather than calculations. Intuition-based decision-making thereby describes the tendency of members of the dominant coalition to make decisions based on "thoughts and preferences that come to mind quickly and without much reflection" (Kahneman, 2003: 697). As such, intuitive decision-making differs from fact-based 'rational,' decision-making, which is based on elaborate market analysis and strategic planning. Decision-makers who rely on their gut feeling rather than calculations when making decisions decide promptly according to the outcome of their interpretation process without justifying these moves with, for instance, calculations and market research data. The following examples illustrate how CEOs of my sampled companies described the nature of their decision-making processes to adapt to the new technologies.

> *[The decision to adopt the innovation] stems from a gut feeling, the feeling that one has to keep options alive. (Alpha Star)*

> *Before you ask: No, this was not tied to any great strategic plans, scenarios, or discussions and so on. [...] It was gut feeling. A feeling that we had to become active, that we had to enter the market. (King's Goods)*

> *I know the business. My father built the business; I took it over from him. Buck up! That's the direction we'll go now. (Anything & More)*

> *[Toys 2000 is characterized by] personalized [as opposed to calculative] decision-making. (Toys 2000)*

> *The senior CEO has a gut feeling and says: 'Yes, this direction could be promising.' And concerning the CEO, the directors ask: 'How can the CEO take this decision? It is doomed to failure. Particularly the numbers guys, like me, calculate the cases and conclude: 'no.' But*

---

[62] After the assessment of the discontinuous technology as 'not pursuable for *Kiddies*,' the CEO quickly decided how to adapt (retrenchment in niche) and initiated the required (incremental) changes.

> *then, a year later, everyone is happy about the CEO's decisions.*
> *(Walter & Colleagues)*

Three companies (*Retail 2000*, the slowest adapter in my sample, *Play & More*, a non-adopter, and *Powerhouse*) deviated from this preference of intuitive as opposed to calculative-rational decision-making. The CEO of *Retail 2000* described the decision-making within his/her firm based on "factual, objective arguments," whereas the CEO of *Play & More* relied on fact-based decisions for any strategic questions. *Powerhouses*' initial decision to enter the market was fact-based. However, as the calculation only relied on rough estimations of potential future revenues and *Powerhouse*'s family owners in the past already showed their willingness to invest in auspicious yet risky new technologies, this assessment process did not significantly delay decision-making. As soon as the implementation had started within *Powerhouse*, decision makers refrained from applying financial measures and rather applied a 'wait and see strategy' without precise capacity planning.

In all these cases, once the decision to adopt the new technology had been taken by the decision makers, the respective CEOs quickly spread those decisions throughout the organizations without engaging in political 'framing contests' (Kaplan, 2011). High levels of command helped the companies to quickly implement the changes, as the following citations describe:

> *One of the advantages of an owner-managed business is that you can*
> *get an idea into your head and just say—given a certain congeniality*
> *and assertiveness—'we'll do it that way.' (King's Goods)*

> *Whenever I was convinced, I decided for it and we subsequently*
> *implemented it. (White & Blue)*

Toys 2000 and *Walter & Colleagues* are theoretical replications (Christensen & Bower, 1996; Yin, 1994) of this phenomenon of a rapid implementation. In both companies, the decision to adapt was promptly made due to low formalization and gut feeling. However, the implementation of the technology took an extended time compared to competitors' reactions. For *Toys 2000*, the reason for this slow implementation lies in the company's striving for high quality (as described in chapter 4.4.1.2.) rather than political resistance or bureaucracy. Fulfilling their own quality requirements coerced the *Toys 2000* to build up deep knowledge and competencies related to the new technology before commercializing the new products. *Walter & Colleagues*, a particularly resource constrained company, initially lacked the competences and capabilities required for adoption of the new technology and thus suffered from some delay during technology implementation

Summarized, intuitive decision and lack of 'framing contests' rooted in high levels of command accelerated decision-making and implementation within the sampled firms. Formally, I propose

> P2-3a. The more decision criteria in a family firm are non-formalized and allow for decisions based on gut feeling rather than precise calculations, the faster the firm decides to adopt a discontinuous technology.
>
> P2-3b. The higher the family CEO's authority and perceived legitimacy in a family firm, the faster the firm implements a discontinuous technology due to low political resistance.

The tendency of decision-makers to rely on few formalized processes as well as gut feeling, and the subsequent fast response are also supported by extant family business literature (Dreux Iv, 1990; Dyer, 1988). Most previous studies have envisaged the tendency of family influenced firms to show low levels of formalization and to refer to intuitive decision-making under continuously changing environments, when stabilized routines and standardized decision-making are beneficial to companies due to the high levels of accountability and cost-efficiency entailed (Arrow, 1974; Cyert & March, 1963; Hannan & Freeman, 1984; Levitt & March, 1988). Thus it is comprehensible that scholars called for more professionalization[63] in family influenced firms (Aronoff, 1998; Craig & Moores, 2005; Dreux Iv, 1990; Dyer, 1989). Technological discontinuities, however, typically come along with high levels of ambiguity and uncertainty (Tushman & Anderson, 1986). As a consequence, market data on such breakthroughs are either unavailable or inconsistent. Ultimately, fact-based decision-making will prolong the decision-making phase (Ocasio, 1995). While such a strategy appears to be relevant in circumstances of continuous change, fact-based decision-making may be highly maladaptive if applied by companies facing discontinuous change.

---

[63] It is important to note that the call for 'professionalism' typically encompasses two components: First, the call for trained, external management talent, and second, the void for formalized and reliable structures. As the lack of talent and the lack of formalization differ fundamentally in their consequences, I argue that a more fine-grained discussion is required.

4.4.2 Intensity

In the following subchapter, I will show that the sampled family firms initially invested low amounts of financial and operational resources into the commercialization of the new technology. This trend was particularly salient for firms that strived for independence and companies that perceived a mismatch between the new technology and their predominant SEW dimension. During the course of the technology's emergence, several companies increased their resource commitment due to an enhanced level of perceived threat.

Intensity of adoption refers to the amount of financial and non-financial resources committed to commercialization of a new technology. In this part of the study, I will rely on relative ($I_{relative}$, see Figure 5) rather than absolute data, as my main focus is on explaining variance in organizational adaptation rather than investigating performance aspects of such organizational moves. I assessed the level of intensity by asking my informants to provide me with quantitative data on financial and operational investments in the discontinuous technology over time and compared those numbers with figures on employee numbers and firm financials (where available). For instance, the CEO of *Alpha Star* (a company employing roughly 2.000 persons in total) informed me that no staff was hired to implement the new technology, whereas *Toys 2000* (a firm with approximately 200 employees) built up a new department consisting of more than ten engineers. Further, I cross-checked the data where possible, for instance, with information gained from press releases indicating new acquisitions or investments.

Most of my sampled companies started their initiatives with low or very low financial and operational investments. (see Table 9, page 142): *Alpha Star*, *Kiddies*[64], *King's Goods*, *Powerhouse*, *Retail 2000*, *Walter & Colleagues*, and *White & Blue*. As shown consistently throughout the interviews, hesitant resource commitment was primarily rooted in the unwillingness to take out loans and in the family businesses' tendency to decrease business risk.

> *I once experienced being dependent on banks. I do not want to experience such a situation ever again. (King's Goods)*
>
> *We invest in new products and new markets [...] but we refrain from risky invests. (Walter & Colleagues)*

---

[64] For *Kiddies*, I coded the investments in incremental product innovations required to realize the company's adaptation by retrenching in a niche segment.

158

Compared to the general tendency in the sample, two companies invested eagerly in the new technology: *Toys 2000* and *Anything &*More. *Toys 2000* entered the market of electronic games and toys with a high initial financial and operational commitment: The company had to purchase tools, hire experienced and knowledgeable staff, and train long-term employees in order to apply an adaptation approach that is in line with the company's goals and intentions as outlined in chapter 4.4.1.2: *Toys 2000*'s main goal is to preserve its reputation as a vendor of high quality toys and games. To fulfill these self-implied standards not only in the established business segments, but also in the emerging market of the new technology, *Toys 2000* had to heavily invest in building up the necessary capabilities. A second reason for *Toys 2000*'s comparably high investment was the organization's high level of perceived threat, as indicated in statements of the CEO referring to "murderous" competition induced by the new technology. Extant literature (Gilbert, 2005) suggests that companies invest more in organizational responses if they perceive a strategic issue as a threat rather than as an opportunity.

Anything & More's early and continuously aggressive commitment to discontinuous technology is also explicable by strong threat perception among the company's executives. At the time the discontinuity emerged, *Anything & More* faced severe financial troubles with declining revenues in the core business and increasing competition. The company felt an immediate hazard to its long-term as well as its short-term goals, and the CEO perceived investment in e-commerce as a promising way to overcome these challenges.

In early phases of the lifecycle of the discontinuous technologies, investment was lowest for those companies that perceived the discontinuous technology as being at odds with the preservation of their most relevant SEW dimension: *Alpha Star*, *Retail 2000*, and *White & Blue*. These three players faced an intriguing dilemma: the preservation of their reputation (i.e. their most important SEW dimension) required rejecting the new technology; and preserving family power and influence (i.e. their second most important SEW dimension) required adopting the new technology. The companies solved this dilemma—mostly after a long interpretation time as described in chapter 4.4.1.2—by finding compromising solutions. Such solutions included, for instance, offering the product with minimum marketing efforts or offering an adapted version of the new technology satisfying most SEW goals at a minimum level. For instance, *Alpha Star*, a player renowned for its generally high marketing budgets, did not actively promote the new product at all. As analyzing of the resource investment patterns shows, these compromising solutions also included low financial and operational commitments to the discontinuous technology. For instance, *Alpha Star* decided on a "silent introduction," according to the CEO; moreover this player did not hire any new employees to commercialize the new technology. This can be explained as follows: These companies somehow felt they had to 'keep their options alive,' however, they did not take full ownership of the new technology due to a perceived mismatch between their SEW and the directions obtruded by the new innovation, and were thus not willing to invest large amounts of money, manpower, or managerial attention. Formally, I propose

> P2-4a. The more family firms strive for financial independence the fewer resources they commit to the discontinuous technology.
>
> P2-4b. The lower the congruence between the family firm's predominant socioemotional wealth dimension and the discontinuous technology, the fewer resources these firms commit to the discontinuous technology.

As Table 9 (p. 142) shows, several firms (*King's Goods*, *Retail 2000*, and *White & Blue*) increased their resource investment as the discontinuous technology further emerged[65]. This can be explained by the increasing framing as 'threat,' both to the business as well as SEW (Chrisman & Patel, 2012). In the mid/late 1990s, e-commerce was mostly perceived as a vague future threat to SEW and, at the same time, as an opportunity to improve the adopters' competitive position as compared to established competitors. In the early/mid 2000s, the threat component began to increase because the German retailing market continued to stagnate, with stationary and e-commerce competition increasing. In line with Gilbert's (2005) predictions, this threat framing relaxed resource rigidity and resulted in intensified investment.

4.4.3 Stamina

In the following subsection, I will provide evidence on how long time planning horizons and CEO commitment enhance a family firm's stamina when adopting a discontinuous technology. Furthermore, I will outline several different decision-making paths and the respective outcomes as observed during my analysis (see Figure 13).

---

[65]*Alpha Star* also started with a low commitment; due to the ongoing success of its new products, there is no need for an increase of commitment for this company. *Walter & Colleagues* and *Powerhouse* invested cautiously in the new technology. As the discontinuous technology of the energy sector is still in an early phase, it is not possible to predict, how framing of the discontinuity and resource commitment of these two companies will develop.

Stamina, defined as 'degree in which established companies continuously reinvest in a new technology following initial investment' (see Chapter 2.1.2.1.3) is an important dimension of adaptation to discontinuous technological change, as discontinuous technologies often emerge over an extended period of time with initially scarce financial success (Anderson & Tushman, 1990).

Showing stamina was or is relevant for all three technological discontinuities examined in this study: In the retailing industry, stamina has been important primarily because, during the early 2000s, a temporary setback in the new market ('burst of the internet bubble') frightened off many investors. Similarly, the toy and gaming industry faced a severe temporary market breakdown in the 1980s. Equally important to German incumbents in the toy and gaming industry were the frequent failures in European-Asian cooperation around the 1980s and early 1990s. Finally, the smart metering industry has not yet experienced substantial setbacks as it only emerged a few years ago. However, due to the market structure (the average period of use of meters exceeds one decade), quick success of the new technology cannot be expected in this industry. As a consequence, this market was, at the time of this study, only attractive for firms willing to invest over an extended period of time before harvesting substantial financial returns.

To assess the level of stamina exhibited by the sampled companies, I coded interview statements referring to previous setbacks, investment time horizons, and success criteria applied for strategic projects[66]. Wherever possible, I triangulated interview data with information gathered from other sources such as historical websites, annual reports, interviews with media, and press releases. Moreover, I analyzed the iterative decision-making processes of each sampled firm, and graphically outlined the main paths in Figure 13. In this illustration I present several decision-making paths regarding the technology adoption as observed for the sampled firms. I indicate the (intermediary) stopping point in the decision tree for each company as of end 2011.

---

[66] For instance, interview codes stating that strategic investments were expected to be profitable in seven years or later were coded as 'long planning horizon,' four to six years were coded as 'medium,' and less than four years as 'short planning horizons.' Moreover, the notion of grace periods exceeding at least one year, for which the new project was not measured against any of the established success criteria was also coded as long planning horizon. In addition, I differentiated between whether organizations stuck to their initial decisions or whether they reversed them after they have turned out wrong.

Figure 13: Iterative Decision-making Tree

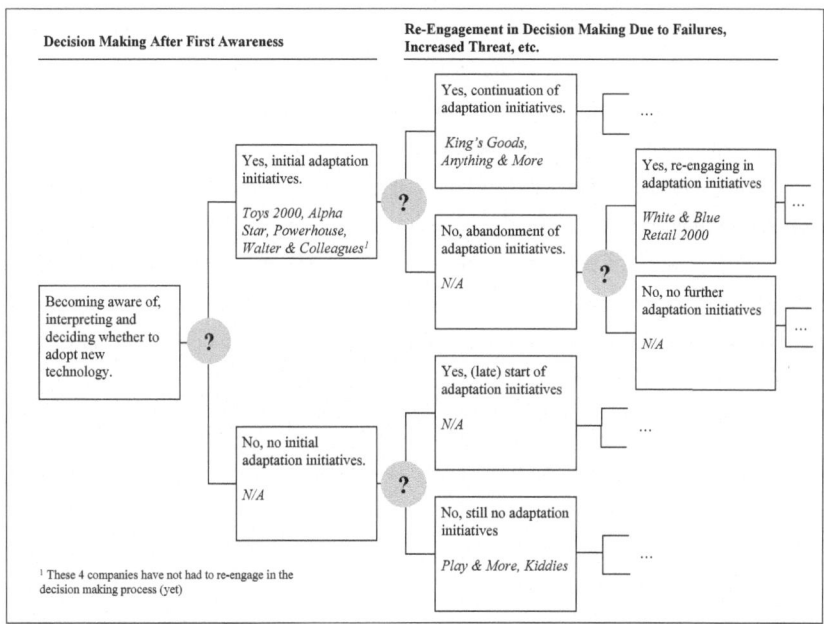

My first analysis, referring to the planning horizons, showed that all but one (*Play & More*) of the family influenced firms in my sample emphasized long-term rather than short-term goals as the following quotations exemplify:

> *Planning horizons for online business are approximately 30 years.*
> *(King's Goods)*

> *This [increased interest of customers] will not be measurable within*
> *two or three years. I believe one has to look thirty years ahead.*
> *(Walter & Colleagues)*

These goals were manifested, for instance, in managerial incentives tied to long-term business performance and giving freedom to employees (*Alpha Star, King's Goods*), the willingness to accept high short-term costs in order to harvest long-term earnings (*Retail 2000*), and, somewhat related, grace periods of several years for any strategic activities (*King's Goods, Anything & More, Powerhouse*). Although nine out of ten companies fulfilled one important pre-condition for stamina—'long-term planning horizons'—as theorized in Chapter 3.8.3, I observed variance in their continuity of investment.

Two companies, *Anything & More* and *King's Goods*, showed high levels of stamina, although they both experienced failure with their first attempts to commercialize online business models. For instance, *Anything &More*'s first internet shops did not attract the interest of customers. However, neither of these companies stalled their investments, and they both maintained a certain level of investment, continued to experiment, and quickly ramped up e-commerce investments again when customer interest increased (*Anything & More*) and profitable business models emerged (*King's Goods*). Both firms stuck to their original online shops and restructured them when required. In addition to its main shop, *Anything & More* possessed a number of e-commerce investments, many of which lacked success in the early years of e-commerce. After the failure of these e-commerce initiatives, *Anything & More* re-invested the capital in online businesses.

To the contrary, *White & Blue* and *Retail 2000*'s levels of stamina were initially low. These organizations' first attempts to enter the e-commerce market failed[67], similar to those of *King's Goods* and *Anything & More*. However, contrary to *King's Goods* and *Anything & More*, the CEOs of *White & Blue* and *Retail 2000* abandoned almost all of the companies' online activities[68] for five to ten years. This temporary withdrawal can be explained by the lack of initial commitment to online business: The CEO of *White & Blue* had not decided to enter the e-commerce market because he/she was convinced about the importance of the topic, but as a reaction to regional activities initiated by *White & Blue*'s employees. The CEO of *Retail 2000* did not perceive e-commerce as an important revenue channel at that time according to the personal interview. At the end of the 2000s, both CEOs re-engaged in the sense and decision-making process. This time they interpreted the new technology as relevant because competition in the retail sector had further increased, e-commerce had meanwhile permeated the German retailing market, and business models had substantially improved. The CEO of *White & Blue* explained:

> *In the beginning, I was very diffident. However, now I see that we have*
> *to go this way.*

Both CEOs ultimately decided on a new start and re-entered the e-commerce market in 2011/2012. This was made possible by the CEOs' open-mindedness, which enabled them to deal with setbacks in a constructive way, which is to reverse decisions that have turned out wrong. One of my informants called this the principle of "allowing for mistakes" (CEO of

---

[67] *Retail 2000* started to engage in some low-effort e-commerce activities around 2004. These attempts to commercialize online services failed.

[68] Online digital photography services have continuously been offered by *White & Blue*.

White & Blue). In a video interview (2011), a next generation family member re-emphasized the "importance of learning from mistakes."

For four companies, *Alpha Star*, *Powerhouse*, *Toys 2000*, and *Walter & Colleagues*, levels of stamina in the context of discontinuous change could not be assessed, as those companies had not yet engaged in a re-interpretation process, either due to ongoing success (*Alpha Star*, *Toys 2000*) or due to the recentness of the reaction (*Powerhouse*, *Walter & Colleagues*). However, all four companies exhibited high levels of stamina in the past. For instance, *Powerhouse* dedicated ten years of development time to a single component in the 1980s and 1990s before it ultimately became profitable.

Two companies, *Play & More* and *Kiddies* initially decided either not to actively react to the discontinuity (*Play & More*) or with retrenchment into a niche (*Kiddies*) without reversing these decisions at a later stage. This can be explained as follows: *Play & More* does not plan on a long-term basis as several interview statements show; rather the CEO adopted a 'wait and see tactic.' The CEO of *Kiddies* openly admitted in the personal interview that sticking to the parent's decision not to adopt the new technology, might have been a mistake; however rather than dealing constructively with such potential misdeterminations, the CEO reacted with resignation—"it is, what it is" belongs to the CEO's most frequently used sentences during the personal interview, indicating a somewhat low level of open-mindedness.

Based on the above-mentioned observations, I conclude

> P2-5a. The more family firms have long- as opposed to short-term planning horizons, the more these organizations will show stamina, i.e. continuously invest in discontinuous technologies.
>
> P2-5b. The more a family firm CEO is initially committed to the decision of adopting a discontinuous technology, the more likely a continuous investment is.
>
> P2-5c. The more open-minded a family firm CEO is, the more likely it is that the company will re-engage in the commercialization of the new technology after temporary suspension of the investments.

These findings are in line with my theorizing (P1-8c, P1-9c, see chapter 3) as well as extant literature in the context of antecedents of escalation of commitment (Woods, Dalziel, & Barton, 2012).

4.4.4 Routine Flexibility

In the next subsection I will show how routine flexibility varied across the sampled firms. While some companies radically adapted their internal routines, others stuck to tried-and-tested processes resulting in 'cramming' of the new technology. Empirical evidence shows that this variation is driven by variation in the match of the discontinuous innovation with the non-economic utility functions of employees responsible for implementing the new routines. While employees who feel that there is a good match are likely to eagerly engage in implementation, those perceiving a mismatch are likely to build up an 'inner resistance' against the technology, resulting in less flexible adoption. This effect is more salient for long-tenured as opposed to short-tenured employees.

Routine flexibility refers to organizations' replacement of established routines with non-paradigmatic ones when adapting to a discontinuous technology (see Chapter 2.1.2.1.4)(Gilbert, 2005). I assessed routine flexibility in a two-fold manner: First, I drew on information provided in the personal interviews regarding firm internal changes in processes and structures while the new technology was being implemented (König, 2009). Second, I analyzed the structure and features of the final products. This second step allowed me to detect 'cramming' (Christensen & Raynor, 2003; Gilbert, 2005), that is, transferring established processes, design, performance criteria, or revenue channels to the new technology without adequately adapting them.

My sample shows large variance regarding the flexibility of routines: Two of the eight adopting companies, *King's Goods* and *Anything & More*, were able to enhance their routine flexibility by adhering to measures well known and described by standard management literature. Both companies set up ambidextrous structures, led by externally hired managers with previous work experience in the new technology domain. Thus these family firms were able to quickly learn about the key pitfalls and relevant success factors which enabled them to effectively offer online services to customers (according to press releases and annual reports). Moreover, the ambidextrous structures eased the implementation of new and decoupled logistics centers, an important structural determinant of success in the e-commerce business. In both cases, the CEOs work closely together with the leader of the new unit in a trust based relationship, which helped the CEO to cede part of the responsibility as stated in the interview by *King's Goods*' CEO.

Alpha Star, *Powerhouse*, *Toys 2000*, and *Walter & Colleagues* moderately adapted their routines by cooperating with externals yet without substantially changing their internal organizational structures. *Alpha Star*'s CEO stated that his/her company did not adopt a new structure due to lack of time. Instead, *Alpha Star* closely worked together with an external partner firm providing the competencies that *Alpha Star* lacked. *Powerhouse* applied a similar implementation strategy by setting up an alliance with a partner firm. Moreover the metering company hired some experts who were placed into the existing development department led by a long-tenured manager. In addition, long-tenured employees were trained in the newly required capabilities. *Toys* 2000 hired new, experienced employees to implement the new technology. Moreover, the established, long-tenured employees working for *Toys 2000* were avid regarding getting trained and advancing the development of the new technology.

White & Blue and *Retail 2000* predominantly stuck to their established routines[69]. Both companies attempted to implement the new technology without involving externals or changing internal structures, and they faced severe difficulties with the alteration of their established routines. The CEO of *White & Blue* traced this back to emotional attachment of the employees who "regret bygone times." Also the CEO of *Play & More*, a non-adopter emphasized the "conservative" attitude and "unwillingness to change" of long-tenured employees as a major inhibitor of change.

Summarizing the results outlined above reveals two important aspects: First and not surprisingly, structural elements such as organizational decoupling and involvement of externals, i.e. factors that have been identified by standard management literature to relax routine rigidity, are also effective for family firms. Second, and more interesting, employees seem to play a crucial yet ambiguous role in adopting discontinuous technologies. Four companies, *Alpha Star*, *Toys 2000*, *Retail 2000*, and *White & Blue* emphasized their focus on *community* and had long-tenured employees strongly identifying themselves with their respective companies. The routine flexibility of those four companies, however, differed substantially. One salient explanation for this phenomenon is variance in the employees' non-economic utility functions. Not only owner-family members feel attached to their businesses,

---

[69] This assessment only refers to the first commercialization attempts of *White & Blue* and *Retail 2000*. As the interviews were conducted a significant period of time before their second market entry in late 2011/early 2012, no information about the routine flexibility of the second attempt is available.

associate positive or negative emotions with the firm, and are sensitive to reputation gains and losses related to the organizations, but also—to a somewhat reduced degree—employees, in particular those with long tenures. Thus I propose that not only firm owners but also employees possess a non-economic utility function related to the organization that employs them. This assumption is in line with research emphasizing that employees do not merely act as 'agents,' but rather 'stewards' (Davis et al., 1997). Due to the specific *community* characteristics of family firms (Miller & Le Breton-Miller, 2005b), it is likely that stewardship behavior is even more salient in family influenced as opposed to non or less family influenced firms.

Similar to family owners, employees possess one dominant non-economic utility dimension. While, to the best of my knowledge, extant literature does not provide a categorization of employees' non-economic utility functions (as is the case for family owners' SEW), my study revealed that dimensions include (but are probably not limited to) the following: preservation of reputation and status (e.g., working for an innovative/high quality/socially responsible firm) and preservation of affects and emotion (e.g., enjoying a good working atmosphere).

Empirical evidence indicates that those companies were specifically able to implement non-paradigmatic routines where there was a match between the employees' predominant dimension of the non-economic utility function and the discontinuous innovation. For instance, employees of *Toys 2000* felt that the strategy imposed by the discontinuous technological change (providing high tech electronic toys) was coherent with their non-economic wealth (working for a high quality, high reputation topic). Somewhat similar, the externally hired employees of *King's Goods* and *Anything & More*, who were already trained in the technology, did not perceive any incongruence between their non-economic goals and the innovation. Employees of *White & Blue*, however, had adopted a non-economic utility component similar to that of the respective family owners—being close to customers. As a consequence they found themselves in difficulties in adopting new, non-paradigmatic routines as associated with setting up a business model that deals with 'anonymous' rather than known, regular customers. As a result, the employees' 'psychological ownership' of the old technology remained high and 'psychological ownership' of the new technology was low, resulting in low flexibility of routines.

Employees of all four above-mentioned companies had above-average firm tenure, identified themselves with their employer and were emotionally attached to the firm (not necessarily the products). Formally, I propose

> P2-6a. Family firms with higher levels of congruence between non-economic utility considerations of employees and the discontinuous technology adopt the new technology in a more flexible way than firms with lower levels of congruence.
>
> P2-6b. Long employee tenures moderate this association.

While first studies examined SEW on a organizational level (Berrone et al., 2010; Chrisman & Patel, 2012; Gómez-Mejía et al., 2007), recent research has referred to the individual level, investigating the SEW of firm owners (Zellweger et al., forthcoming-b). With my study, I go one step further and propose that not only family members but also employees will gain non-economic wealth by working for a family influenced firm with an inclusive, community-based culture for a long time, and in turn, strive to preserve that non-economic wealth.

4.5   Concluding Remarks

4.5.1   Summary and Contribution

This study aims to shed light on the organizational adaptation pattern of family businesses in times of discontinuous technological change. As one of the first investigations in the family business literature, this study adopts the lens of incumbent inertia theory—a broad, widely recognized, strand of research in the non-family business literature. To summarize the empirical evidence of my multi-case study, specific family related drivers entailed a characteristic response pattern of family influenced firms when challenged with discontinuous technological change (see Figure 14). Most companies adopted the innovation in a fast and persistent manner, yet with low initial resource commitment and different levels of routine flexibility. Fast adoption was mainly driven by intense environmental scanning, which built on the frequent use of semi-private weak ties, non-formalized processes resulting in intuitive rather than rational decisions, and low levels of political resistance within the organization due to high levels of CEO authority and perceived legitimacy.

Depending on the initial match of the family firm's dominant SEW dimension with the anticipated future state induced by the new technology, the duration of the dominant coalition's interpretation time varied. This is because family firms strive to implement strategies that allow for SEW preservation (Gómez-Mejía et al., 2007). Any mismatch of SEW intentions and available strategic options—as observed particularly for firms focusing on reputation and status as well as emotional and affective aspects—will trigger either a long interpretation and discussion process until a comprising strategic move has been identified or trigger non-adoption.

Family firm's initially low resource commitment is rooted in their striving for financial independence. This tendency is enhanced for family firms with a mismatch between SEW preservation and adaptation: Disconnections between the family goals, as manifested in SEW, and the anticipated strategic paths imposed by the discontinuous technology, lower family CEOs' psychological ownership of and commitment to the innovation.

Long-term planning horizons and constructive handling of setbacks allow for continuous investment in the commercialization of the innovation—given a certain level of initial commitment. Somewhat similar to the argumentation regarding speed of interpretation, the routine flexibility within family firms is driven by the match of the employees' non-economic utility function with the innovation. This relationship is moderated by employee tenure. Figure 14 summarizes these findings. Besides the above-mentioned associations, two further relationships are highly conceivable, yet did not emerge from my empirical data (both shown with dotted lines in the figure). First, congruence of innovation and predominant SEW dimension is likely to not only influence speed and intensity, but also stamina. When the investment into the new technology is congruent with the family's predominant SEW dimension, the decision makers are less inclined to abandon such activities after temporary setbacks and thus exhibit higher levels of stamina. Besides this direct effect, I also expect an indirect effect, pointing to a potential relationship between SEW congruence with the discontinuity and CEO commitment to the discontinuity: As family firm owners are tightly linked to their business and as family owners' goals tend to converge with the firms' goals (Berrone et al., 2010), the match of SEW and technology is also likely to enhance the decision makers' commitment to invest into the discontinuous innovation.

Figure 14: Family Influence and Adaptation to Discontinuous Technological Change (Empirical Evidence)

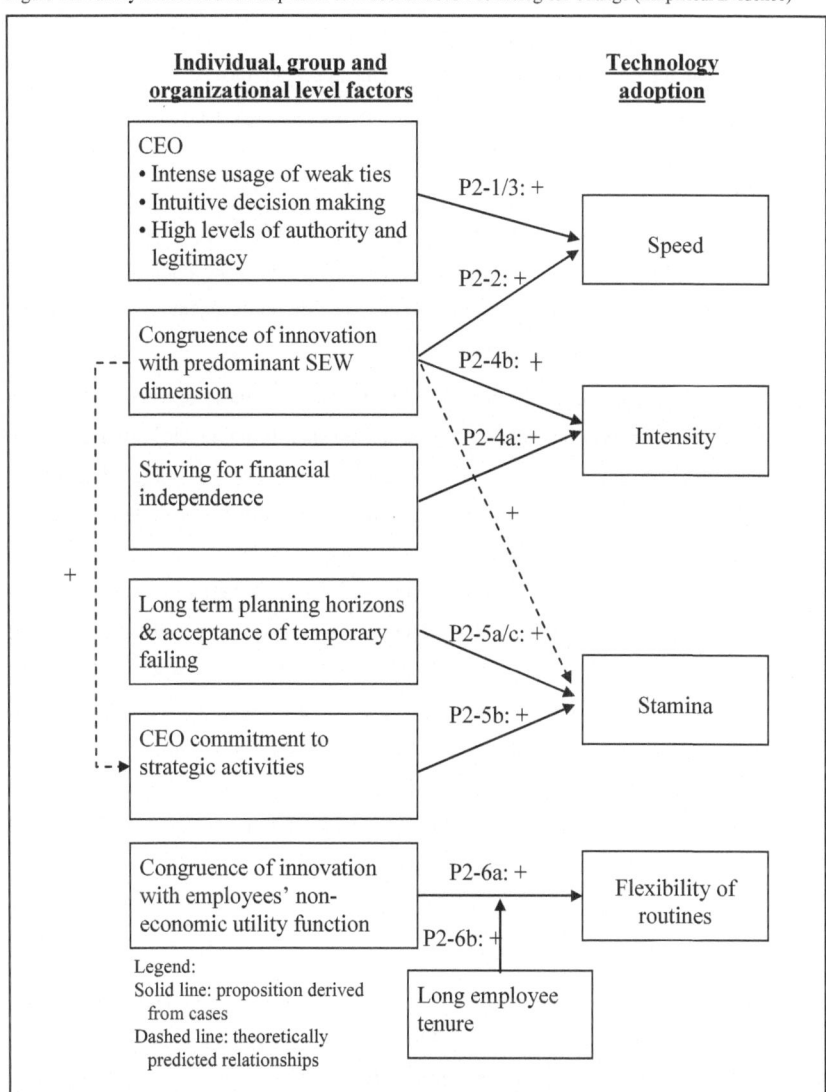

These findings contribute to literature on discontinuous technological changes as well as family business research. First, the analyzed cases exemplify some of the propositions conceptually deduced in chapter 3 and provide first hints of the most salient connections between family influence and determinants of organizational adaptation. This is relevant information for researchers aiming to set up a large-scale study investigating the family influence-response relationship quantitatively. Next and more importantly, the findings of this study reveal the role of socioemotional wealth considerations as well as the role of employees' non-economic utility functions for the speed, intensity, and routine flexibility of the respective organization's adaptation. Research on discontinuous technologies has investigated the role of various related non-economic determinants of organizational adaptation such as rigid mental models or organizational identity. To the best of my knowledge, the influence of non-economic goals on organizational adaptation has remained untackled by research so far. My results show, that specific non-economic goals can either promote or hamper adaptation to discontinuous change.

My findings also contribute to the field of family business research. Predicting organizational response to technological shifts is fundamental for both theory and practice. In this context, the distinction between continuous and discontinuous technological change is particularly salient because predictions of family business responses to change are likely to differ fundamentally depending on the type of change. Thus, the disentanglement of discontinuous technological change from continuous change may help to reconcile some of the contradictory findings in family business literature, thereby leading to the delineation of clearer practical implications for family businesses. For instance, family firms faced with discontinuous technologies are not advised to rely on their long-term and experienced employees to implement discontinuous innovations, particularly if they counteract the employees' non-economic utility function. Moreover, family firm owners and managers should be aware of the positive impact low levels of formalization and intuitive-based decision-making can have in times of environmental turmoil.

4.5.2 Limitations and Further Research

As with any empirical work, this study comes along with several limitations, particularly regarding its internal and external validity. First, as common for most case study research, interview statements referring to past decision-making and interpretation processes might be affected by retrospective bias (Gavetti & Rivkin, 2007; Gilbert, 2005). To alleviate this risk of distortion I took two methodological precautions: When defining the research setting, I deliberately selected industries with mostly recent technological changes, to minimize retrospective bias of the respondents. Moreover, I draw on extensive longitudinal primary and secondary material such as content of historical websites and press releases of a 15-year time period, with specific focus on CEO interviews with media, to triangulate my data.

Hence I expect that my results are stable and largely unaffected by retrospective bias. A comparison of (potentially biased) interview statements with (unbiased) historic press releases, annual reports and website content[70] supports this assumption.

Next, one could argue that 'single respondent bias' affects my results, since for most cases I relied exclusively on interviews with the CEO rather than collecting information from various organizational members. This focus, however, is rooted in the nature of my research questions. My main goal was to understand the decision makers' interpretation of the discontinuous technology and the respective decision-making processes within my sampled firms when faced with the discontinuous technology. Given the organizational and hierarchical structures present for my sampled firms, the details of these processes are best and in many cases only known by the CEOs—my informants[71]. Again, wherever available, I triangulated interview data with information from other sources such as press releases.

Alternative explanations of the observed phenomena (Yin, 1994) cannot be unerringly precluded. In particular, the research setting is confined to three German industries, thus rendering cultural and institutional idiosyncrasies possible. For instance, it is possible that in different cultural contexts, family influence manifests itself in different firm characteristics than those presented in this study. I therefore encourage other researchers to replicate my study considering different cultural settings to gain an even more nuanced understanding of interpretation and decision-making processes of family firms under environmental turmoil. Moreover, future research needs to test the propositions induced in this study by applying empirical, quantitative methods, including highly family influenced as well as non or less family influenced businesses. Content analysis of historical documents appears to be a specially fruitful avenue for doing so (Berrone et al., forthcoming).

To conclude, my results show that adoption of incumbent inertia research as theoretical perspective opens a host of new research avenues for scholars examining the behavior of family businesses in times of radical change. I hope that my findings will inform other researchers in their approaches to develop clearer models that explain the responses of family businesses to change.

---

[70] Comparisons conducted in all cases for which data from both sources was available.
[71] In two cases (*Anything & More*, *Powerhouse*) it was not possible to gain direct access to the family CEO. Consequently, I interviewed top management members who a) closely collaborated with the family CEO and b) were deeply involved in large fractions of the adaptation process, thus being likely to know his/her attitude and interpretation of the discontinuous technology.

5 Empirical Evidence on how Organizational Identity affects Organizational Adaptation to Discontinuous Technological Change

5.1 Introduction

Throughout the previous chapters of this thesis, one important determinant of organizational adaptation to discontinuous change has remained unheeded: organizational identity. However, various interview statements, in particular those outlined in conjunction with SEW considerations indicate that the self-definitions and self-perceptions of organizations substantially shape their organizational behavior.

In the field of incumbent inertia, organizational identity—the perception of "who we are as an organization" (Nag, Corley, & Gioia, 2007: 824)—has recently been identified as an important cognitive-emotional barrier to adaptation by researchers (Tripsas, 2009). Although existing research has generated a wealth of insights, our picture of the role of organizational identity in the context of organizational adaptation is still coarse and equivocal (Clark, Gioia, Ketchen, & Thomas, 2010; Livengood & Reger, 2010). In particular, contrary to the dominant view, recent empirical work (Hatum, Pettigrew, & Michelini, 2010) shows that organizations' identities can also *promote* adaptation to radical environmental turmoil. These authors empirically reveal strong organizational identity based on "ideals favouring change, innovation and risk-taking" (Hatum & Pettigrew, 2006: 131-132) as an important enabler for major organizational transformations within family firms. Given these ambiguous results, rooted in the nascence of the research field, as well as the tremendous impact of organizational identity on organizational adaptation (Kogut & Zander, 1996; Nag et al., 2007; Oliver, 1997), I chose to dedicate a full chapter of my thesis to better understand the mechanisms underlying the family-identity-response relationship.

Organizational identity of family influenced firms is much shaped by their history (Greiner, 1997) and their dominant coalitions through an overlap of the identities of the family, the business and the individuals (Dyer & Whetten, 2006). The owner-family itself constitutes a central, distinctive, and enduring element of the organization (Memili, Eddleston, Kellermanns, Zellweger, & Barnett, 2010) and as such substantially contributes to the organization's identity. Moreover, the image of the founder as well as the family members' vision and values affect organizational identity (Memili et al., 2010). Given the idiosyncrasy of family firms' goals and values and hence organizational identity (Milton,

2008), variance in family influence is likely to cause variance in adaptation to discontinuous change. In the following exploratory study, I therefore aim to answer the following research questions:

(1) *What are elements of organizational identity that exacerbate organizational inertia, and what are constituents of organizational identity that enable organizations to respond adequately to environmental changes?*

(2) *Once these factors are known, how do they exactly affect organizational adaptation?*

(3) *How are these factors linked to family influence?*

In a first step, guided by extant literature on organizational identity (Albert & Whetten, 1985; Gioia & Thomas, 1996; Labianca, Fairbank, Thomas, Gioia, & Umphress, 2001), I collect a rich body of interview and archival data to first investigate how the identities of organizations facing a discontinuous technological change differ with regard to various dimensions. I then aggregate this data to distill two specific, mutually exclusive dimensions of identity, which I label *focus* and *locus*. Focus of organizational identity relates to the degree to which organizational members define their competitive area *inclusively* or *exclusively*. An inclusive organizational identity is anchored in a higher hierarchical cognitive level (Porac, Thomas, & Baden-Fuller, 1989). Correspondingly, members of an inclusive organization perceive themselves as belonging to a broader competitive group and conjunctively draw connections between various fields of competition and technologies (Miles & Snow, 1978). In contrast, members of an exclusive organization define their "home turf" (Livengood & Reger, 2010: 49) more narrowly and highlight elements that differentiate their own domain from other domains. Locus of organizational identity denotes to what extent members of a given organization predominantly refer to the organizational self (e.g., organizational members, narratives, and shared values) or to the environment (e.g., customers, suppliers, and other stakeholders) when legitimizing strategic actions (Gioia & Thomas, 1996; Labianca et al., 2001).

In the second step, I build on the distinction between locus and focus of identity to investigate how these dimensions of organizational identity individually affect adaptation to technological discontinuities (Ford & Baucus, 1987). My analysis reveals that focus of identity is most tightly linked to the degree to which companies adopt the new technology: highly exclusive organizations tend to stick more to their established business model and are less likely to attempt to commercialize the new technology than inclusive organizations. In contrast, locus of identity is most strongly related to the speed of the response to a discontinuity: environment-focused organizations react relatively promptly, while self-focused organizations seem to wait longer until they consider a response.

In the third step, I explore the interactive effects of the two identity dimensions in the context of discontinuous technological change. This analysis provides a central contribution of my paper: a framework that proposes how the four generic focus/locus combinations of organizational identities entail idiosyncratic variation in organizations' adaptation behavior: Type 1 (inclusive/environment, I-N) organizations respond to discontinuous technological change immediately and actively, particularly by attempting to create new domains using the innovation, however, with somewhat rigid routines. Type 2 (inclusive/self, I-S) organizations adopt technologies somewhat later than Type 1, however more flexibly. Type 3 (exclusive/environment, E-N) organizations are the only players that respond passively to discontinuities, particularly with denial and resignation. Finally, Type 4 organizations (exclusive/self, E-S) respond actively, however, by deliberately sticking to, and defending, the established domain.

In the fourth and final step, I investigate the link between organizational identity types and family influence. First, my data shows that family influenced businesses in particular are inclined to own inclusively focused identities that result in passive or defensive responses to change. Second, I provide empirical evidence on how family idiosyncratic factors such as the founder's legacy (Ogbonna & Harris, 2001), organizational history, and family members' goals shape the respective organization's identity.

My research provides a nuanced picture of the impact of organizational identity in the realm of organizational adaptation to discontinuous technological change. Previous research (Livengood & Reger, 2010; Tripsas, 2009), which has envisaged organizational identity as a focal ingredient to organizations' enactment of their environment and to competitive response, has left the precise differences that identity causes in organizational outcomes unspecific. My inquiry is unique as it is the first to explore how two specific dimensions of identity, which have been previously discussed but never explored in combination, can entail diverse responses of incumbents within a single geographically focused industry. As such, my study may serve as a basis to resolve a critical puzzle addressed in recent management research: heterogeneity of incumbent response patterns when faced with discontinuous technological change. I also add to research on organizational identity by revealing and conceptualizing two mutually exclusive and highly influential dimensions that together form the identity domains of social actors. Moreover, my study contributes to family business literature by enhancing knowledge on the links between family influence and strategic behavior of firms. Likewise, I provide empirical evidence on how identities of family influenced firms differ from those of non-family influenced firms.

5.2 Theoretical Background

5.2.1 Organizational Identity and Organizational Adaptation to Discontinuous Technological Change

More than 50 years ago, Selznick (1957) was one of the first to emphasize the importance of organizational members' collective self-definition for their respective organization's capabilities (Nag et al., 2007). In a groundbreaking article in 1985, Albert and Whetten defined organizational identity as the "central, distinctive and enduring" elements of an organization (Albert & Whetten, 1985: 265). This definition was later broadened by researchers to the perception of "who we are as an organization" (Nag et al., 2007: 824) as a collective and self-referential understanding of the "essence" of an organization (Corley et al., 2006). Organizational identity is related to yet different from constructs such as organizational image, organizational culture, and corporate identity (Corley et al., 2006).

Incorporating aspects of prior work by scholars in the organizational identity field, such as David Whetten and Denny Gioia, I define 'organizational identity' as follows:

Definition 3: Organizational Identity

> *'Collective belief, shared among organizational members across all hierarchies and departments, about what constitutes the central, distinctive and enduring elements of the respective organization.'*

As such, organizational identity sets the "cognitive frame or perceptual lens that provides a basis for sense-making" (Cornelissen, Haslam, & Balmer, 2007: 9; Dutton & Dukerich, 1991;

Gioia & Thomas, 1996). In other words, identity represents the underlying collective values that determine whether managers and employees can identify with certain strategic options and, in turn, view these options as legitimate (Albert & Whetten, 1985; Dutton & Dukerich, 1991). Typically, identity is shaped by the historical and social contexts of an organization and, therefore, is profoundly influenced by an organization's concept of creating and capturing value (Dutton & Dukerich, 1991). Organizational identity is mostly envisaged as a rather stable concept (Albert & Whetten, 1985; Whetten & Mackey, 2002), however, several authors have emphasized that 'destabilization' of organizational identity as a result of attempts at strategic change, can in turn induce changes in organizational identity (Dutton & Dukerich, 1991; Gioia & Chittipeddi, 1991; Nag et al., 2007). Several authors showed how organizations such as US post offices (Biggart, 1977) or high tech companies (Fiol, 2002) mastered changing their identities.

A growing stream of literature has suggested that organizational identity is not only influential in strategic decision-making in general (Elsbach & Kramer, 1996; Glynn, 2000; Golden-Biddle & Rao, 1997), but is a critical cognitive ingredient of organizational adaptation (Clark et al., 2010) and, thus, can contribute to explaining the substantial differences in incumbents' responses to technological discontinuities. Several authors have recently illustrated that organizational identity can restrain decision makers and organizational members from adopting non-paradigmatic innovations (Kets de Vries, 2006; Livengood & Reger, 2010; Tripsas, 2009). Tripsas (2009) uses the case of a company in the photography industry to show that because discontinuous technologies challenge deeply embedded norms of creating and capturing value, decision makers often find it difficult to adopt such identity-breaking innovations. Nag and colleagues (2007) provided empirical evidence based on a single case study, of how the organizational transformation of a high tech company from a 'technology-push' to a 'market pull' identity failed due to routine rigidity.

In particular, scholars propose that organizations develop 'identity domains' (Livengood & Reger, 2010), i.e. socially constructed boundaries of competitive fields (Porac et al., 1989), which, in turn, affect organizational attention and the organizational moves implemented by decision makers in response to strategic issues (Barney et al., 1998; Ocasio, 1997). If discontinuous technological change emerges outside of a firm's identity domain and is thus incompatible with the prevalent identity scheme within this domain, decision makers are unlikely to become attentive to this issue and, if they do notice, may tend to consider more rigid, paradigm-consistent organizational responses (Livengood & Reger, 2010; Tripsas, 2009). Despite these recent conceptual and empirical advancements, it still remains unclear which elements define identity domains and how these elements are linked to organizational adaptation.

Existing literature provides only vague answers to this question. Porac et al. (1989) and studies stemming from this work (e.g., Abrahamson & Fombrun, 1994) argue that decision makers in incumbent organizations develop homogeneous sets of fundamental beliefs and values within their ecosystem as a result of repeated transactions (Kaplan, 2011). However, homogeneous identities within a group of incumbents imply homogeneous cognition and adaptation behaviors and therefore cannot explain the phenomenon of response heterogeneity (Livengood & Reger, 2010).

The question whether distinct identity dimensions attributable to a broad set of organizations exist has not yet been conclusively answered (Corley et al., 2006). While proponents of an emic approach deny the existence of universally applicable categorizations, researchers favoring an etic perspective suggest that there are certain dimensions of identity that shape organizational action (Corley et al., 2006). For instance, Miles and Snow's (1978) classic typology of strategic types, although it does not directly refer to organizational identity, suggests that companies differ with regard to how narrowly or broadly they define their domain, and ultimately show variance in their strategic behavior. In another taxonomy, Albert and Whetten (1985) differentiate between 'utilitarian' organizations, whose main stimuli stem from external changes in market conditions and whose goal is to maximize performance, and 'normative' organizations, which are driven by internal values and ideologies and are therefore interested in "upholding traditional values and ideologies" (Martins, 2005: 707). In a similar vein, Nag et al. (2007) investigate the change of a high-tech company from a 'technology-push' towards a 'market-pull' organizational identity, thereby focusing on knowledge grafting processes.

However, there is a void in research to scrutinize and integrate these findings to understand how identity dimensions combine into types of organizational identities (Corley et al., 2006), and how these identity types differ in their responses to discontinuous technological change. This research gap is critical because unless scholars have explored the underlying mechanisms that explain how and why and under which circumstances dimensions of identity interactively play out in the context of technological turmoil, it is impossible to precisely theorize on the association between radical environmental shifts and organizational adaptation.

5.2.2 Organizational Identity and Family Influence

Generally, owners and managers exert significant influence on their companies' identities (Arregle et al., 2007) and, in turn, on strategic activities (Benner, 2007; Fiss & Zajac, 2004). In family businesses, this influence is particularly strong, because the 'family system' in family businesses (Habbershon & Williams, 1999) is typically nurtured by strong, historically consistent, and clearly defined values and norms (Gersick et al., 1997). Thus, Memili, Eddleston, Kellermanns, and Zellweger conclude (2010: 201): "For family firms, organizational identity is unique in that the family is a distinct, central, and enduring component of the firm."

However, the interplay between organizational identity and family influence and its implications in times of discontinuous change are so far largely underexplored. Although family business research has long studied constructs related to organizational identity (Arregle et al., 2007) such as the cultural idiosyncrasies (Vallejo, 2008) or specific goals and values (Zellweger & Astrachan, 2008) of family firms, literature explicitly investigating the role of identity in the context of family firms remains surprisingly scarce. A search for papers containing 'identity' in their titles in the two most relevant family business specific scientific outlets, revealed no hits for *Family Business Review*, and only eleven hits for *Entrepreneurship Theory and Practice*, all of them having been published in 2008 or later[72].

Most research published in *Entrepreneurship Theory and Practice* related to identity, however, relates to entrepreneurial identity (Farmer, Yao, & Kung-Mcintyre, 2011), social identity (Miller & Le Breton-Miller, 2011), or meta identity (Reay, 2009; Shepherd & Haynie, 2009). Authors discussing social identity mostly argue on group level, proposing that the overlap of and interaction between family and business identity has ponderous implications, for instance regarding conflicts among family members (Shepherd & Haynie, 2009). Other researchers (Arregle et al., 2007; Foreman & Whetten, 2002) point out the hybrid identities prevalent in most family influenced firms.

The sparse research available on organizational identity of family firms points out that owning families shape their respective organizations' identities (Memili et al., 2010) by adding a normative component (Albert & Whetten, 1985) to the utilitarian identity of the business (Zachary, McKenny, Short, & Payne, 2011). Arregle and colleagues (2007: 80) theorize that "family members [...] transmit [family social capital's] main characteristics

---

[72] Search conducted with googlescholar.com on April 8, 2012. No further filters despite title and journal, were set. Most of the identified papers referred either to entrepreneurs rather than family firms or studied group level rather than organizational level identity.

(norms, values, narratives, etc) to the firm through two specific constructs: the firm's organizational identity and rationality." Besides the influence of the active generation, the "founder's shadow" (Short, Payne, Brigham, Lumpkin, & Broberg, 2009: 10) is likely to shape a family firm's organizational identity and subsequently its strategic behavior (Davis & Harveston, 1999; Kelly, Athanassiou, & Crittenden, 2000).

5.3 Methodology

5.3.1 Research Design and Setting

Given the limited knowledge about the dimensions of organizational identity, their impact on incumbents' responses to technological discontinuities, and the interplay with family influence, I adopted a theory-informed, inductive multiple-case research strategy (Eisenhardt, 1989; Yin, 1994). Building on multiple cases allowed me to collect comparative data within a replication logic and thus to derive a more generalizable theory than by analyzing a single case (Eisenhardt & Graebner, 2007).

The research setting for this study was the organizational response of established firms in the German publishing industry to digitized products and services from the mid-1990s to the end of 2011. This setting is appropriate for several reasons: First, the complexly intertwined socio-technological shifts associated with content digitization and subsequent shifts in the value chain render extant knowledge bases and other assets of traditional publishers obsolete (Ronte, 2001). For instance, in digital publishing, IT skills replace knowledge about traditional layout and printing techniques. Moreover, new performance metrics, such as interactivity and ubiquitous availability, have largely dispelled established performance criteria, particularly haptic attributes. Digitization in publishing therefore represents an archetypal case of technological discontinuity (Tushman & Anderson, 1986). Second, organizational identity is likely to play an important role in the response behaviors of incumbent publishers to these changes. The technological innovations that have evolved over the past decade are significantly at odds with the traditional assumptions and values shared by most established publishers. In addition, German publishing houses have historically enjoyed high social status and have substantially influenced the development of Germany's culture and ultimately, its values, ideologies, and beliefs. Thus, these companies can, on average, be expected to be relatively cognizant of their own values and identities (Meyer & Scott, 1983).

Third, there is high variance among the established German publishing houses regarding their adaptation to digitization.[73] Fourth, the timeframe of this study is sufficiently distant to allow for the emergence of longitudinal patterns, whilst at the same time being recent enough to permit detailed data collection with limited retrospective bias.

Another advantage of the German publishing industry is that it consists of numerous businesses with varying characteristics, which allowed me to compare polar cases, in order to scrutinize and enrich my theory development (Eisenhardt, 1989; Harris & Sutton, 1986). In this vein, I selected a diverse sample of 14 of the top 100 businesses[74] in the German publishing industry that are active in the segments of producing and selling books and/or periodicals, but differ with respect to aspects such as size and age (see Table 11). 'Providing high quality products'[75] was a key concern for all 14 companies. All of the selected companies employed 50 or more employees when the discontinuity started to emerge. Thus, these firms are all formalized businesses with multi-level resource allocation processes (Bower, 1970). Moreover, the companies I investigated are non-diversified publishing houses, which have been active in publishing for at least 25 years. Hence, all these firms qualify as incumbent players.

Table 11: Organizational Identity and Adaptation—Overview of Sample

Disguised Publishing House[a]	Primary Market Segments	Number of Emp-loyees[b]	Age of Organi-zation	Family influence[c]
Reader's Finest	Books—fiction, non-fiction (reference books)	~50	>150	+
Rocket Book	Books—non-fiction (reference, workbooks), periodicals—formerly: customer magazines	~50	~50	+
Ars Legendi	Books—fiction, non-fiction (reference books)	~50	~50	+++
Books & More	Books—non-fiction (reference, textbooks)	~75	~50	o

---

[73]For instance, in the field of journal publishing, Springer was one of the pioneers in the German market, offering electronic journal access to its customers as early as 1996. In contrast, Franz Steiner still refused to grant readers electronic access to its journal publications in 2011. In the book market, Hermann Simon started to sell ebooks as early as 1998. In contrast, the ebooks of the Thienemann Verlag were not available until 2010.

[74] Ranked according to annual revenues, published by the German publishing association.

[75] As revealed by word count analysis.

Disguised Publishing House[a]	Primary Market Segments	Number of Employees[b]	Age of Organization	Family influence[c]
House of Books	Books—non-fiction (reference, text, workbooks), periodicals—journals, trade magazines, newsletters	~75	>150	++
Star Print	Books—non-fiction (text, workbooks)	~75	>150	++++
Secret Books	Books—fiction, non-fiction (reference books), periodicals—customer magazines	~100	~50	o
TopPress	Books—fiction, non-fiction (reference, textbooks), periodicals—journals, trade magazines, customer magazines	~150	>100	++++
Reader's Publisher	Books—non-fiction (reference books), periodicals—trade magazines, newsletters	~200	>150	+++
Peter's Publishing House	Books—non-fiction (reference, textbooks), periodicals—journals, newsletters	~250	>150	++
Yellow Books	Books—fiction, non-fiction (reference, textbooks), periodicals—trade magazines, newsletters	~300	~50	+++
Book 2020	Books—non-fiction (reference books), periodicals—trade magazines, newsletters	~400	~50	+
Superbooks	Books—non-fiction (reference, textbooks), periodicals—trade magazines, newsletters	>500	~50	o
Arthur & Sons	Books—non-fiction (reference, textbooks), periodicals—trade magazines, formerly customer magazines	>500	>100	++
Median	**n/a**	**~125**	**n/a**	**++**

[a] ordered according to size (small to large), family influence (low to high), and alphabetically
[b] in 2009.
[c] measured based on adapted F-PEC scale (Astrachan et al., 2002).

5.3.2 Data Collection

Over a period of 15 months, I collected data from three main sources:

> (1) semi-structured interviews with members of the focal firms and industry experts;
>
> (2) archival documents including historical websites, business publications, and internal documents; and
>
> (3) additional data.

An overview of my data is presented in Table 12. The core objective of my semi-structured interviews was to gain a deep understanding of the companies' respective identities and to explore the mechanisms that linked firms' identities and their adaptation behaviors. In this regard, I first adopted a top-management focus, assuming that upper echelons influence organizational identity most strongly (Barney et al., 1998; Hambrick & Mason, 1984). Accordingly, I conducted semi-structured interviews (minimum length 60 minutes) with the CEO and/or owner of each of the companies[76]. I then expanded the data to a broader range of hierarchical levels and functions by interviewing additional employees and managers across hierarchies, departments, and tenures (80 percent in-person; average length 60 minutes). In total, I conducted 63 case-specific interviews and enhanced my understanding through seven extensive, more general conversations with industry experts and the founders of new entrants, i.e. companies that recently entered the publishing market with digitized products and services.

This approach allowed me to obtain a variety of perspectives. The interview guidelines for my conversations were primarily informed by the identity literature (Labianca et al., 2001) and literature on organizational adaptation (Gilbert, 2005). Although I continually adapted the interview guide to the particular interview situations (Glaser & Strauss, 1967), all interviews included three main sections that were composed of a total of approximately 40 mostly open-ended questions. The first section aimed to gain an enhanced understanding of the business and the respective owner-family's influence. The second part encouraged the informants to provide a broad overview of their company's vision and goals, core products, strengths and weaknesses, positioning among the competitors, self-image and history. Information gathered from this section was crucial for identifying the company's

---

[76] Nine of the 14 CEOs/owners were either already CEOs/owners of their company when the new technology emerged (8) or worked in the company in a leading position (1). The other five CEOs I interviewed had joined their respective companies as external CEOs later during the technology evolution. I therefore paid particular attention to interviewing other managers with longer tenures when collecting data from these organizations.

identity and for scrutinizing identity strength and homogeneity. The third part of the interview aimed to answer the questions of 'When and how did organizational members recognize and interpret the discontinuous technology, and how did the organization decide on the response to this shift?' Besides this, I intended to understand in depth, how companies that had responded actively to the new technology had implemented these responses. All interviews were recorded and afterwards transcribed within one week. In the fourth and last part, I aimed to probe some initial hypotheses with my informants.[77]

---

[77] For the detailed interview guide, see Appendix.

Table 12: Organizational Identity and Adaptation—Overview of Data Sources

Publishing House	Interviews		Archival (Primary and Secondary) Documents			Additional Data	
	Num-ber	Informants	Number	Examples of Business Publications and Internal Documents	First Historical Website[a]	Industry Fair Ob-servation	Sur-veys[d]
Reader's Finest	6	CEO, 2 production assistants, sales assistant, editor, industry expert	18	Internal workflows, brochures, press articles, financial statements	1999	yes	4
Rocket Book	3	CEO, editor, industry expert	4	Financial statement	1996	yes	2
Ars Legendi	3	CEO, VP production, sales assistant	49	Brochures, press articles, financial statements	1998	yes	3
Books & More	3	CEO, VP production[b], industry expert	4	Brochures, press articles, financial statements	1998	no	0
House of Books	4	Owner, VP sales, VP production[b], supplier	17	Brochures, press articles, financial statements	1998	yes	0
Star Print	3	CEO, editor, industry expert	5	Brochures, financial statements	2001	yes	2
Secret Books	4	CEO, VP production, VP sales, manager digitization[b]	76	Brochures, press articles, financial statements	1999	yes	1
TopPress	4	CEO, VP production, manager digitization[b], industry expert	17	Press articles, financial statements, CEO interview with media	1998	yes	3

Reader's Publisher	5	CEO, VP purchasing, 2 managers, editor	59	Press articles, financial statements	1998	no	3
Peter's Publishing House	7	CEO, 2 production assistants[b], sales assistant, head of work council, editor, industry expert	39	Company chronicle, internal organization charts press articles, brochures, financial statements, CEO interview with media	1996	no	4
Yellow Books	8	CEO, manager digitization[b], editor, 2 sales assistants, executive assistant, product owner, industry expert	7	Brochures, financial statements, CEO interview with media	1999	no	2
Book 2020	5	Owner, CEO, head of online marketing, manager online marketing; industry expert	47	Brochures, press articles, financial statements, CEO interview with media	1999	no	1
Superbooks	4	CEO, product owner[b], project team member, industry expert	84	Press articles, financial statements	1996	yes	0
Arthur & Sons	4	CEO, VP production[b], editor, industry expert	16	Company chronicle, internal organization charts, press articles, brochures, financial statements	1996	no	2
Not case specific[c]	7	2 founder-owners of new entrants, 5 industry experts	41	Academic books on the publishing industry, newspaper articles, blogs, websites, market research studies	n/a	n/a	n/a
Total	**70**	**n/a**	**483**	**n/a**	**n/a**	**8**	**27**

[a] Historical Websites were retrieved in 2-year steps until end of 2011. [b] Person responsible for digitized products.
[c] These interviews focused on general issues concerning the response of publishing houses to digitization.
[d] Numbers of interviews that included a short, systematic, structured questionnaire.

In addition to the interviews, I was able to assemble a large amount of *archival data*, including press clippings and company presentations. Some organizations granted me access to confidential archival documents. One particularly interesting source of archival data was historical websites dating back to the mid-1990s, which I accessed via web.archive.org. I analyzed all available subsections of the website data in two-year increments, resulting in an average of eight data points per organization.

Additional data included approximately 15 informal follow-up conversations and 27 systematic complementary surveys on the company's relevant network and market screening activities, which I used to validate my findings. Moreover, during my interview visits, I systematically documented on-site observations, which included aspects such as body language of the informant, the atmosphere of the conversation and summaries of casual conversations with other employees, which were subsequently used for triangulation (Bogdan & Biklen, 1982). Furthermore, I was able to observe members of the focal organizations unobtrusively at industry conferences and industry fairs.

5.3.3 Data Analysis

I began the data analysis process with a computer-assisted (Krippendorff, 2004) grounded analysis of the individual cases through the lens of my research question (Eisenhardt, 1989; Glaser & Strauss, 1967). Given my explorative goals, these single case studies aimed to allow "the insiders' point of view to [become] the foundation of the analysis" (Clark et al., 2010: 403). I therefore adopted van Maanen's (1979b) two-step process and first extracted first-order concepts from the collected documents and interview transcripts. In a subsequent step, I interpreted the respondents' cognition in the light of the respective context and I compared my interpretations with findings in extant literature. This allowed me to distill abstract, second-order concepts and to induce grounded theory. Afterwards, I began the cross-case analysis, looking for common patterns and themes (Eisenhardt & Graebner, 2007). In doing so, I iteratively analyzed the data whilst enfolding extant literature (Eisenhardt, 1989) and steadily refined emerging themes and patterns by revisiting the single cases.

In particular, I first entered all textual data relating to the cases into my online case study database[78]. In parallel, I created an initial set of coding guidelines that was informed by

---

[78] http://www.hidrive.strato.com

extant literature. These coding guidelines consisted of the overall concepts I aimed to investigate (called 'nodes' in the computer program NVivo; for instance 'type of adaptation to discontinuous change') and their mutually exclusive and comprehensively extensive components (called 'sub-nodes,' such as 'domain creation,' 'domain offense,' 'other active reactions,' and 'passive reactions'). Each node and sub-node was based on a clear definition and linked to relevant literature (see A.A-3 for an example). When adequate and required, I added specific coding rules (such as 'Activities were coded as other active domain reaction if the decision makers were aware of the change and deliberately took actions to face the challenge, but activities were neither domain creation nor offense.') to provide additional details on how to codify interview statements.

I began my single case analyses by reading and coding the respective interviews (first CEO interviews, then interviews with managers and employees). All textual passages that matched the definitions and rules given in the coding guidelines were attributed to the respective sub-node. Interview statements that appeared insightful and important but did not match any of the defined sub-nodes were coded as so called 'free nodes.' On a regular base (in average after completion of 10-20 interviews) I adapted the coding guidelines rendering them increasingly precise (Miles & Huberman, 1994) and adding other relevant nodes (based on salient free nodes). I re-coded the interviews whenever required, so that in the end all interviews were coded according to the final coding guidelines (as portrayed in Appendix A.A-3). After completion of the interview coding, I re-narrated the adaptation process (as well as the firm's background and identity) for each organization in a bullet point form in order to create a holistic understanding of each single case. Based on this data, I abstracted first order concepts to second order concepts. For instance, I interpreted *Ars Legendi*'s signing on of new authors for print books as domain offense activity in the given contextual setting.

Moreover, I exported the NVivo summary into an Excel file showing how many times I categorized interview statements of each case into each of the sub-nodes. I used this matrix-type overview to identify first, preliminary cross-case patterns (e.g., observation that companies that defined their focus narrowly refrained from adopting the technology). After identifying these potential associations, I returned to the transcripts searching for statements indicating causal relationships (e.g., statements that publishers did not commercialize ebooks because they felt that this was out of their comfort zone, they felt unable to produce those products, etc.)

In a subsequent step I compared my preliminary results, i.e. the Excel matrix and the first set of propositions, with those of four student coders, who conducted the above described analysis process at the same time as I did[79]. The coders had diverse, but related, educational backgrounds (two with university degrees in management and two with university degrees in book studies) in order to obtain a broad variety of perspectives. Involving coders with a background in book studies was particularly helpful in this study to draw on tacit knowledge about the book publishing industry. In two joint full-day workshops and several subsequent phone conferences, I crosschecked my findings with those of the student coders. I regularly performed intercoder agreement checks (intercoder correspondence of final coding guidelines measured by Cohen's Kappa >0.6, thus indicating a substantial agreement of the coders[80]). Discrepancies were debated until consensus was reached. This part of the data analysis process terminated with a shared understanding of each single case, existing cross-case patterns, and propositions as well as a link to extant theory.

In parallel to discussing discrepancies with the other coders, I followed the call of other scholars (Jick, 1979), and systematically triangulated data from multiple sources and on multiple occasions to enhance the accuracy of my findings. To do so, I read through all of the non-interview-based, textual data (e.g., press releases, annual reports, web sites, brochures,…) and compared the relevant statements—in particular dates and numbers—to those included in my single cases in note form. I added missing data and noted whether interview data were confirmed or contradicted. In case of still missing or conflicting data, I also followed up with emails to my informants, in line with suggestions of Lincoln and Guba (1985)[81]. For more complex concepts (such as locus and focus of identity) that involved abundant textual evidence, I created Excel sheets, in which I counted evidence for each sub-node (e.g., how many statements on historic websites referred to environment-centric as opposed to self-centric *foci*). Moreover, I roughly cross-checked whether the findings of my textual sources

---

[79] In order to increase the reliability of the results, we did not communicate within the coding team on our preliminary findings throughout the initial coding processes. We did, however, align on the initial coding guidelines and timely communicate any change relating to those, in order to improve comparability of our results.
[80] Cohen's Kappa is a statistical measure that indicates the quality of raters' agreement (Cohen, 1960). It is calculated as: (relative agreement among raters – hypothetical probability of chance agreement) / (1 – hypothetical probability of chance agreement). Other than merely counting the ratio of agreements vs. non-agreements Cohen's Kappa takes agreements by chance into account. Researchers (Landis & Koch, 1977) categorize Kappa < 0 as poor agreement, 0 < Kappa < 0.2 as slight agreement, 0.2 < Kappa < 0.4 as fair agreement, 0.4 < Kappa < 0.6 as moderate agreement, 0.6 < Kappa < 0.8 as substantial agreement and 0.8 < Kappa < 1 as almost perfect agreement. One major critics of Cohen's Kappa is that due to the specific way to calculate its value, it is not always possible to reach Kappa = 1, even in cases of perfect inter-rater agreement.
[81] However, due to non-responses it was not possible to fill all missing details and solve all data conflicts. All such cases were indicated as such in the respective subsections of Chapter 5.4.

190

matched with the observations I (and the other coders) made at the industry fair and during the interviews. This was particularly helpful to identify potential 'extenuations,' e.g., due to social desirability bias, by our informants. Whenever discrepancies between the interview data and other data emerged, I brought those issues back into the discussions with the other coders.

As soon as I had a strong match between theory and data, I created a set of propositions that explained the cross-case patterns observed for my sampled firms. Moreover, I re-engaged in communication with my informants and experts to validate my observations and to discuss alternative explanations. For instance, I gathered feedback from industry experts and publishers when presenting my results at the international conference of book scientists.

5.4   Detailed Case Description

In line with Sutton and Staw's (1995) call for more detailed case description, I will next provide information on each of the publishing houses included in this study. Due to anonymity concerns, information regarding, for instance, the concise product portfolio, dates of strategic actions, etc. cannot be revealed in this section. Moreover, I will disguise the gender of the interviewed persons in order to further address anonymity concerns. Similar to the case descriptions in Chapter 4.3, German quotations were translated to English by myself and then back-translated by students to ensure correctness.

The intention of this section is to expose single case studies in a descriptive and neutral manner. Data provided here will largely represent ordered sequences of first-order concepts (van Maanen, 1979b) as provided by the informants as well as data obtained from documents. The abstraction of second-order concepts and the following interpretation will take place in a second step. The information provided in the following subsection mainly stems from historic websites, personal interviews, and personal observations. Data sources are indicated where required. Please note that due to spatial constraints, only a selection of data can be provided in this thesis. However, all collected data (incl. codings of all interviews and all secondary data) was used for the categorization of the single cases and for the interpretation of the results. A summary of this holistic data interpretation process is shown in Table 15. Each case description will follow the same structure:

(1) I first provide background information regarding the respective company's size[82], products, its ownership structure, recent strategic moves such as acquisitions, and its level of international activities.

(2) Next I provide information regarding the company's culture and identity.[83]

(3) Third, and last, I describe the company's adaptation to digitization as a discontinuous technology over time. I will include statements revealing the underlying interpretation that caused specific activities wherever appropriate.

5.4.1 Reader's Finest

5.4.1.1 Background Information

Reader's Finest is a medium-sized, family-owned organization publishing fiction books as well as reference books, and it is managed by a family-external CEO. After decades of independence, the company was sold to another, large-sized family-owned business several years ago but even today it still acts rather independently. Throughout the last three decades and particularly during the last ten years, *Reader's Finest* has founded and acquired a total of four publishing houses and ultimately divested two of them. Products of *Reader's Finest* can be purchased globally via licensing partners.

5.4.1.2 Culture and Identity

Reader's Finest is seen as a "traditional business" (CEO, personal interview) which emphasizes its "long tradition" (website, 1999). This is congruent with the equipment of the physical publishing house (displaying the organization's development over time to the visitor) and also the graphical content of all historic websites (until a major relaunch in 2011), which, for instance, prominently displayed the logo of the company and other elements referring to the publisher's history. Employees describe the publisher as "reliable, traditional, a perpetual organization one can trust" (personal interview).

---

[82] Definition according to IfM Bonn (2012; http://www.ifm-bonn.org); small companies employ up to 9 employees with annual revenues less than 1 million EUR. Medium sized refers to those companies with 10—499 employees and annual revenues larger than 1 and smaller than 50 million EUR. Companies with more employees and annual revenues than medium-sized firms are labeled 'large.'

[83] As these two constructs are closely related and to some level interdependent, I chose to provide information on both of them to in order to amplify the reader's understanding of each of the single cases.

The company is very focused on its printed books. In the personal interview, the CEO emphasized: "Books, print books [are our core products]. And this won't change." Supporting this view, historical websites of *Reader's Finest* showed photographs of the first book produced by the publisher and the central meeting room of the publishing house contains a selection of all 'key' print books the company has developed over the most recent decades. One assistant described *Reader's Finest*'s employees as "very book-affine. We like the haptics, we like the smell, we like the sound [of turning pages]."

Despite the stability of *Reader's Finest* core products, the CEO expects some future changes, particularly in the product portfolio. "Innovation," referring for instance to new writing styles and topics, is seen as important. Anything not directly tied to the core business, however, is outsourced. This is the case for any non-hardcover and non-book product—those products are licensed to third party contractors and are not mentioned at all (except in short press releases) on the company's website. Moreover, acquired firms are not fully integrated into the organization. Again, the acquisition deals were only briefly noted in *Reader's Finest*'s press releases. Nine other established publishers, active in the same segment as *Reader's Finest*, are seen as main competitors (CEO, personal interview). Contacts to firms outside the book industry are perceived as "highly irrelevant" and contacts to book-related startups as "mostly irrelevant, as they first have to become established" (CEO, personal interview).

While the website did not and does not contain the word 'customer' at all, there is an extensive focus on the organization itself and its contracted authors. The website content is all about the company's history, its products, and—in particular—awards the company, its products, or authors have won. Each of the company's authors gets introduced in detail in a dedicated section of the historic websites (until 2011) including a photo of him or her. In the personal interview, the CEO explained that new innovations were typically introduced without any direct feedback from customers; rather it belongs to the publisher's "principles to generate ideas internally" (employee, personal interview).

Moreover, "independence" is a core theme for *Reader's Finest* (website, 1999). Interestingly, nowhere did the publisher notify of its acquisition by another family business (as mentioned in Chapter 5.4.1.1) but just dropped the phrase "independence" on its website, in a paragraph relating to its self-description in the respective year of the sale. Only since 2011, after the major website relaunch, can one find information regarding the publisher's current owner. Because of "decentralized structures" organizational members rather have a

"[*Reader's Finest*] feeling than a [*name of the mother company*] feeling" (CEO, personal interview). Employees and managers strongly identify with the publisher, its "history and its core products" (employee, personal interview).

5.4.1.3 Adaptation to Discontinuous Technological Change

In 1999, *Reader's Finest* first offered its customers the option to order books online. Up to 2011, all innovations were related to print products. For instance, in 2004, the company launched new series of books in non-German languages, new editions of 'classic' books, and also amended its product portfolio by publishing new genres. According to the CEO, it was a deliberate decision among the top managers to be "very cautious" regarding activities related to new media. "We never said: 'That will be the market of the future.' Maybe our books [of the core segment] will become more precious, more expensive" (CEO, personal interview). This perception is shared throughout the organization. An employee remarked in the interview: "I believe, that the hype about ebooks and apps will cool down again. Printed books will not disappear."

In the personal interview, the CEO and one employee stated that they perceived digitization as a radical change in other segments of the book industry that, however, left *Reader's Finest* unaffected. Moreover, the CEO assessed the "mediating role of the publisher between shops and authors as out of any question" and ultimately disintermediation as non-existent. In mid-2011, the publisher introduced its first ten ebooks (pdf, epub format) and also offered DVDs as free supplementary material for selected books. As described in the personal interviews, these innovations were forced by the parent company and were intended to "be prepared [for the potential development that] inexpensive reading matter might be distributed electronically in the future" (CEO, personal interview). This underlying motivation might explain the fact that electronic products are not actively promoted on the homepage. Moreover, the structure of the website is designed in such a way that all electronic books are categorized into a special section at the very bottom of the website. At the end of 2011, the number of ebooks had increased to slightly above 100. Interestingly, only books of relatively 'new editions,' those launched in 2004 for example, were offered in an electronic format. The 'classic' books were still only available in print, and, according to interview statements will not be offered electronically in the future. Since 2011, two apps have existed that were developed and commercialized by a licensing partner. When I asked the person representing *Reader's Finest* at the 2011 industry fair, he/she could not give me any information on those products. Asked about potential internal changes related to digitization, the CEO revealed that *Reader's Finest* had not hired any externals to work on new media development and commercialization (personal interview).

5.4.2 Rocket Book

5.4.2.1 Background Information

Rocket Book is a medium-sized, family-owned publisher that focuses on reference and workbooks. Formerly, the business was also active in producing and selling customer magazines. Several years ago, the family-owned company was acquired by another large German family business, but is still able to act relatively independently. It is managed by a family-external manager. *Rocket Book*'s international activities are constrained to licensing products to third parties. The mother company, however, is internationally active via its other subsidiaries.

5.4.2.2 Culture and Identity

The *CEO* of *Rocket Book* describes his/her company's core product as "the book—supported by other media" (personal interview). Moreover, several press releases emphasized the importance of the "classical book series" (2004). The company's culture and identity is centered on the publisher's core brand, which was also one of its first products and the first product that turned out to be successful. The logo of this brand has been central to the website throughout the years. Several non-book products related to that brand exist and were proactively promoted on the company's website as the company aims to offer a "complete" spectrum (website, 2004), "independent of the medium" (CEO, personal interview), and thus has long engaged in a "diversification strategy" (CEO, personal interview). However, non-publishing products of the core brand were strictly outsourced to third party contractors who were responsible for commercialization. This is because an important philosophy of *Rocket Book* is to "keep one's hands off activities which organizational members are not capable of doing" (CEO, personal interview). In contrast to other outsourcing models (see e.g., Chapter 5.4.1), *Rocket Book* strives to build up *some* competence internally in order to be able to control the quality provided by the partner and thereby protect the core brands (CEO, personal interview). In another part of the interview, the CEO remarked: "We are specialists, we do not offer a broad portfolio. We know exactly what we are good at."

One established German publisher is perceived to be the main competitor (CEO, personal interview). In 1996, one could read on the company's website: "In the beginning [of the company's history], our new [products] were not accepted. There were objections [against them]." The former owner-CEO explained why he/she continued despite the initial criticism (website, 2006): "I felt the famous gap." A press release (2004) characterized the publishing house as "creative and innovative." As the CEO explained, many changes in the company's recent history were introduced by employees. For instance, two employees of the graphics department proactively attended classes in photography and subsequently were able to broaden *Rocket Book*'s product portfolio. The CEO was not informed about the plans of the two employees before their training was completed (CEO, personal interview). Somewhat bluntly, the employee and the CEO summarized the working culture in the organization as: "Everyone does what he/she is capable of and likes to do."

For more than 10 years, the company did not mention its sale to another family business on its website. And even when they ultimately revealed their ownership structure, they only did so on the English subsection of the company website that referred to international licensing partners.

The company strongly focuses on "high quality" and "good value" and desires to offer a "concentrate of the best" to its customers by engaging "the best authors" and creating a "sophisticated program" (website statements). The company's website refers to numerous prizes that the publisher's products have won.

5.4.2.3 Organizational Adaptation to Discontinuous Technological Change

In a *2011* press release, the company's adaptation strategy is summarized: "Printed books are intended to remain the core of our publishing house. [...] Some genres could become even more successful in the future, when we offer digital products as supplements to books."

In 1996, the company offered CD-ROMs, developed by an external partner, as substitutes for books. From 1997 onwards, an 'online corner' of the website offered free supplementary information, however, this data was not updated for a long time after 1997. In 2002, for instance, all new product launches were print products—interestingly, the topics of several of them were related to 'computers' and 'internet.' In 2003, *Rocket Book* launched a new CD series in cooperation with a contracting partner (also a publisher). Customers were only able to order those products via the contractor's website, it was not possible to order the CDs directly from *Rocket Book*. At the time of the study, *Rocket Book* only offered print books. However, the publisher actively announced and promoted electronic products, commercialized by a licensing partner, such as CD-ROMs (since 1996), DVDs (since 2006), videos (since 2006), audio books (since 2006), ebooks (since 2011) and also apps (since 2011). In 2011, premium 'online only' content for specific customer target groups was implemented in cooperation with a partner. Interestingly, at the 2011 industry fair, *Rocket Book* was one of only four[84] publishers to present smartphone applications. *Rocket Book* had hired a small number (roughly four or five) of employees specifically for tasks related to new media, e.g., managing content in an electronic database. In 2011, 15% of the revenues was generated by (licensed) digital content.

5.4.3   Ars Legendi

5.4.3.1 Background Information

Ars Legendi is a medium-sized, family-owned and -managed publishing house with high family influence. The publisher's focus lies on fiction and reference books. Several years ago the company divested two of its original four business units and founded a new one. Hence, it now consists of three business units that are structurally separated from each other. The publisher is internationally active through various licensing contracts.

5.4.3.2 Culture and Identity

When asked to describe the organization, the CEO answered: "We are maybe the most renowned publishing houses of the post-war era in [our segment]. And we are one of the few pure family businesses in this sector." Quality is crucial to *Ars Legendi* and hence the publishing house "would not produce junk, just to survive" (CEO, personal interview). Three publishing houses, all of them 60 years or older, are perceived as the main competitors by the CEO and the interviewed employees. The culture of *Ars Legendi* is "hierarchical and patriarchic" (CEO, personal interview), however, the CEO strives to establish a more democratic way of working within the organization. *Ars Legendi*'s main goal is "to produce pretty, formidable books, that [the organizational members] also like to position in bookshops and sell" (employee, personal interview). According to the same employee, the core segments of *Ars Legendi*, will not change over the next decade, and *Ars Legendi* "will remain

---

[84] New entrants focusing on ebooks excluded. Interestingly, two of these four electronic devices did not work or had even been removed by the time I visited the industry fair.

a publishing house that produces books." The "long history and [*Ars Legendi*'s] brand" is what makes the publisher unique (employee, personal interview). A manager described *Ars Legendi*'s main goal as "to continue to produce the books, which we have produced in the past, also in today's rough economic times." The "print medium per se" is *Ars Legendi*'s core product (manager, *Ars Legendi*). The publisher is very closely tied to print books as its core products. Supporting the informants' statements on the importance of the medium 'printed book,' information on the historic websites (available from 2000 until 2010) emphasized the average weight of paper that the publisher and its employees made use of per year. Moreover, the company ran a distinct section announcing "particularly precious books" on its website (from 2000 onwards).

In the 1990s and 2000s, *Ars Legendi* broadened its product portfolio of print books. This was made possible by the "great freedom in decision-making," rooted in family ownership and influence (CEO, personal interview). However, the company struggled with recognizing and making use of the synergies between the various book segments. The strict internal separation of the business segments was mentioned in two of the personal interviews. Moreover, in 1998, the company informed its readers via a text on its website that its organizational members found it difficult to see the similarities between two of the business units that produced books on different, yet related topics.

In recent years, the publisher's website has been focusing on the publisher's employees as opposed to its customers. Employees are "proud to work for [*Ars Legendi*]," identify themselves "exceedingly" with the "beautiful and great products," and "often forget that [*Ars Legendi* is] a commercial business" (CEO, personal interview). This results in a problematic "ivory tower" perspective (CEO, personal interview). A manager described working for *Ars Legendi* as a "privilege," as he/she "likes working with the product 'book,' […] identifies [him-/herself] with the product 'book,' [… and] enjoy[s] reading books, also privately." All interviewees of *Ars Legendi* confirmed their high level of identification with the publishing house. Moreover, they repeatedly emphasized that they would not deviate from their quality standards just to increase revenues.

According to the CEO and one manager, reacting to soft-cover publishing was one of the biggest challenges the publisher had to master in its past. After some experimentation, the company ultimately decided to retrench into a high quality niche as a consequence of this "threat."

5.4.3.3 Organizational Adaptation to Discontinuous Technological Change

In the personal interview (mid 2011), the CEO acknowledged that "radical changes induced by new media" were ongoing in the publishing industry. At the same time, the CEO assessed the impact of those changes on *Ars Legendi* as "not as dramatic as it is often portrayed. […] I really do not believe that there will be nothing but ebooks [10 years from now]." Moreover, the CEO believes that "ebooks are overrated." [… *Ars Legendi* does] not earn any money with 'that stuff,' money is earned with print" (CEO, personal interview). The CEO further elaborated: "There are many hundreds of thousands of people who enjoy reading books. There are more of those than ever. That is statistically proven. And those people are our main customers—not some nerds surfing on the internet." The CEO assumes major dissimilarities between the book publishing and music industry, both challenged with digitization, as reading books and turning pages is assumed to be a "sensual delight," whereas the recording medium of music is "irrelevant." The CEO perceives print books as "a product that is not easily replaceable by digitized products." In a similar vein, the CEO rationalized the legitimization of printed books as follows: "you cannot digitize everything. […] Maybe babies will one day look at teddies displayed on an iPhone. I cannot imagine that."

In 2002, the company engaged in domain offense activities and specifically announced light books as an important innovation. The company's first ebooks, produced by an external partner (employee, personal interview) were available in 2010 (historic webpage, personal interviews). These ebooks were associated with the newly acquired business unit. In 2011, no ebooks or other digitized media were available for products of *Ars Legendi*'s 'core' segments. Ebooks had only been launched within the newly founded divisions—as "one of the last publishers in that segment" (manager, personal interview). In total, roughly 10% of the books were available electronically at that point of time. Since 2011, one smartphone app related on one of *Ars Legendi*'s recent books has been available via the shopping portal *App store*. Interestingly, this app was not mentioned or even promoted on the company's website. The CEO evaluated *Ars Legendi*'s level of digitization as "backward compared to [international] competitors. We are unable to cope with them." The employee confirmed this evaluation.

Internally, *Ars Legendi* has not hired any new employees dedicated to the commercialization of the new media (CEO, manager, personal interviews). Moreover, *Ars Legendi* has not installed any incentive systems for employees coupled with the commercialization of new media. The manager acknowledged in the interview, that customers occasionally asked for ebooks of older products, but *Ars Legendi* was not able to meet these requirements due to the early stage of digitization within the organization.

5.4.4 Books & More

5.4.4.1 Background Information

Books & More is a medium-sized non-family business that publishes non-fiction books with focus on reference and textbooks. The firm originates from two distinct family-owned publishers that merged into *Books & More* several years ago. Nowadays, *Books & More* is owned by an international holding company with focus on publishing. Although the proprietary company provides strategic guidelines and sets strategic targets, *Books & More* can act quite independently on the German market. Several years ago, the two main pillars of *Books & More* were supplemented by a new business unit, which soon became a core element of the company. Moreover, the company has entered the training and consulting market in its recent history.

5.4.4.2 Culture and Identity

Books & More perceives itself to be a high-quality, flexible provider of content that is embedded in an international context. One manager described the publisher as an "international company that strives to advance the world's knowledge." When asked to describe his/her company in a few words, the CEO provided the following answer: "focus on quality and content [...]; global [...] and very much in the process of changing." With 'global' he/she refers to the frequent exchange with international companies belonging to the same proprietary company, particularly those active on the US market. With 'process of changing' he/she refers to the "transformation of the business from a publisher to a service provider" as well as the portfolio restructuring described in Chapter 5.4.4.1 that was initiated in order to address the challenge of growing complexity. Brochures of the company (2006) emphasize that *Books & More* is a provider "of the entire spectrum [of the education market]" that interprets "textbooks [...] as an active service package."

   Books & More acts as a self-confident publisher who wants to shape the publishing industry. In a recent, publically available document, *Books & More* states: "We adhere to our basic principles. [...]. We self-confidently state what we believe is right and we have the heart to strike new paths." While this has been a constant claim over the past roughly ten years (website 2004: "We shape the 21st century's literature," brochure 2006: "Our principle—always one step ahead"), it has not always been like that: Before their merger into *Books & More*, one of the two founding companies (the larger one) mainly reacted to customer demands rather than acting proactively: The word 'customer' was frequently used on the historic website, the importance of the publisher's sales department was overly emphasized as compared to other publishers and 'reacting to customer demands' belonged to the self-conception of this publisher. On the contrary, the historic websites of the second founding company (the smaller one) do not refer to 'customers' at all. In the first years after the merger, the historic websites of *Books & More* referred to customers ("market-driven, customer-oriented strategy") whereas press releases already emphasized the driving force of the employees: "We owe our success to our employees [whom we provide] space for pioneering spirit." After a few years of coexistence of the two claims, the notion of 'reacting to customer demands' became less prevalent, whereas the number of statements relating to the 'shaping character' of *Books & More* increased in press releases, brochures, and on the historic websites. This does not imply that *Books & More* neglects the customer when developing new products. However, the source of innovation and the motivation to launch products changed. This change process started when a new business unit was created that soon became the core of the company. At the time of this study, contact to end customers was scarce; innovations were either driven by the employees or adapted from similar activities started by other international subsidiaries of the holding company (manager, personal interview). The same manager summarized: "The firm philosophy is highly relevant [for all product innovations. ...] In any case, it needs to fit the company."

The culture of *Books & More* is described as highly creative and the level of employees' identification with the company is high. The CEO explained: "I believe that people here care much about what they do. Let me phrase it this way: there is a high level of intrinsic motivation, and employees identify with the contents they work on, the authors they work with [...] I think that is quite cool. And it also entails that we experiment [with new products] quite a lot." Public data confirms this view: one of the company's slogans, the company's goals and also press releases (2000) explicitly contain the term 'creativity'.

Books & More's organizational identity is rather highly pronounced. Besides the above-mentioned commentary of the CEO on the employees' identification with the firm and the product, and the fact that the central, distinctive, and enduring element (Albert & Whetten, 1985) is clearly defined and shared among employees ('providing high quality content to the [...] market'), some additional observations support this finding: Chronicles of the different business units of *Books & More* are highly visible on the internet. Despite the above-mentioned creativity and flexibility, the CEO stated, "This is an incredibly traditional firm." The meeting room, where the semi-structured interview with the CEO took place, was equipped with posters displaying the company's products and slogans.

5.4.4.3 Adaptation to Discontinuous Technology

In 1998 (and repeated, with different words, in 2000), *Books & More* stated on its company website that not the medium but the content is crucial. In 2000, *Books & More* founded a startup providing a platform for online sales of publishing products. In the same year, *Books & More* went into the print on demand market. In 2002 (press release), *Books & More* denominated ebooks and e-learning as "present state" and advanced solutions such as online books and enhanced media as "future state." The publisher launched its first digitized products (ebooks) in 2002 and thus was one of the earliest adopters amongst the group of book publishers. The first products were ebooks, implemented as pdf-documents. In 2002, ebooks were—in contrast to products of many other publishers—not offered for free but were priced comparable to print books (with a discount for the reduced material cost). Since 2004, *Books & More* has offered interactive supplementary material for its textbooks offered online. The CEO described the adaptation pattern as a three-step process ranging from 'We are here' (first contact with digitized content: launch of website with supplementary material) over 'digitization as sales-booster' (free supplementary online material provided to customers who had bought the print book) to 'digitization as a core segment' (offer of purely digitized products, that in many cases are not related to any print products). Today, *Books & More* offers a variety of digitized and also highly innovative products and services, including: learning software, computer games, print on demand, ebooks, platform solutions, e-learning, online books (specific way of providing textual content for devices such as tablets and smartphones), electronic books with interaction of author and reader already during its development process, video trainings and applications. In 2011, more than 20% of the revenue was already generated through digitized products. To achieve its digitization goals, *Books & More* engaged in cooperation with software developing businesses early, but implemented the products by themselves.

202

5.4.5 House of Books

5.4.5.1 Background Information

House of Books is a medium-sized, family-owned and -managed publisher with focus on reference books, textbook, workbooks, journals, trade magazines, and newsletters. In the past, *House of Books* acquired two independent publishers, one of them comparatively traditional and conservative, and a second one which presents itself as an innovative publishing house. *House of Books* owns subsidiaries in several European countries. Moreover, the company is active in the licensing business. Printing is outsourced to external partners.

5.4.5.2 Culture and Identity

House of Books' main goal is "maximizing profit, besides other, subordinate goals such as ethical goals" (manager, personal interview). The organization constitutes a "traditional and very social family business keeping up many of the old values and for which employees and their experience are of high importance" (manager, personal interview). "We attach great importance to the fact that our core products continue to exist [in their traditional form]" (manager, personal interview). *House of Books* has two core aspects—"content and customers" (manager, personal interview). One of the next generation owners describes *House of Books* as "a publisher that offers any information [the customers] need. [...] absolutely everything." Several established German publishing houses are perceived as main competitors (manager, personal interview). "Capability to abstract," is perceived as a key driver for future success (manager, personal interview). This broad and integrated perception of *House of Books'* core products, however, is not shared by all organizational members. One manager emphasized that the focus on "traditional print businesses" is anchored in the minds of numerous employees and managers.

"Commitment to the customer" and "closeness to customers" have been important characteristics of the publisher for at least the last two decades (two managers, personal interview). Organizational members "have always looked for potential market extensions [...] in order to enhance customer retention" (manager, personal interview). Customer behavior, in particular conservative attitudes of specific customer groups, was repeatedly named as a legitimization for *House of Books'* strategic behavior (manager, personal interview). "We have the ear at the customer and we do what he wants" (manager, personal interview). Communication with the customer takes place via various channels, such as surveys, conferences, and bulletin boards. *House of Books*' direct sales channels further facilitate exchange with customers. "Doing the right things for the market" is seen as *House of Books*' formula for success (manager, personal interview). Applied to new media, "[*House of Books*] offers all existing media forms that the customers require" (manager, personal interview).

The owners of *House of Books* are particularly open to change which was one of the main reasons for one of my informants joining the company several years ago (personal interview). Since a major relaunch of the website in 2002, the publisher's website has changed into an online sales platform dedicated to selling products to customers rather than presenting the organization.

5.4.5.3 Organizational Adaptation to Discontinuous Technological Change

House of Books' first digitized product was a database, created in the early 1980s and still available in 2011 (manager, personal interview). First contracts related to digitized products with international partners were signed in 1990 (manager, personal interview). In 1998[85], the company offered roughly 20 software programs, videos and also CD-based games. Customers were able to gain online access to one of the publisher's journals in mid/late 1990s. In 2000, one of the trade magazines went online and offered some extra features for customers buying the electronic version. In the same year, the first online database was launched. Two years later, in 2002, *House of Books* offered print-CD packages as well as the first product offered for download only. In the same year, *House of Books* acquired an online platform offering download products for customers. An analysis of the product portfolio in 2002, as displayed on the historic website, suggested that most of the publisher's magazines were still print-focused at that time with only few of them offering (premium) online access. In 2004, the company created its first product catalogue focusing exclusively on software programs (roughly 40 programs included). Two years later, the first smartphone application became available. In the same year, *House of Books* ceded to offer free online content related to its journals; rather, journal content was made electronically available for print customers. Since 2008, *House of Books* has launched several product innovations in the audio and service segment.

---

[85] As there is no reliable information available regarding the point in time when *House of Books* started to commercialize its database, I take '1998' as date of adoption.

In the personal interview, a manager described *House of Books*' business model as follows: "Most journal content is available online. Print subscribers can access the online content—only print subscribers." Content of the online journal transcends print content and offers additional interviews and videos (manager, personal interview). In 2011, roughly 20% of *House of Books*' revenues were generated with digitized products (manager, personal interview).

Internally, *House of Books* hired several new employees dedicated to new media. Those employees were structurally integrated in the established organizational setup. The owner commented on the adaptation process as follows: "Our customers are slowly adapting to digitization. That means we also have time to change slowly." The conceptual development of digitized products is carried out in-house, whereas the programming is executed by external partners (owner, personal interview).

5.4.6 Star Print

5.4.6.1 Background Information

Star Print is a medium-sized, family-owned and -managed publisher of textbooks and workbooks. The publishing house is not active outside Germany and does not have any licensing contracts. It has not acquired other companies or divested any business units in its past. The company has experienced some changes related to the topics of products offered during the last decades.

5.4.6.2 Culture and Identity

"Sustaining independence" is *Star Print*'s most important goal (CEO, manager, personal interviews). *Star Print*'s CEO described his/her company's core values and goals as "consistency, continuity, and long-term orientation" (open letter, 2011). The company's presentation on its website is focused on topics such as "the graphic design of books" (chronicle portrayed on website, 2001). In the website's 'about us' section, *Star Print* presents itself as an organization covering the value chain from "development of books up to sales" (2006); events such as the purchase of new book storage equipment are extensively celebrated with all employees. Despite the fact that *Star Print* does not possess a printing press, the CEO and the interviewed manager perceive "books [in a certain segment]" as the core product, and the manager further elaborated (personal interviews): "We publish books.

Everything else needs to go alongside." The CEO believes that this focus will remain unchanged in the future and the manager hopes that "things will remain the same throughout the next years." Three traditional German publishers, active in the same market segment as *Star Print* are perceived to be the main competitors (CEO, manager, personal interviews).

Contact to its customers is central to the publisher. *Star Print* actively presents itself at numerous industry fairs (as stated on the website, 2001). Roughly 15% of the organization's employees belong to the field crew. The manager explained in the personal interview: "Frequently, customer demands reach us [...] and when something new happens, we have to react. When the customers decide that they need new product innovations, then we have to react. [...] We are customer-driven." Despite some intention to 'influence' the market segment (manager, personal interview), *Star Print* perceives its own level of control as somewhat limited. In 2001, the family CEO wrote in an open letter: "[We are] dependent on external conditions!" In another letter, in 2004, the CEO further complains about external forces threatening his/her business. This is in line with statements of the interview such as "No one knows in which direction the market will develop," or "I am not able to predict where the journey [of the market segment] will lead to" (CEO, personal interview). When asked about the strengths of *Star Print*, the manager explained: "We select certain topics and we focus on our competitors. This allows us to survive." Before commencing new strategic activities, the organization seeks advice from external service providers and subsequently engages in a "trial and error working mode" in cooperation with external partners (CEO, manager, personal interviews). During commercialization of new products, the CEO meticulously controls the implementation process as he/she "strongly believes in human laziness" (CEO, personal interview). Under certain conditions, *Star Print* also engages in non-profitable activities, e.g., for the purpose of "doing an author a favor" (manager, personal interview).

"Ability to exert solidarity" is one of the core principles illustrating the company's culture (CEO, open letter, 2011). Referring to the publisher's identity, the CEO stated in an open letter (2005): "It has always been important to us to transfer the consciousness of a common identity as well as civil work ethic to the successors. The family pledged questions, such as 'Where do we come from?' 'Who are we?' and 'Where do we want to go?' [...] Someone who has defended independence over the last decades, does not frivolously follow the caprices and current consulting trends." Interestingly, the entire letter focused on a retrospective rather than an outlook. *Star Print*'s employees strongly identify themselves with the publishing house and also its products (CEO, personal interview). The manager summarized: "I like the reading experience, I like the haptics of those books." The culture and working atmosphere within the organization is described as "trustful, cordial, humane and social" (manager, personal interview).

5.4.6.3 Organizational Adaptation to Discontinuous Technological Change

According to the CEO, *Star Print*'s top management team became aware of digitization at the end of the 1990s, when some of the competitors started to initiate discussion rounds referring to that topic. In 2000, *Star Print* included one software program (available on CD-ROM) in its product portfolio (according to the company's history; website 2004). Although this was award-winning software, *Star Print* did not announce or even actively promote it on its main website (historic websites, 2000, 2002, 2004). In the following years, *Star Print*, focused on expanding its product portfolio by including new book series. In 2010, *Star Print* started to offer free newsletter services, video clips, and also premium news-tickers. The latter ones were implemented in close cooperation with a partner. At the time of the study, *Star Print* offered premium downloads and audio CDs. In December 2011, *Star Print* launched a premium webportal developed by an external partner. However, *Star Print*'s promotion of this product was parsimonious. As most of those products were offered either at low prices or for free , the company strived to "find balance between the economically acceptable and the innovatively required" (CEO, personal interview). The motivation to offer those digitized products is that they are required "to support the sales of books," "if you do not offer digitized materials, customers won't buy your products any more" (manager, personal interview). Revenue generated with digitized content amounted to several thousands of EUR in 2011.

In 2011, *Star Print* was already aware of virtual 'clouds' as potential new digital innovation as the CEO revealed in an open letter: "Despite innovations from a different, 'multimedia' world, such as enhanced ebooks, interactive apps and clouds for the community, selling books still guaranteed stability [in the previous accounting year]." The CEO perceives *Star Print*'s achievements regarding digitization compared to those of competitors as follows: "We are lagging behind them. But I don't care."

Any activities related to new media were carried out by long-tenured employees. *Star Print* did not hire any externals in order to develop and commercialize new media (CEO, personal interview).

5.4.7 Secret Books

5.4.7.1 Background Information

Secret Books is a medium-sized, non-family publisher owned by an institution. The company produces and sells fiction books, reference books, and customer magazines. It is solely active in Germany. Several years ago, the publisher abandoned one of its core sales channels. Besides this eschewal, the business has remained largely constant over the years without major acquisitions or divestitures or any other major organizational changes.

5.4.7.2 Culture and Identity

Information on *Secret Books'* culture and identity mostly relies on interview data and my own observations in the publishing house and at the 2011 industry fair. Contrary to the internet presence of all other sampled publishers, the *Secret Books* website is relatively scarce, neither revealing information on the publishing house itself, nor offering its customers the opportunity to purchase products online. The only relevant information obtained from this data source is from historic press releases that mainly referred to new product launches.

"Fulfilling [its] own quality standards," is the most important goal of *Secret Books* (CEO, personal interview). In order to adhere to its standards, *Secret Books* invests substantial financial and human resources, for instance on rechecking the quality of work provided by external partners (CEO, personal interview). In many cases, this intensive quality control, that has resulted in numerous award-winning products, "does not pay off—customers will not pay for it" (CEO, personal interview). Besides quality, being profitable is another important goal of *Secret Books*, however, maximizing profit is not (CEO, personal interview).

As one manager repeatedly remarked, the organization identifies itself via the content it provides, not the medium: "We are not a publishing house" (personal interview, repeated three times). *Secret Books* perceives a broad range of companies as (potential) competitors, ranging from established publishers to "small technology startups" (manager, personal interview).

Whenever a new innovation—such as audio books, for instance—emerges, the organizational decision-makers engage in long discussions on whether the new product is in accordance with the organization's core values and beliefs: "When our experts ultimately agree on a match or mismatch, you frequently do not need the product anymore," is how a manager described the interpretation process (personal interview). *Secret Books* does not engage in "any commercial activities, regardless of the medium, that do not suit [*Secret Books*]. For instance, we would not launch an internet platform related to cooking just because we could earn money with it. [...] We only do things that suit us." The CEO goes further to speak about the "pride" of organizational members when launching new innovations. *Secret Books* aims to offer products "which the industry sector has not seen before" and is reluctant to introduce products that are "nothing new." The CEO described the organization's strategy as "cautious."

Ideas for new product innovations, e.g., the award-winning app described in the next subchapter, frequently occur in discussions between employees and long-established partners. The publisher "actively offers new products to the customers, as there is no direct pull" (manager, personal interview). The manager continued to describe: "It does not happen that our customers say: 'Wow, I want to have that product digitized as an app.' That is not the case. One has to drum for it, to actively offer it, that is by no means an item that sells itself." Those 'drumming' activities are, however, outsourced to an external partner, as *Secret Books* has no sales force.

Nevertheless, the publishing house is "customer oriented" (manager, personal interview) in the sense that in the end, the customers need to buy *Secret Books*' products to guarantee economic survival of the company. Hence, the publisher is concerned about the profitability of its product and only launches innovations that suit the company and that are likely to reach break-even within a year. Customer orientation at *Secret Books*, however, transcends the mere fulfillment of warranted customer requirements. Indeed, *Secret Books* also reacts to requests of non-customers and helps in un-warranted situations in order to generate "satisfaction." A manager commented: "This is a crucial part of our organizational culture."

According to the CEO (personal interview), not all employees are equally aware that *Secret Books* "is a company that needs to be profitable." The culture of *Secret Books* is "open," "cooperative," and marked by "individual responsibility" (CEO, managers, personal interviews). Whether *Secret Books*' employees "enjoy working with a new innovation" or not, is an important driver for success of the new product.

5.4.7.3 Organizational Adaptation to Discontinuous Technological Change

Internally, *Secret Books* began discussing digitization in 2003. In 2005, *Secret Books* started to offer one of its customer magazines in an electronic version including supplementary material such as video clips. As stated in a press release (2005) the electronic version was intended as a substitute, not as a complement to the print version. In 2006, the ratio of customers reading the electronic version versus customers reading the print version was roughly 0.3% (press release, 2006). In 2007, a new, free internet portal of *Secret Books* went online according to a press release (2007). In the same year, the publisher launched an award-winning DVD. In 2008, the publisher introduced two further internet portals as well as their own web-based TV channel according to historic websites. *Secret Books* does not offer any ebooks. A manager explained in the personal interview: "We will not produce classic ebooks. Reading books as an ebook is boring. Especially when it's black and white without color. What's interesting is offering apps, real applications." Since 2011, *Secret Books* has offered apps, which were internally developed with the support of an external technology partner. One of those apps was among the three top-selling German apps for several months and number one within its segment. At the 2011 industry fair, *Secret Books* was one of only four publishing houses that actively promoted their digitized products. Besides the commercialization of digitized products under the flag of *Secret Books*, the company also actively licenses its digitized contents to other players. The company is selective as to which of its products are offered in a digitized version (manager, personal interview). Several new employees and (experienced) managers have been hired within the last years to build up the department dealing with new media.

5.4.8 TopPress

5.4.8.1 Background Information

TopPress is a medium-sized, family owned publisher who focuses on reference books, textbooks, journals, trade and customer magazines, and fiction books that are related to specific topics. A family CEO manages the business. In its early years, the publisher solely offered books and broadened its portfolio to journals and magazines over the years. Meanwhile, the company has entered the international market via fully dependent subsidiaries in selected countries. In 2003, *TopPress* started to expand through acquisitions and new subsidiaries.

5.4.8.2 Culture and Identity

Organizational members perceive *TopPress* as a "service provider" (manager, personal interview). The CEO described the company's characteristics as "innovative products based on a traditional anchoring within the customer group and on competence" (personal interview). "Trial and error" and "evolutionary development" further characterize the publisher (CEO, personal interview). *TopPress* is deeply rooted in its two classical program segments referring to two content-related topics. During the last years, *TopPress* expanded into "fringe areas related to those classical segments that were inconceivable in the past" (manager, personal interview). According to the same manager, *TopPress'* unique selling feature is "its broad product portfolio that covers everything [within its market niche]." The organization's most important goal is to "continue the organization's tradition and to offer products and services to the target customers" (manager, personal interview). Members of the organization foresee *TopPress'* future as "new things will be added, other things will be abandoned" (manager, personal interview). *TopPress'* main goal is "stability [...] to guarantee survival of the family business by providing literature on specific topics."

Asked about *TopPress'* core product(s), a manager replied in the personal interview: "We do not have a core product. We have target customers whom we serve. [...] We have to ask ourselves: What does the customer group require?" 'Competitors' are broadly defined: According to a manager, not only incumbent publishing houses but also startups and even *TopPress'* authors are perceived as (potential) future competitors.

Contacts with non-publishing businesses play an important role for *TopPress* as they are perceived as a "source of information across the boundaries of the publishing sector" (manager, personal interview). The CEO is "deeply embedded in external networks [...] and bring[s] ideas [he/she has] detected into the organization and initiate[s] activities" (personal interview). At the same time, "creativity" among employees is important to the CEO. Moreover, customers play a crucial role in *TopPress'* identity as indicated by its slogan "Being Really Close" as well as a self-description on the website (2000): "[*TopPress*] lives on its customer focus." *TopPress* is strongly embedded in the industry associations related to its program segments. To fulfill the needs of specific industry customers, *TopPress* is willing to engage in some unprofitable projects (manager, personal interview). The strong customer focus is also reflected in the company's product catalogues (2002): For each (digital) product, *TopPress* provides a detailed overview of the customer groups addressed. *TopPress* focuses its activities on those areas where there is pull and "resonance" from the customers (manager, personal interview).

TopPress' culture can be described as a "balancing act between tradition and innovation" (two managers, personal interview) and as "creative and down-to-earth" (manager, personal interview). Employees identify themselves with the publisher to a high degree, in particular with the core topics of the program segments (manager, personal interview).

5.4.8.3 Adaptation to Discontinuous Technological Change

In 1994, the CEO became aware of the discontinuity during discussions with (future) employees (CEO, personal interview). The first CD-ROMs were offered at the end of the 20th century as valuable 'add ons' for print books. Moreover, in 1998, the first magazine went online and was immediately presented at the book fair. In 2000, the historical website of TopPress stated: "We perceive ourselves as an information provider with books, periodicals, and electronic media." In the same year, one can read the following statement propagating a coexistence of printed and digitized media on the company website: "Our CD-ROMs cannot replace books but they will complement them." Moreover, in 2000, TopPress began to enthusiastically promote its digitized products (historic website). Latest in 2001[86], TopPress offered standalone CD-ROMs with databases as content. A dedicated webshop was created in the same year to sell digitized products. In 2000 and 2007, TopPress launched internet based books in cooperation with partners, including a website where pdf-documents related to the online books could be downloaded for free. The website offered additional material such as interviews with the authors, lexica, and bulletin boards. Customers were able to order additional products via that website. In 2004, TopPress offered its first non-free, electronic-only products by selling articles in pdf-format. Since 2006, TopPress has offered chargeable access to online databases. In 2011, TopPress sold CD-ROMs with software programs and databases, more than 500 ebooks (as compared to more than 1500 print books), online access to the most important journals including premium content, online databases, 24 apps (some of them offered for free), and online services for small and medium sized enterprises. Most of these digitized products relied on print-pdfs converted to ebooks without any adaptations. Moreover TopPress uses advertisements of third parties on twelve topic-specific webportals as an additional source of revenue. The organization is active on the web2.0 platforms

---

[86] Between 1998 and 2001—more detailed information not available.

Facebook and Twitter and offers a blog as well as newsletters. In 2011, roughly 15 to 20% of *TopPress*' revenues were generated through digitized products (manager, personal interview).

5.4.9 Reader's Publisher

5.4.9.1 Background Information

Reader's Publisher is a medium-sized, family-owned and managed publisher with a focus on reference books, trade magazines and newsletters. Moreover the company provides training services and owns a printing house. *Reader's Publisher* is active in several European countries via co-subsidiaries.

5.4.9.2 Culture and Identity

Website statements characterize *Reader's Publisher* as a company that "serves customer media-independently and in a target-group orientated way" (press release, 2000) and that "provides any type of information, from classic print media to electronic offline and online media." The CEO explained, "Our history is rooted in printed words. [...] Looking into the future, I would say, it does not matter in which medial form the customer wants to perceive the information." However, later in the interview, the CEO started to refer to the "advantages of print books" such as "haptics [and the] opportunity to turn the pages." Interview statements of several managers and employees confirm that until several years ago the CEO was narrowly focused on print products and that this perception is still widely spread among certain groups of employees.

Reader's Publisher perceives itself as mainly reacting to external forces. For instance, when re-labeling one of its periodicals, the company stated: "We wanted to emphasize that the content had been adapted to changing market conditions." (press release, 2002). In another press release of the same year, *Reader's Publisher* stresses the challenges going along with market changes. In order to get in closer contact with externals, management positions were filled by those people whom top management believed were best suited to strengthen the ties with customers (press release, 2002). Moreover, the publisher frequently engages in customer surveys to learn about how they use the company's products (CEO, personal interview). When developing new products, the publishing houses engages in a close feedback loop with customers, asking them: "Is that what you need? What else could we add?" (CEO, personal interview). In addition, conservative customer behavior was used as an explanation for *Reader's Publisher*'s cautiousness in launching new innovative products (editor, personal interview). *Reader's Publisher* strives to alter business routines and build more on employees' creativity. In this context, an innovation award was recently introduced to tell the employees, "be creative. Maybe you are already creative and you just do not know it. Communicate it" (editor, personal interview). Only recently, the company commenced to refer to internal aspects when legitimizing activities: "Even though our customer groups don't come to us, knock on our doors, and say: 'listen, I want [a new product innovation];'—we just do it and we have to actively market it" (CEO, personal interview).

"Guaranteed quality" is a key concern of *Reader's Publisher*. The company's website describes the publisher as "dynamic and traditional" without mentioning any family influence. Similar to another sampled firm, *Superbooks*, *Reader's Publisher* refers to: "Nothing is as stable as change" as one of its slogans. According to the detailed, 18-page, chronicle on the company's website, the company has changed on a yearly basis, including numerous acquisitions and divestitures.

5.4.9.3 Organizational Adaptation to Discontinuous Technological Change

Reader's Publisher's CEO first became aware of digitization at the end of the 1980's/early1990's. In 1995, *Reader's Publisher* launched its first CD-ROM (according to the company's chronicle as displayed on the website 2011). In 1998, *Reader's Publisher* amended its product portfolio by free and premium software and acquired a premium webportal in cooperation with a partner. Press releases from 1998 refer to the company's investment in internet activities. Interestingly, job offers for "media consultants" as published on the internet job market of *Reader's Publisher* in the late 2000's did not require any knowledge of or experience with digital products from the applicants. However, in the mid-2000's *Reader's Publisher* hired an external who was made responsible for coordinating internet activities. The main responsibility of this person is to serve as an advisor for business units requiring support regarding new media (editor, personal interview).

In the first decade of the 21st century, *Reader's Publisher* launched a series of webportals, some of them for free, some of them with premium content. Premium content is frequently offered for free to subscribers of the print product and contains access to databases and downloads. At the time of the study, *Reader's Publisher* also offered ebooks for sale and had created its first apps. The core business segment of *Reader's Publisher*, periodicals, had not undertaken any digital activities. An editor explained in the personal interview, that most online activities had started in 2011. The majority of the technical tasks were outsourced to external service providers.

Digitized products are seen as a "meaningful addendum" (website, 2002) to print products and online shops were long assessed as something that would not replace physical book-sellers in the short term (press release, 2001). An editor termed *Reader's Publisher*'s digital products as "free goodies for print subscribers, intended to strengthen the customers' loyalty."

The success of *Reader's Publisher*'s attempts to adopt digitization is seen ambiguously. One editor stated in a personal interview (end of 2011): "Publishing the multimedia way is still difficult, we have not succeeded in doing so thus far. It is an ongoing project with which we are not really happy." Moreover, the editor lamented about the 'cramming:' online texts were mainly based on print material, and also graphics were merely re-used for digitized publishing, without generating native digitized content. The CEO commented on new media as follows: "I am not a digital native. And now, I have to deal with something I don't completely understand. And I won't do certain things anymore [which I did before], I don't understand everything there [related to product digitalization], and in my opinion I don't have to." The CEO continued: "On the one hand, product digitalization is a curse. On the other hand, it is a blessing. But I can't complain about it every day, I can't change it anymore. Digitalization will move on. [...] But I have to accept what our customers demand. [...] We said that we needed to do something about it [product digitalization]. Our customers expect that from us." One manager described the adaptation process as follows: "I would say that we [employees] pushed that decision through [to face product digitalization]. [...] We pushed top-management to take new media on board. Well, it wasn't like top-management initiated that process. We pushed it through." Another employee supported this view: "Well, we [employees] had to do a lot of persuading. But with the initial investment [in product digitalization] we really experienced a profound change at our company. [...] Today, we've got a clearly different understanding about new media compared to the past. We turned ourselves inward;—we started to deal with our contents again, with our structure, because all that implies a profound change. And how can I say it? We had to go through a mental change process here, so that we can now say, 'I don't think in print anymore, I think in generating content.'"

5.4.10 Peter's Publishing House

5.4.10.1 Background Information

Peter's Publishing House is a medium-sized, family-owned publisher with focus on reference books, textbooks, journals, and newsletters. The company has been managed by an external CEO for the last decade, but the family owners still influence strategic decision-making through the supervisory board. *Peter's Publishing House* owns international subsidiaries. The company has acquired several companies in the past, however, according to information obtained from interviews with long-time employees, these companies were fully integrated in *Peter's Publishing House.*

5.4.10.2 Culture and Identity

Peter's Publishing House perceives itself to be a "service provider" that "combines advantages of traditional books and multimedia concepts" (website, 2004). According to its CEO, *Peter's Publishing House* takes the role of a "communicator" (personal interview). Even editing assistants with long tenures (30 years and more) and education and experience in the publishing sector admit—despite some emotional attachment—that electronic documents are a "priceless advantage [...] that one has to strengthen and develop" (personal interview).

According to the CEO (personal interview), the "owners have a strong emotional attachment to the publishing house and want to continue the founder's heritage. [But they] are not emotionally attached to a specific product. [...] That implies that there are no 'emotional reservations' that are taboo for any change attempts." However, according to the current CEO, his/her predecessor mainly tried to "keep the status quo." Industry experts, however, did not support this assessment.

In the past—including the majority of time of adaptation to discontinuous technological change—the publisher launched activities regardless of specific customer needs. The organization itself served as source of and driver for new ideas. The editing assistant stated in the personal interview: "[As a customer] I would ask [when a specific product was launched only in electronic form]: Why should I spend money on this? You are crazy!" (Despite those potential reactions, the organization launched various products, electronic only). The editing assistant went on, "we work a lot based on ideology," and "we determine the directions of the [strategic] paths." When asked to elaborate on that topic in more detail, the interviewee continued: "The market had never been a central topic. To ask:

What does the market demand? [...] We never thought about this in the past. My previous boss enjoyed some nice dinners with authors during which they generated great ideas." The CEO confirmed that many of the company's activities were "driven by supply" rather than demand. However, he/she personally tried to make the publisher become more customer focused, since he/she believed that "each book that many customers like is a good book." To achieve this goal, the CEO had recently implemented customer boards in the company. However, the focus on customer needs did not fundamentally change the source and legitimization of innovation. "Sometimes—but on very rare occasions—we say: They are all crazy, we'll do it the way we like. However, making decisions contrary to the customer demands does not happen often nowadays."

The personal interviews mirror a conflict-ridden atmosphere in the publishing house. Those conflicts implied a comparatively high degree of fluctuation throughout the last years, including a replacement of roughly 50% of management staff. Referring to the company's goals, maintaining its economic independence is the main intent of the publishing house (CEO, personal interview).

Peter's Publishing House is "committed to highest quality standards" (production manager, personal interview) and "known for quality" (sales staff, personal interview). The production manager stated in the personal interview, that his/her "heart is attached to the beautiful quality." The identification of employees with the company is particularly high, and employees are "proud" of working for *Peter's Publishing House* (personal interviews, CEO, editing assistant).

5.4.10.3 Organizational Adaptation to Discontinuous Technological Change

In the mid-1990s, *Peter's Publishing House* first offered content on CD-ROM and in the following years, the publisher amended its CD-based product portfolio. In 1996, the company's website indicated two employees who were responsible for electronic products. At that time, roughly 7% of the periodicals were available electronically, however, all of them for free. In 2007, *Peter's Publishing House* announced its plan to make each of its books, including those on the backlist[87], electronically available. Experience with new media was a key requirement for any job applicants at the turn of the century. In 2003, most of the company's periodicals offered online access. Most of them were only accessible to

---

[87] Backlist refers to the comprehensive list of books, magazines, etc. that have been published by the company since its foundation.

subscribers of print periodicals; some could also be viewed on a pay-per-view basis. In 2006, the company launched its first smartphone application. Almost all periodicals offered electronic pay-per-view access at that time. The company's product portfolio comprised a large number of book/CD-packages. In the subsequent year, in 2007, *Peter's Publishing House* launched its first ebooks (according to information available on the website; according to interview data this launch happened already in 2004/2005). Since 2008, the publisher has allowed buying of selected chapters instead of entire books (press release). In the same year, *Peter's Publishing House* introduced new innovative webportals and online services. Nowadays, the company's digital product portfolio encompasses online journal portals, pay-per-view periodicals, ebooks, various databases, CD-ROMs, apps, customer-specific books, and various combination packages. At the time of the study, *Peter's Publishing House* generated roughly 30% of its revenues through electronic products (CEO, personal interview). The company thereby relies on sophisticated pricing models, in particular for print/electronic packages. The CEO emphasizes: "It has always been important for me, that each employee perceives digital products as his or her core business."

5.4.11 Yellow Books

5.4.11.1 Background Information

Yellow Books is a medium-sized, family-owned and -managed publisher with a focus on fiction, reference books, textbooks, trade magazines, and newsletters. The publishing company is solely active in Germany and has neither acquired nor divested firms in its past. However, the company has changed its products, in particular the medium (from periodicals to newsletters).

5.4.11.2 Culture and Identity

Yellow Books perceives itself to be a "traditional, medium-sized family business" (website, 2010), a "holistic" provider of "products and services" (brochure, 2004) offering a "broad product portfolio, [...], innovative solutions [...], multifold services" (website, 2010). The company aims to unify opposites, "established and new products" (website, 2010) and is proud of having gone the way from "steam engine to major publisher" (website, 2010). The CEO described the company as follows: "We have left saying 'we sell content' far behind. No, we rather provide services and products [for specific tasks]." Competitors of *Yellow*

Books, as perceived by the CEO (personal interview) encompass a broad spectrum of firms, ranging from established publishing houses to software and IT-providers.

Yellow Books is characterized by customer orientation and trust based relations with customers (website, 2004). However, despite this focus on customer needs, customers are not seen as the primary source of innovation. A manager explained: "When you ask the customers 'what do you want?,' the mostly vague answers you receive are always based on something you have already seen or something that is not innovative at all. But when you have your own ideas that are fundamentally different from your competitors' ideas, then you have to give them a shot." In the same vein, another manager stated in the interview: "We don't just want to be in a market. We want to offer value." This view is also shared by the CEO, who noted: "We want to decide on our own what we do and what is right for our customers." Competitors are seen as a non-relevant source of information and the CEO is "not very active in observing competitors" (CEO, personal interview). Hence, "energy and creativity of [its] employees" distinguish *Yellow Books* from other companies (website 2010). To harvest the employees' creativity, *Yellow Books* has, in contrast to many of its competitors, its own R&D department (manager, personal interview).

"High degrees of personal responsibility, long-term thinking and independence" (website, 2004) belong to Yellow Book's core characteristics. Products of *Yellow Books* "transport innovation and quality" (website, 2006). Employees perceive the organization as "young, dynamic, and committed" (manager, personal interview).

5.4.11.3   Organizational Adaptation to Discontinuous Technological Change

In 1999, *Yellow Books* offered lexica and databases on CD-ROM. In 2000, premium access to online databases was offered. A specific contingency made this product launch possible at that point of time: One of *Yellow Books*' employees had a personal affinity to IT and electronics, based on his/her educational background. Perceiving the future potential of databases, this employee proposed a respective project—the digitization of content into a comprehensive database—to the CEO in the 1980s. The CEO was not fully convinced of the idea, but due to the inclusive and trust based organizational culture, he/she provided some time and resources for the employee's 'pet project.' The employee worked hard on the digitization with high levels of personal commitment for more than a decade. As a consequence, when *Yellow Books*' decision makers, after a long time of interpretation and cogitation whether and how digitization fitted the company's vision and value, ultimately decided to act, the basis for all digitized products—the database—was almost ready to use.

In contrast to many other publishers, even subscribers of *Yellow Books*' print products had to pay a fee for using the online database. In the following years, *Yellow Books* steadily increased its number of CD-ROMs and online databases. In 2005, the first online books became available. In 2006, *Yellow Books* launched premium information for download (pdf format). This download was not merely a copy of print newsletters, but went through various finishing processes before being published (manager, personal interview). According to a press release, *Yellow Books* launched its own social community in 2008. At the same time, the organization started to offer audio CDs and online training. Since 2007, the company has won innovation prizes on a regular base. At the time of this study. *Yellow Books*' database was "the most widely spread database" in its segment. Moreover, *Yellow Books*' digital product portfolio comprised CD-ROMs, online databases, subscriptions to audio files, electronic premium newsletters, software with various types of contents, internet TV, ebooks, apps, and online seminars (since 2010).

Ostentatious for the visitor of the company's website, *Yellow Books* extensively listed and somewhat enthusiastically explained advantages of electronic products as compared to print products. Each electronic product is graphically marked to immediately indicate to the website's visitor that it is a digital product.

One manager described the role of the customers and the organization itself in the adaptation process as follows: "Our customers respond very slowly and in a conservative manner to the emergence of new media. It took us forever to make them use DVDs rather than CD-ROMs." Another manager explained: "We thoroughly reflected on our options and asked ourselves: 'What is going on there?' 'Is our current way of developing products right?' 'What kinds of technologies are really interesting for us?' Acknowledging that we are a forward-looking organization, we wanted to make sure that we would not develop solutions based on the status quo rather than on the future opportunities."

5.4.12 Book 2020

5.4.12.1   Background information

Book 2020 is a medium-sized, founder-owned and externally managed publisher of journals, loose-leaf services, and reference books on several selected topics. Moreover, *Book 2020*

220

organizes seminars for its customers. *Book 2020* co-owns subsidiaries around Europe. Worldwide, products of *Book 2020* are offered through various licensing contracts. Besides its subsidiaries and licensing partners, *Book 2020* is well embedded in an international network. Information gained from these contacts, especially regarding new innovations launched in the US market, is seen as a major driver of *Book 2020*'s growth.

5.4.12.2   Culture and Identity

Book 2020 perceives itself as a "medial consulting firm" (website, 2004) with "printed 'consultants'" and "periodicals" (website, 2002) as core products. Historical websites (e.g., 2000) emphasized the focus on "substance and quality" as well as the firm's "long-term experience." A member of the online marketing department stated in the interview: "Our core products are our print products, even though they will certainly have to fight in future." *Book 2020* aims to focus on its core competencies, as the CEO noted in a press release (2006): "Specialization defeats generalization," or somewhat differently phrased: "[Our employees] concentrate on their core competencies" (chronicle, 2011). Thereby, the company frequently refers to its history (website, 2002) when talking about the company's core competencies.

The customer is seen as the focus of the company's strategic actions. The publisher's goal is "to make customers successful [by providing information]" (website, 2002) and "to fully concentrate on the customers' demands and to quickly react to changes" (website, 2002). In the yearly review (2001) the CEO stated: "Due to our customer proximity, we have a broad and differentiated portfolio." In 2002, one could read on the company's website: "Each customer reaction, positive or negative, directly impacts on the results of the [business units]. [Our] organizational setup is a further warrantor that ensures that our customers always get high quality and up to date information."

The strong customer focus of *Book 2020* manifests itself in the fact that customers are seen as major drivers of innovation: "Ultimately [product development] is simply a question of: What does the customer want? What does his or her consumer behavior look like? [… The launch of new products] is always a reaction to what the customer demands. [...] This is why we want to know what the customer says, what he/she wants, what he/she desires" (CEO, personal interview). After the first product launch of CD-ROMs, *Book 2020* stated on its homepage: "To satisfy the trends, we launched the first electronic product." Only one employee of the internet department was more circumspect regarding the customer's role in the innovation process, remarking "the implementation of customers' ideas as real life products is a step that has to be taken by us [in contrast to the customer]."

In order to gain a deep knowledge on the customers, their demands and behavior, *Book 2020* engages in various activities, such as discussion panels with customers, customer invitations to seminars, customers contacts at industry fairs, surveys, special events for target customer groups, and domiciliary visits. Moreover, *Book 2020* uses information generated by the number of how many individuals newly subscribe free electronic newsletters on specific topics. In 2004, *Book 2020* stated on its homepage: "Newsletters are a particularly appropriate format for such requirements [*comment author: to identify customers' interests*]. Thus, for instance, we were able to react to the steep increasing demand for [*name of the topic sanitized*] by launching three new titles within one year only." Asked about how he/she typically became aware of new trends, the owner described in the personal interview: "We listen to customers, we listen to suppliers, we listen to employees. We are permanently engaged in market research." The owner observes the (potential) market and (potential) customers during all his/her daily activities such as travelling by train, visiting other companies and so on.

Book 2020's culture is "open […], characterized by flat hierarchies, and decisive" (CEO, personal interview). *Book 2020* is deeply engaged in supporting social activities, sponsoring sports events, and financing entrepreneurs.

5.4.12.3   Adaptation to Discontinuous Technological Change

In the personal interview, the CEO stated, "[Whether digitized media has a future] depends on the form in which it is offered. I think that a pure transfer of knowledge is difficult. [The customer is not willing to pay for] solely providing textual elements and some small tools." This view is more or less repeated in the publisher's sustainability report (2011): "Despite digitization, the strong position of the [print products] will remain untouched. Reading during a lunch break or after work—no one is willing to abstain from that."

The first CD-ROM of *Book 2020* was brought to the market in 1997. In 2000, *Book 2020* founded an on online marketing department which has, since then, been responsible for harvesting internet based sales channel for *Book 2020*'s core products. In 2002, *Book 2020* stated on its website: "The printed knowledge of our information services is flanked by multimedia contents on CD-ROMs and our websites." In 2011, the publisher's portfolio of electronic and digitized products encompassed roughly 15 CD-ROMs (most of them as supplements of print books), and premium electronic newsletters. Ten of the loose-leaf services were combined with online portals, all with free access for customers. According to the product catalogues and the online shop, *Book 2020* did not sell any ebooks in 2011. However, they offered one or two ebooks as free giveaways to new customers. In recent years, *Book 2020* launched a series of webportals. The underlying business model of these platforms that offer free content to customers is threefold: First *Book 2020* promotes its own products on these websites. Second, customers have to register to get access to certain content and *Book 2020*, in turn, sends newsletters offering new products to all registered users. Thus, those platforms are intended to increase sales within the existing domain (established products, same customer group). Third and last, businesses can advertise their own products on *Book 2020*'s websites using banners and advertisements. The current company website lists its product innovations from its foundation up to today: The last product innovation (CDs and software) was brought to the market in early 2000.

In the mid-2000s, *Book 2020* claimed that its main growth traces back to publishing new titles (amending the established product portfolio) and international activities (year end retrospective 2004, website). Since 2010, the company's revenues have declined steadily (figures given on website) entailing divestiture of some products and abandonment of several international activities.

Members of the publishing house assess *Book 2020*'s reaction to digitization as follows: "So far, revenues with digitized products do not play a central role for us. Thank goodness, we have not made the mistake of trying to mercilessly change everything here to digital so our core business is still based on print products." A historic website statement supports this notion: "As it is our principle to work in a sound manner, we deliberately were not among the first [companies] on the internet." (2002). The owner believes that neither ebooks nor apps will "have a future" (personal interview).

5.4.13 Superbooks

5.4.13.1   Background Information

Superbooks is a large-sized, holding-owned publisher that grew historically by acquiring independent publishing houses. *Superbooks* focuses on reference books, textbooks, trade magazines, and newsletters. International exchange takes place through other subsidiaries of the holding company.

5.4.13.2   Culture and Identity

Superbooks describes itself as a supplier of "holistic, thorough, and high quality information" (website 2001) and as a "provider of knowledge and information with core competencies on [several special topics]" (website, 2007). In 2006, one could read on the company's website in the job offers section: "There are hardly any limits to the multitude of media and services [existing at *Superbooks*]." Offering "integrative solutions" (manager, personal interview) is one of *Superbooks'* unique selling propositions.

Superbooks is a customer-oriented publishing house that perceives customers as the major driver of its innovations. A press release (2006) summarizes it like this: "Our customers interests are core to us." In the same year, the publisher states on the website: "Concrete demands and requirements of our customers are our starting point. To benefit our customers, we are open for cooperation and strategic alliances and to work jointly with the book sellers." The company's website had long been structured along the specific customer groups (website, 2006) and *Superbooks'* internal strategy is aligned to customer demands (website, 2001). The CEO described the underlying rationale for *Superbooks* to adopt digitized products (as described in Chapter 5.4.13.3) as follows: "This trend is not driven by us, it is driven by the customer." An analogous explanation was provided by an employee describing why there are still business units of *Superbooks* offering only print products (as described in Chapter 5.4.13.3): "The market is still conservative and sedate in that segment. That means that there are relatively few stimuli that force us into any reaction." However, the same manager admitted for his/her segment: "We realized that we did not know what the customer really wanted [...] because the people we talked to were not the people we should have been talking to." To get input from the customers regarding their demands and requirements and to also gather ideas for innovation, managers of *Superbooks* are obliged to spend some time with customers, every week. Moreover, the publisher relies on measures such as customer surveys, focus groups, usability labs, organizing congresses, and also a large field crew to generate information on customer demands (CEO, personal interview). To ensure implementation of those ideas into commercial products, internal processes such as prizes for commercializing ideas were introduced. Besides customer demands, a second driver of innovation is the "close cooperation with authors" (website, 2001).

One of the company's slogans is: "nothing is as stable as change" (website, 2011). The CEO explained in the interview: "Our employees live change. Daily. This, for instance, is the reason, why there is no chart displaying *Superbooks*' organizational structure." One manager describes the employer as an "innovative and dynamic publishing house."

According to website information, the organization's values comprise "innovation," "integrity," "value creation for customers, and owners," "teamwork" and "responsibility." Compared to other companies of this sample, fluctuation of employees and managers was high in this organization. On the website, the company relies heavily on its "long standing experience within the publishing industry."

5.4.13.3 Organizational Adaptation to Discontinuous Technological Change

The organization's framing of discontinuous technological change is best summarized in one of the CEO's statements in the personal interview: "We sell information to target customers. […] In the pre-electronic era, we had no other media at hand to fulfill this task."

Periodicals were first offered in an electronic way in 1999. In 2000, *Superbooks*, in cooperation with e-learning experts, started an online project encouraging customers to jointly write an ebook. In 2001, the company invested heavily in new media, as indicated by the *Superbooks*' online job market. In the subsequent year, the publisher acquired a software-developing house and in 2002, it bought an online portal providing premium information to its customers. Around 2004, *Superbooks* offered a series of CDs. Most initial online projects of *Superbooks* failed to meet the organization's revenue expectations and were thus relinquished one or two years after their implementation (according to software employee, personal interview). As a consequence, *Superbooks* changed its strategy to offer more integrated rather than stand-alone products in the mid-2000s. The underlying motivation for the investments in digitized products was a decline in the customer base, and subsequently in revenues (manager, personal interview). From 2004 onwards, *Superbooks*, on average, announced one major electronic product launch per year. Since 2007, the company has been engaged in a major cooperation with a leading technology firm. At the time of the study, *Superbooks* offered CD-ROMs, webportals, online learning, and software to customers, however, no ebooks or apps (according to website data and confirmed by the manager in the personal interview).

More employees have an educational background in computer science as compared to those being trained in traditional book publishing. In 2011, *Superbooks* generated more than 50% of its revenues based on electronic media. In order to achieve these contributions of digitized products, the company "had acquired businesses, mainly active in the online and software business and divested substantial parts of the print portfolio" (manager, personal interview). Despite the financial success, the transformation process is not yet perceived to be finished (manager, personal interview). Roughly 15 to 20% of the products are still offered in print only, with no intentions from the management side to change this status quo, since those products are targeted at specific, conservative target groups and *Superbooks* decided to apply a 'cash cow strategy.'

Most of the work related to new media is done in-house, as "this is a substantial strategic core competence that each publisher requires on its way to becoming a service provider" (manager, personal interview). To achieve its growth goals, *Superbooks* installed ambidextrous structures several years ago.

5.4.14 Arthur & Sons

5.4.14.1 Background Information

Arthur & Sons is a large, family-owned publisher, managed by an external CEO. The owner-family still exerts influence on major strategic decisions via positions on the supervisory board. In particular, when the publisher discussed investments in digitization, the senior owner played a crucial role by supporting that strategic move as he/she clearly framed the situation as an opportunity. The product portfolio of *Arthur & Sons* comprises reference books, textbooks, and trade magazines on a variety of topics. Besides publishing, *Arthur & Sons* also offers seminars, printing, and other media services. *Arthur & Sons* is internationally active via co-subsidiaries and licensing contracts.

Throughout its history, *Arthur & Sons* has undergone some major changes, for instance regarding the target groups or the thematic segments they are active in. The core business, however, has remained constant over the past decades.

5.4.14.2 Culture and Identity

Arthur & Sons perceives itself as a "multimedia partner offering full services" (brochure, 2011). For the publisher it is particularly crucial that its products and services are "holistic and integrated" (website, 2006; manifold mentions). The core product of *Arthur & Sons* is a "clear and practice-oriented transfer of knowledge" (website, 2002). In the personal interviews, particularly with the production manager, synonyms of "geared," "integrated," "become indistinct," and "bring together" are frequently mentioned when describing the products and strategic activities of the company.

Moreover, members of *Arthur & Sons* perceive the organization as a publisher that "serves society" (website, 1996) with a "main focus on customers" (website, 2002) and the goal "to make customers even more successful" (website, 2006). The CEO stated in the personal interview: "Our most important goal is to survive [...], to remain independent, and of course to deliver value to customers." Put differently, "[c]ompetence, independence, and focus on readers are the basics of [*Arthur & Sons'*] success" (website, 2002). When advertising new products on the website, *Arthur & Sons*, meticulously describes the target groups these new products aim at. *Arthur & Sons* has some ambitions to influence their environment as stated on the website (2006): "We want to contribute to the advancement of our sector." In 2011, this is re-emphasized in a brochure: "*Arthur & Sons* accompanies people and markets and shapes the future together with them" (brochure, 2011). However, somewhat contrary to these claims, many of the recent innovations of *Arthur & Sons* were not driven by the publisher itself but by the environment. Asked why routines have changed during the recent past, the production manager answered in the personal interview, "because the market required changes." He/she continued, "The change [in our company] is in principle based on massive, external factors that force us to change. Completely new and different market conditions. [...] Thus we aim to become a service provider who covers all the services demanded by the target groups." In particular when integrating new media, "we did a lot of market research and we also started a close cooperation with a market research institute and a consulting firm."

To get further input and feedback, managers of *Arthur & Sons* are strongly engaged in association work. "It is important for the publisher to get input from externals [...] to see what others are doing and to be able to assess our own activities" (production manager, personal interview).

The culture and leadership style of *Arthur & Sons* are described as "cooperative" (CEO, production manager, personal interviews) due to intensive "intellectual exchange with employees" (production manager, personal interview). Moreover *Arthur & Sons* is "highly influenced by the owners" and constitutes a "traditional company with a major emphasis on family aspects" (production manager, personal interview).

Employees and managers have a strong identification with the publisher, yet not necessarily with its products. The production manager stated in a personal interview: "Sure, I absolutely identify myself with the publisher! From a private perspective—honestly speaking—I am not interested at all in the products. From a professional perspective I identify myself with the products in such a manner that I say: Ok, I have to bring them along. But there is no emotional attachment to the products." The CEO emphasized the "pride" many employees sense regarding their company. To further foster this identification, *Arthur & Sons* is deeply engaged in supporting social activities.

5.4.14.3   Adaptation to Discontinuous Technological Change

In 1996, *Arthur & Sons* stated on its website: "We serve our society by providing information […] through printed and electronic media." In the same year, CDs were included in the images presenting the company on the internet. The production manager stated in the personal interview: "We became aware of digitization during the 1990s. […] However, in the beginning, our expectations were far too high." *Arthur & Sons'* reaction to digitization happened in three waves:

(1) First reaction including electronic media on CD-ROMs in the mid-1990s.

(2) Intense investment in internet activities at the turn of the millennium.

(3) Re-investment in the mid-2000s.

In 1996, *Arthur & Sons* offered various (>13) CD-ROMs and at least one CD-ROM subscription. Two years later, in 1998, the first periodicals went online. In 2000, an independent online department was founded. In the same year, one of the key magazines of *Arthur & Sons* launched a new premium online portal. In the subsequent year, *Arthur & Sons'* own online shop went live and a cross-media team with eight employees started to offer digitization services. In 2002, *Arthur & Sons* announced two internet-based co-subsidiaries (including a specialized online shop) as well as a cooperation agreement with an e-learning company. In that year, roughly 11% of *Arthur & Sons'* reference books were available electronically (as ebooks, based on historical website information). *Arthur & Sons* was responsible for 17 webportals, generating revenues through a large number of advertisement possibilities for business customers, and also offered software tools for media-independent publishing. In 2006, roughly one third of the products was available electronically (as CD-ROMs). Roughly at the same time, a two-digit number of new employees were hired to support *Arthur & Sons'* online activities. According to the production manager (personal interview), all products were available in a media-independent form at the time of the interview (2011). Moreover, a mobile media group had been set up to develop products for mobile devices such as apps for smartphones. The publisher is active on Youtube, Facebook, Xing, and Twitter.

5.5 Results

5.5.1 Focus and Locus as Dimensions of Organizational Identity

Which are the main dimensions of organizational identity? Given a vague guidance only in literature to answer this question, I adopted an exploratory, grounded approach and iteratively distilled the core aspects of identity from the various data in a set of revelatory cases (Glaser & Strauss, 1967; van Maanen, 1979b). Two discriminant dimensions, which I later termed 'focus' and 'locus,' turned out to be robust and influential. The definitions of 'focus' and 'locus' can be formally stated as follows.

Definition 4: Focus of Organizational Identity

> *'The degree to which organizational members define their competitive area either inclusively or exclusively (Livengood & Reger, 2010; Porac et al., 1989).'*

Definition 5: Locus of Organizational Identity

> *'The extent to which members of a given organization predominantly refer to the organizational self or to the environment when legitimizing strategic actions (Gioia & Thomas, 1996; Labianca et al., 2001) and when describing the main source of stimuli for organizational attention and action (Rotter, 1966).'*

In a second step, I developed concise coding guidelines, which I used to confirm my taxonomy with data from the remaining cases. In this section, I briefly describe the two identity dimensions and I discuss how I attempted to reliably assess their level for each case at each point in time during the period under investigation. The definition of the two dimensions, their operationalization, as well as exemplary quotes from interviews and historical websites are summarized in Table 13 and Table 14. Before describing the two dimensions in detail, I formally propose

> P3-1. In the context of adaptation to discontinuous technological change, focus and locus of identity represent two mutually exclusive and influential dimensions of organizational identity.

5.5.1.1 Focus of Identity

I defined an organization's 'identity focus' as the degree to which the members of an organization define their competitive arena *inclusively* or *exclusively*. An inclusive identity is broader and less bound to specific pre-existing identity domains than an exclusive identity, and thus somewhat similar to Miles and Snow's (1978) 'prospector' stereotype. Members of an inclusive organization think inclusively, as they draw connections between various fields of competition and technologies. In contrast, members of an exclusive organization define their "home turf" (Livengood & Reger, 2010: 49) more narrowly (like Miles and Snow's 'defender') and highlight elements that differentiate their own domain from others.[88]

To allow for a reliable characterization of my 14 firms, the final coding guidelines for interview transcriptions and archival data contained exemplary codings as well as dictionaries for each of the two categories of identity focus, which were informed by extant literature.[89] As I also compared identities within the organizations, I included a node for hybrid, ideographic identities (Albert & Whetten, 1985; Corley et al., 2006), i.e., identity discrepancies between certain hierarchical levels or functional departments of the organization. After finalizing the coding, I created ratios of textual passages that referred to the *inclusive* or *exclusive* focus for each case[90] and I compared frequencies within each organization.

My data shows that focus of identity is relatively stable, as it remained constant for all sampled organizations throughout the assessed timeframe[91]. Seven companies showed a

---

[88] For instance, when studying Scottish knitwear manufacturers, Porac, Thomas, & Baden-Fuller (1989) observed that these companies defined their competitive area rather exclusively, both on a geographical level (perceived competitors only included other Scottish manufacturers, as compared to other manufacturers from the UK or even from around the world) and on a technological level ('high quality knitwear' rather than 'textiles').

[89] For instance, I coded self-portrayals such as 'communicators and mediators of knowledge' or 'providers of integrative and holistic solutions' as indicators of an inclusive focus, whereas self-portrayals such as 'producer and seller of books' were coded as exclusive. The complete guidelines are available in the appendix.

[90] Identities were categorized as 'inclusive' if ≥60 percent of codes were labeled as 'inclusive.' Identities were categorized as 'exclusive' if ≥60 percent of the codes were labeled as exclusive.

[91] As the interview statements (outlined in Chapter 5.4) indicate, *Reader's Publisher* might be about to change its identity from exclusive to inclusive. This change was triggered by conflicts and framing inconsistencies as they emerged due to the development of digitized products.

coherently *inclusive* focus,[92] five companies a coherently *exclusive* focus.[93] Companies with an inclusive focus emphasized the similarities between their products and services and those of related yet different markets. For instance, the CEO of the textbook publisher *Books & More* stated:

> We actually do not feel obliged to produce content. We always feel
> free to say: Let's build and run a school! Or: Let's start consulting.[94]

In contrast, respondents from organizations with an *exclusive* focus strongly identified their company's activities with the medium of books and stressed the differences between their perceived identity domain and others. The CEO of *Ars Legendi* emphasized that, in his/her opinion:

> [Companies of other related domains] simply do not pursue the same
> high quality standards as those that publish books.

Two of the cases[95] were characterized by hybrid, ideographic identities (Corley et al., 2006). In both cases, frequency counts of the individual interviews revealed substantial discrepancies between the perceived focus of the CEO, top managers, and long tenured employees (*exclusive*) and that of lower hierarchy managers and new and mostly younger staff (*inclusive*).

Table 13 provides an overview of the operationalization and exemplary quotes used to determine the organizations' *foci* of identity. Information on the categorization of each of the sampled firms is also provided. A more detailed list used to categorize firms as either exclusive or inclusive can be found in the Appendix A.A-4.i.

---

[92] *Arthur & Sons* (~75%), *Books & More* (~90%), *Peter's Publishing House* (~65%), *Secret Books* (~80%), *Superbooks* (~80%), *TopPress* (~70%), and *Yellow Books* (~75%).
[93] *Ars Legendi* (~70%), *Book 2020* (~60%), *Reader's Finest* (~75%), *Rocket Book* (~60%), and *Star Print* (~90%).
[94] Similar to the procedure described in the previous chapters, German quotations were translated to English and back-translated to German in order to fulfill quality standards.
[95] *Reader's Publisher* and *House of Books*.

Table 13: Focus of Organizational Identity

Specification	Operationalization	Exemplary Quotations	Firms in Sample
Inclusive: Organization focuses on similarities of own and other identity domains	Frequency counts of interviews and archival material based on dictionary containing words such as 'broad,' 'integrative,' 'holistic,' 'from…to,' 'communicator,' 'mediator' (for a detailed dictionary, see Appendix A.A-4.i) Ratio of sum of inclusive textual elements as opposed to exclusive elements > 60%.	"We offer integrated solutions." *(CEO, Superbooks)* "We offer the whole spectrum." *(Website image brochure, Books & More)* "We see ourselves as service provider." *(Website, TopPress)* "There are hardly any limits." *(Website, Superbooks)* "We want to be a mediator that organizes communication processes." *(CEO, Arthur & Sons)*	*Arthur & Sons, Books & More, Peter's Publishing House, Secret Books, Superbooks, TopPress, Yellow Books* *Hybrids:* *Reader's Publisher,* *House of Books*
Exclusive: Organization focuses on differences between own and other identity domains	Number of textual elements in interviews and websites that refer to a narrow definition of the core business or clear separation between own and other identity domains. (For a detailed dictionary, see Appendix A.A-4.i) Ratio of sum of exclusive textual elements opposed to inclusive elements > 60%.	"Our core products are books in high quality." *(CEO, Ars Legendi)* "Initiatives outside of [the book] business would be impossible for us." *(CEO, Reader's Finest)* "Our standards have to remain the same—in that, we are absolutely uncompromising." *(CEO, Ars Legendi)*	*Ars Legendi, Book 2012, Reader's Finest, Rocket Book, Star Print*

5.5.1.2 Locus of Identity

'Locus of organizational identity' is the extent to which an organization's identity is centered on the organizational self or the environment. Members of a *self-centric* organization predominantly refer to the 'organizational self' (Scott & Lane, 2000: 45), whereas members of an *environment-centric* organization refer to the environment of their organization, when legitimizing strategic actions (Gioia & Thomas, 1996; Labianca et al., 2001) and when describing the main source of stimuli for organizational attention and action.. As such, locus of identity is comparable to a sustained perception of 'perceived control' (Brockner et al., 2004), which encompasses the two notions of 'self versus environment-determination' (DeCharms, 1968) and 'impact' (Rotter, 1966) and is a stable organizational trait rather than an ad hoc appraisal of a concrete environmental stimulus. Locus of identity is also coherent with Nag et al.'s (2007) categorization of 'technology-push' (similar to 'self-centric') versus 'market-pull' (similar to 'environment-centric'). However, these authors' classification is based on what organizations *actually do*[96], whereas locus is related to *why* organizations engage in activities. Moreover, Nag et al.'s categorization is bound to technology, rather than a broader set of organization-internal aspects such as values, goals, or shared history.

Similar to the analysis of identity focus, I created a precise coding protocol to characterize the locus of each of the 14 companies and applied it to all interview and archival data[97]. The assessment of locus was cross-validated by interview responses to slightly adapted questions from Labianca et al.'s (2001) survey on identity types, which referred to the company's mission, performance goals, engagement in marketing campaigns, and the nature and characteristics of the key decision criteria that are applied for strategic activities.

As profit-seeking organizations that compete in an open market, each of the selected businesses inherently paid attention to environmental stimuli, at least to some degree. Nevertheless, I found ample variance in my sample regarding the companies' locus. Six of the cases showed high levels of *self-centricity*[98], whereas eight were coded as environment-

---

[96] Compare Nag et al.'s (2007: 825) definition: "a pure technology-push mentality is characterized by internal creative processes focused on designing and developing cutting-edge technology (often without an obvious commercial application); this technology is then pushed into the market-place regardless of current market interest."

[97] The terms I associated with an environment-centric identity included 'customers' (as opposed to 'employees' or 'values'), 'market [or environment] as driver of action,' 'react' (as opposed to 'act' or 'shape'), and statements such as 'in the past we were coerced to...' or 'consultants told us to....' The full guidelines are available from the first author.

[98] *Ars Legendi* (~90%), *Reader's Finest* (~85%), *Rocket Book* (~70%), *Peter's Publishing House* (~70%), *Secret Books* (~75%), and *Yellow Books* (~65%).

centric[99]. My longitudinal analysis also revealed that the locus of all but one (*Books & More*[100]) remained constant over the previous one and a half decades. For instance, companies with an environment-centric locus consistently emphasized that "it is the market that dictates the terms" (CEO of *Star Print*) and saw themselves as executors who "do what the customer wants [them] to do" (CEO of *Star Print*). In contrast, organizations with a self-centric locus referred to themselves, their history, values, and beliefs when legitimizing actions, both ex ante and ex-post. For example, the CEO of *Peter's Publishing House* stated:

> *When we at the board are unsure about future strategic moves, we always ask ourselves: 'What would Peter [the founder] have done?'*

Furthermore, self-centric organizations believed in their ability to have an impact on their environment. For example, the CEO of *Books & More* noted:

> *We shape the [...] publishing of the 21st century.*

Table 14 provides an overview of the operationalization and exemplary quotes used to determine the organizations' *foci* of identity. Also information on the categorization of each of the sampled firms is provided. A more detailed list used to categorize firms as either self or environment-centric can be found in the Appendix A.A-4.ii.

---

[99]*Arthur & Sons* (~65%), *Book 2020* (~85%), *Books & More* (until 2003: ~75%, 2003—2011: ~ 20%), *House of Books* (~70%), *Reader's Publisher* (~80%), *Star Print* (~85%), *Superbooks* (~85%), and *TopPress* (~75%).
[100]*Books & More* evolved from a merger of two German publishing houses (one with external and one with internal locus) at the end of the 1990s. The quotations from the early years of *Books & More* reflect a mostly environment-centric locus, but data indicates a change of identity in the early 2000s: At that time, a third 'brand' with a distinct product portfolio and a clearly self-centric mission statement and values was founded, which instantly became a main component of *Books & More*. Since then, the publishing house has been primarily self-centric (see detailed case description in Chapter 5.4 for more information).

Table 14: Locus of Organizational Identity

Specification	Operationa-lization	Exemplary Quotations	Firms in Sample
Self-centric organizations evaluate issues and answers along their own values and norms. They intend to shape the market	Coding of interviews/websites: Ratio of sum of textual elements related to the organizational self/values/norms as opposed to customers or other external stakeholders. (For a detailed dictionary, see Appendix A.A-4.ii)	"We shape [...] publishing of the 21st century." *(CEO, Books & More)* "[As a customer,] I would ask: 'Why should I spend money on this. You are crazy!' [...] We don't care." *(Editing assistant, Peter's Publishing House)* "We don't just want to be in a market. We want to offer value." *(Manager, Yellow Books)* "We only do things that suit us." *(CEO, Secret Books)*	*Ars Legendi, Peter's Publishing House, Reader's Finest, Rocket Book, Secret Books, Yellow Books* Since 2003: *Books &More*
Environment-centric organizations evaluate innovations along existing demands of established or potential customers	Coding of interviews/websites: Ratio of sum of textual elements related to customers or other external stakeholders as opposed to the organizational self/values/norms. Frequency codes of 'customers' and 'react' (as opposed to 'act') (For a detailed dictionary, see Appendix A.A-4.ii)	"It is the market that dictates the terms" *(CEO, Star Print)* "We listen carefully to what our customers say. And we do what he wants us to do." *(CEO, House of Books)* "New developments are not driven by us, the customer drives them." *(CEO, Superbooks)*	Up to 2002: *Books & More* *Arthur & Sons, Book 2020, House of Books, Reader's Publisher, Star Print, Superbooks, TopPress*

5.5.1.3 Strength of Identity

To rule out potential alternative explanations, I also evaluated the strength of identity—the depth in which core values and concepts are embedded in an organization's cognition and action (Labianca et al., 2001). I applied several measures such as counting how often identity-related issues were mentioned overall. I also extensively referred to additional observations, expert impressions, and the organization's narrative and symbolic self-portrayal in archival company documents (e.g., how extensively the company's history was elaborated upon on the company's website, position and size of the firm logo) to corroborate my interpretations. The identities of all my companies had medium to high levels of strength.

5.5.2 Organizational Identity Dimensions and Adaptation

My work is embedded in the literature on cognition and strategy (Kaplan, 2011), which studies how organizational interpretation (Daft & Weick, 1984) affects organizational responses to discontinuous technologies (Kaplan & Tripsas, 2008). I first analyzed how the two focal identity dimensions I had dissected affected organizational attention to and interpretation of the involved 'issues and answers' (Ocasio, 1997; see also Gilbert, 2005; Livengood & Reger, 2010; Weick, 2001).

In a second step, I studied how the identity dimensions—mediated through interpretation—influenced two relevant dimensions of adaptation that have been frequently discussed by other scholars: timing and type. 'Adaptation timing' refers to the swiftness with which organizations recognize discontinuities, interpret these changes, and implement competitive responses (Miller & Friesen, 1980). Although there is no *a priori* association between the temporal structure and the success of adaptation, response timing is pivotal to surviving and leveraging change (Anderson & Tushman, 1990; Suarez & Lanzolla, 2007).

'Adaptation type' refers to the mix of various dimensions of organizational response (Ford & Baucus, 1987). Organizational response can be either *passive* or *active*. Passive responses "do not attempt to respond to [the change] and are evidenced where decision makers ignore, become angry, deny or create illusions about [the discontinuity]" (Ford & Baucus, 1987: 371-372). In contrast, active responses include all deliberate strategic steps an organization takes to adapt. Active responses can be differentiated into decisions to adopt the technology and the deliberate decision not to adopt the technology. Decisions to adopt a new technology are labeled as 'domain creation.' Decisions not to adopt the new technology include three possibilities: 'domain defense' encompasses moves to "preserve or restore legitimacy [of the established technology]" (Ford & Baucus, 1987: 372), for instance by publically slandering the new technology as harmful to cultural, ethical, or political norms and values (König et al., forthcoming); 'domain offense' attempts to preserve the old technology through strategic business activities, for instance, investments into the improvement of the established technology and 'bold retreat' into remaining market niches (Adner & Snow, 2010); 'domain abandonment' involves the exit from the technological field through divestiture.[101]

I assessed the timing of organizational adaptation by using data on product launch dates and information on the timing of internal routine changes gathered from my various sources. For instance, I tracked at which point in time each company started to offer electronic books, to provide electronic access to magazines, or to present smartphone applications, online databases and platforms. I used all the information on strategic initiatives to categorize the sample firms as 'early,' 'average,' or 'late' with regard to recognition time and all four active dimensions of active response. Particularly important in this regard were decisions to create new domains by adopting the new technology[102]. I found considerable variance in my sample: six companies were 'early,' one 'average' and five 'late' domain creators or domain offenders. Two of the fourteen organizations have not adopted the new technology to the present day.

To distinguish between 'active' and 'passive' responses and to determine the precise pattern of organizational adaptation, I relied on interview statements, counted the number of revenue-generating digitalized products, and assessed the degree to which these deviated from the traditional print business paradigm. Four companies were classified as passive or predominantly passive responders (*Book 2020, House of Books, Reader's Publisher*, and *Star Print*)[103]. All of the ten active responders engaged in some level of domain defense.[104] However, they were clearly classifiable as either 'domain creators' or 'domain offenders'

---

[101] Note that domain creation, defense, and offense are continuous dimensions and that companies can engage in all three response patterns simultaneously. I coded the responses to identify the organizations' predominant response behaviors. As highlighted by Ford and Baucus (1987), the activity level of responses is also influenced by the degree to which external moves are accompanied by internal changes.

[102] Within the magazine market I classified electronic product launches before 2000 as 'early' and launches since 2003 as 'late.' The electronic book launches were categorized as 'early' when they took place before 2004 and as 'late' when they occurred in 2007 or later. I focused on real digitized media such as internet content, pdf and ebooks, etc. when determining the time of adoption, this excluding any launch of CDs/DVDs. I derived this classification from some of my non-case specific interviews with industry experts.

[103] For the passive organizations, I coded interviews and archival material to identify the specific type of passive response ('denial,' 'anger,' or 'resignation;' (Ford & Baucus, 1987). My data shows that denial and resignation are most common among our passive responders.

[104] For instance, the CEO of *TopPress* stated: "There is nothing more delightful than burying oneself in a physical book."

(none of the companies abandoned publishing). I classified reactions as domain offense when the company refused to develop and publish digital content (also reflected by a low ratio of digital products to physical products mentioned as revenue sources in press releases and company reports) and simultaneously engaged in strengthening and extending the traditional print business.

I categorized strategies as domain creation when interviewees emphasized (and other data corroborated) that their companies had actively invented new digitization-based revenue sources, such as online access to electronic magazines via subscriptions and pay-per-article, customized ebooks, smartphone applications and fee-based learning platforms.[105]

The coding scheme for organizational adaptation allowed me to study which type of response to discontinuous technological change results from variance in organizational identity. My data provides two insights: First, focus and locus of identity shape two particular dimensions of response; and second, the interaction between the two identity dimensions engenders heterogeneity of incumbent response, which explains phenomena that run counter to standard theory. Below, I describe my observations and reflect them in the light of prior theory.

5.5.2.1 The Effect of Focus on Adaptation

Although focus affects multiple response dimensions, it is most tightly linked to the degree to which companies engage in domain creation: exclusive organizations tend to stick to their established business model and are less likely to commercialize the new technology than inclusive organizations. In fact, my sample suggests that, ceteris paribus, an inclusive focus is required and sufficient for domain creation. Only, and all of, the seven inclusive companies committedly adopted digitization-based business models over the observed timeframe.

Statements of the respondents indicated that an inclusive focus lowered the 'identity challenge' (Tripsas, 2009) involved in accepting digitization as a viable and valuable business. For instance, interviewees at *Books & More* suggested that their company felt no obligation to stick to books as the main medium and was able to identify with the new technology as a result of the broadness of their organizational self-definition. This inclusiveness was also reflected in an image brochure published by this firm in 2006, which

---

[105] With regard to these activities, I captured both the level of resource commitment to the new technology and the extent to which companies adopted paradigmatically different business routines compared to attempts to 'cram' (Christensen & Raynor, 2003; Gilbert, 2005).

stated:

> *We offer [this digitized product] because we consider [...] books [to be] an active service package instead of a static stack of paper.*

In contrast, respondents at exclusive companies explained the unwillingness of their company to invest in the new technology with their perception that the new domain was not their 'home turf' (Livengood & Reger, 2010). The CEO of *Star Print* stated,

> *[Digitization] is something that is clearly out of our industrial sector.*

Additionally, one middle manager from the same company noted:

> *One of the leitmotifs here is 'Cobbler, stick to your trade.'*

This finding echoes assumptions developed in recent literature. As Livengood and Reger (2010) conceptualized, the focus of an identity determines the scope of 'issues and answers' (Ocasio, 1997) that are perceived to be relevant and legitimate to take into consideration. The more inclusive the identity, the more issues and answers enter the organizational repertoire. In addition, decision makers of inclusive organizations were more willing to 'connect the dots' rather than differentiate themselves and their organizations from developments at the periphery of their traditional domains[106].

P3-2. The more inclusive (as opposed to exclusive) the focus of a company's identity, the more likely this organization is to adopt a discontinuous technology to create new domains.

5.5.2.2 The Effect of Locus on Adaptation

I found the locus of identity to be most strongly related to the speed of the response to discontinuous technological change, particularly the speed of technology adoption. Overall, environment-focused organizations responded relatively promptly, whereas self-focused organizations seemed to wait longer before considering a response. All five 'late' responders had self-centric identities, whereas the identities of the six publishers that were categorized as 'early' responders were environment-centric, at least at the time of their first response initiative. One company was categorized as 'average' responder. Two companies responded passively, therefore I did not capture any level of response speed for these firms.

---

[106] This definition is coherent with Miles and Snow's description of 'prospectors' (1978).

According to my data, environment-centric organizations feel obliged and legitimated to respond to external changes early. For example, the CEO of *Arthur & Sons* stated that his/her company never had a choice other than to respond faster than others:

> *[The world around us] coerces us to adapt.*

Similarly, the CEO of *Superbooks* stated:

> *[The trend to digitization] is not driven by us, it is driven by the customer.*

In contrast, regardless of the point in time when they became aware of the change, decision makers in self-centric companies took more time to decide how to respond, particularly because they pondered longer over question of what kind of response would be consistent with their firm's identity.

My findings resonate with and combine, prior theory. Albert and Whetten's (1985) notion of utilitarian, as compared to normative, organizations implies that organizations that feel more bound to core beliefs and values have more difficulties in assimilating to external change[107]. However, locus of identity also includes a strong facet of inherent and stable control perception (Brockner et al., 2004) and implies that environment-centric organizations respond to changes in the environment earlier, while self-centric organizations wait and only act when their threat perception reaches a threshold level (Gilbert, 2005; Tversky & Kahneman, 1974) or when they see an opportunity to shape the environment by taking initiative.

> P3-3. The more environment-centric (as opposed to self-centric) an organization's locus of identity, the earlier this organization will respond actively to a technological discontinuity.

---

[107] Note that 'utilitarian' is not equal to self-centric. Albert and Whetten's (1985) distinction between utilitarian and normative also refers to the level by which the respective organizations are driven by profit. My categorization of self- vs. environment-centric firms does not include that characteristic. Indeed, as profit-seeking organizations, all firms in my sample were acting towards (some) financial goals. However, independent of their locus of identity, many firms also engaged in non-profitable activities based on varying motivations: Self-centric companies pursue 'pet-projects' aligned with their organizational shared values, whereas environment-centric companies do favors for their suppliers (here: authors) or customers.

5.5.2.3 Interactive Effects of Focus and Locus on Responses to Discontinuous Technologies

In the subsequent analysis, I proceeded to ask: 'How do identity dimensions *interactively* influence organizational adaptation?' To do so, I clustered the sampled firms into the four generic categories of identities that can be constructed by the combinations of inclusive/exclusive focus and environmental-centric/self-centric locus. My analysis revealed idiosyncratic response patterns for all four identity types, which were either literally or theoretically replicated by each single case study. Below, I present my findings (see also Table 15).

Table 15: Organizational Identity and Adaptation—Empirical Findings

Publishing House	Identity Type[a]	First Adoption[b]	Type of Reaction[c]	No. of New Revenue Sources[d]	Description of Adaptation Behavior	Match with Theory[e]
1. Congruent cases						
Superbooks	I-N	Early (p: 1999, b: ~ 2000)	*Active (creation)*	4	Complete change of business portfolio from published products to service/IT provider. Currently offers services along the entire value chain.	L
Arthur & Sons	I-N	Early (p: 1998 b: 2002)	*Active (creation)*	5	Conservative customer groups. As suggested by an external consultant, *Arthur & Sons* started development of new (non print-related) business segments including the trade of whitepapers and webinars.	L
TopPress	I-N	Early (p: 1998 b: 2000)	*Active (creation)*	6	Early switch to media-independent data editing and launch of various new products: online access for magazines, ebooks, smartphone apps.	L
Secret Books	I-S	Late (p: 2005)	*Active (creation)*	4	Long-lasting internal discussions about digital products. Recently launched selected, innovative products (apps but no ebooks).	L
Yellow Books	I-S	Average (p: 2000 b: 2005)	*Active (creation)*	7	First online product in 2000. Afterward long internal discussions about new innovation. Since 2007 development of innovative digital, award-winning products with complex, tailored business models.	T

Peter's Publishing House	I-S	Late (p: 2003)	*Active (creation)*	11	Long-lasting internal discussions since 1996 about digital products. Launch of a broad variety of digitized products, seen as new core product.	L
Book 2020	E-N	n/a	*Passive (denial)*	1	Denies need for digitized products. Restricts activities to online distribution of physical books and magazines; premium pdf-newsletters. Strives to maximize profits for print magazines and loose-leaf editions.	L
Star Print	E-N	n/a	*Passive (denial and resignation)*	1	Sees overall trend to digitization, however denies shift to digitized products in its own market segment. Recently launched a very limited number of digital products, mostly as "goodies for free" (manager) on request of customers (no additional revenues). One interactive webportal in cooperation with external partner.	L
Rocket Book	E-S	Late (b: 2011)	*Active (offense and defense)*	1	Strong focus on main product. Licensing of any non-print products. 2011: launch of one smartphone app.	L
Reader's Finest	E-S	Late (b: 2010)	*Active (offense and defense)*	1	Some level of denial. Attempt to strengthen print market share by sustaining innovations. Licensing of non-print products.	L
Ars Legendi	E-S	Late (b: 2010)	*Active (offense and defense)*	1	Attempt to strengthen print market share by contracting new promising authors.	L

2. Incongruent / hybrid cases

Case	Type	Time[b]	Active/Passive	Count[d]	Description	
Books & More	I-N before 2003, then IS	Early (b:2002)	Active (creation)	9	Development of online, interactive learning platforms, ebooks and online books, online and video trainings.	T
Reader's Publisher	Hybrid[f]: E-N / I-N	Early (p: 1998)	Mostly passive (denial at top management level)	3	Company started adoption early, launches various online initiatives. However, none of these affect the core business. Most digitized products are for free, few additional revenues.	T
House of Books	Hybrid[g]: E-N / I-N	Early (p: before 1998)	Core segments: Passive (denial at top management level)	3	Online portals as new revenue source. Core products (magazines and ebooks); however, so far exclusively print.	T

[a] I-N: inclusive/environment-centric (Type 1); I-S: inclusive/self-centric (Type 2); E-N: exclusive/environment-centric (Type 3); E-S: exclusive/self-centric (Type 4).

[b] Point in time of first adoption. "p" refers to periodicals, "b" refers to books.

[c] Categorized according to Ford & Baucus (1987) hierarchically into 2 categories: (1) active vs. passive and (2), for active responses, domain creation/offense/defense/abandonment-

[d] Count based on information provided in interviews and current websites. Count excludes CD-ROMs and usage of internet as direct sales channel for books (as compared to distribution for digital content).

[e] L: literal replication, T: theoretical replication.

[f] Idiographically hybrid (Corley et al., 2006): Owner-manager exclusive, top managers primarily inclusive.

[g] Idiographically hybrid (Corley et al., 2006): Acting owner-manager and long-time employees exclusive. Future owner-manager highly inclusive.

Type 1: Inclusive and environment-centric (I-N). Three companies—*Arthur & Sons, Superbooks*, and *TopPress*—were classified as Type 1 (Inclusive-eNvironment; I-N). Their reaction pattern was characterized by early attention to the change and only minor hesitation to legitimize adoptive responses, which ultimately resulted in quick domain creation and only low levels of domain defense and offense.

Due to their tendency to be driven by the environment, all I-N-organizations were engaged in intensive market observation. At the same time, these companies defined their environment very broadly, which is why the number and breadth of factors they took into account when screening the environment was high. For instance, the CEO of *Arthur & Sons* stated:

> *Some of our competitors are, of course, classical publishing houses.*
> *However, new competitors, specifically mediators of content such as*
> *Google, Xing, internet newsgroups, and even industry fairs, have*
> *become even more important.*

Two out of three I-N companies were active in major industry associations to build up social capital, which helped them to gain access to and exchange information on market developments with peers. Moreover, all I-N-companies actively engaged in market observation, both by commissioning classic market research and by using advanced social media tools. Keeping themselves 'up-to-date' formed a fundamental part of these organizations' identities. Driven by their low level of self-centricity and their general openness, I-N-companies were also willing to hire external consultants to reflect strategies from an outside perspective.

Their high level of market sensitivity allowed the I-N-companies to become attentive to the digitization very early. More importantly, the decision makers within these companies also interpreted the discontinuity as a legitimate issue and felt the necessity to develop competitive answers (Ocasio, 1997). For example, already in 1991, the CEO of *Arthur & Sons* urged other publishers in a speech, to watch out for 'the advance of digitization' (source: archival document).

The responses of I-N-organizations were also shaped by their specific identities. In contrast to my observations at other companies, I found almost no evidence for extensive discussions and 'framing contests' (Kaplan, 2008b) with regard to the interpretation of the discontinuity at the top management level, which could have protracted decision-making. To the contrary, decisions were made promptly. Decision speed was reinforced by the inclusive focus of I-N-companies. For instance, as early as the 1990s *Arthur & Sons* described its mission as the "transfer of knowledge that is relevant to practice" (historical website), which is why, from very early on, "media-neutrality" (CEO) was at the top of the strategic agenda.

Intriguingly though, as a result of sense of environment determination, the I-N-companies' competitive moves were also limited in their proactivity. In fact, my data shows that these organizations rather tended to imitate other companies, particularly new entrants, in their implementation of new businesses instead of developing their own business designs. As one editor at *Arthur & Sons* noted:

[We] lacked entrepreneurial spirit and good ideas for products.

Moreover, *Arthur & Sons* felt that the environment coerced the organization into the abandonment of the online business in 2002 after the burst of the internet bubble. According to the managers, customers had become disillusioned. Only in 2005, when customer interest in online services had resumed, did the digital business at *Arthur & Sons* revive. This time, a strategy consultancy developed a plan for the new business (20 million Euro invest), which was fully implemented and is still in service. However, respondents noted that the internal, structural changes related to the new business were still limited. While the new online unit was established to be responsible for the technical maintenance, no new, online-focused editors were hired. As a result, "A majority of the online content is reuse of print content" (editor, *Arthur & Sons*).

As shown in Table 15, the pattern described for *Arthur & Sons* was replicated by all other Type 1 (I-N) cases. Overall, the identity-related idiosyncrasies of I-N-organizations affect the various elements of these organizations' attention systems (Ocasio, 1997). In particular, the environment-centric attention structures and the broad definition of the competitive domain in an I-N-company cause this organization to recognize discontinuities fast and coerce the firm into adopting innovations early. However, even though their portfolio of competitive answers is relatively broad, the implementation routines of such companies often lack flexibility, as they are imitative, as opposed to proactive.

P3-4. Organizations with an inclusive and environment-centric identity will engage in early, yet somewhat rigid, domain creation, and low levels of domain offense.

Type 2: Inclusive and self-centric (I-S). I classified three of the organizations studied— *Peter's Publishing House*, *Secret Books*, and *Yellow Books*—as belonging to this type. While, similar to the I-N-organizations described above, I-S-organizations adopt new technologies to create new domains, I observed that they did so later and, at the same time, in a more flexible way. Furthermore, I-N-organizations commit more operational, as opposed to financial resources, when implementing new technologies.

As I-S-organizations are more self-centric, they are less engaged in market observation and it ultimately takes them longer to become aware of discontinuities. For instance the CEO of *Yellow Books* noted that his/her company recognized the implications of digitization only in the late 1990s, and explained: "I admit, I am not very active in observing new competitors."

While delayed attention is somewhat predictive for a later response, the most important reason why I-S-organizations show a delayed response to technological discontinuities is rooted in the 'impetus' and 'commitment' phases (Bower, 1970) of the resource allocation process. As a consequence of the inclusive identity, decision makers in the I-S-organizations that I observed, did not reject the innovation in itself as identity inconsistent. However, they found it challenging to invent a way to adopt the technology in a manner that suited the company's identity: to managers at I-S-organizations, it is important not to be determined by the environment. As the CEO of *Yellow Books* explained:

> *We want to decide on our own what we do and what is right for our*
> *customers, [which is why we hesitated for so long].*

Intriguingly though, once they have decided to adopt a new technology, I-S-organizations seem to commit more operational resources, i.e., more to managerial attention, enthusiasm, and time to domain creation than other companies. Apparently, members of I-S-organizations take more ownership for their innovations due to their self-centricity. For instance, I compared the enthusiasm that the companies showed when describing their innovations on websites or in company brochures and observed that I-S-companies describe and promote digitized products up-front and highlight these innovations visually. As such, these companies attempted to persuade customers of the uniqueness of internally developed innovations, rather than following homogeneous demands of customers. The manager of online publishing at *Yellow Books* stated:

> *Our goal is more than simply offering an iPad App. That is not*

enough for us. Rather, we ask ourselves how we can create a product
that really offers new value. We are looking for an advanced solution.

At all the I-S-organizations I studied, higher levels of operational resource-commitment and identification with new self-developed technologies resulted in more flexible routines and path-divergent products. For instance, *Yellow Books* has won numerous innovation awards for its digital products. As another example, of all the companies on my sample, I-S-companies created, on average, the largest number of new revenue sources (7.75 as opposed to 5 for I-N-companies, which rank second on this scale).[108]

In total, their inclusive focus and determination to shape the environment interactively form a unique and highly creative response pattern of I-S organizations. While organizational attention structures in these organizations are sensitive to a wide spectrum of issues, their repertoire of answers (Ocasio, 1997) is reduced to those organizational moves that meet these companies' craving for independence and impact.

P3-5. Organizations with an inclusive and self-centric identity will engage in late, yet flexible domain creation, and low levels of domain offense.

Type 3: Exclusive and environment-centric (E-N). Two cases were categorized as E-N-organizations: *Book 2020* and *Star Print*. E-N-organizations recognize change relatively early as their point of reference is located in the environment. Nevertheless, the fact that the focus of their identity is narrow and bound to an established low-level business paradigm (Porac et al., 1989) causes them to initially respond passively with cognitive dissonance and denial. Furthermore, E-N organizations generically show relatively low levels of perceived control, which, once they recognize the change as important, can result in resignation.

High levels of internal framing incongruence and passivity characterized the

---

[108] Note that at *Yellow Books*, the timeliness of the adoption ('average') did not match the prediction of a late adoption response. Managers of this company became aware of the discontinuity relatively late (end of 1990s) and spent long months on internal discussions. However, due to serendipity, one specifically motivated employee had, out of personal interest, worked on a digital database in the previous years. Thus, the first digitized product of the company was in fact almost ready-to-launch when top managers finally decided to become active. As this happenstance reduced product development time by years, *Yellow Books* was already able to offer digitized content in 2000. I thus classified this response as a theoretical replication. (For details see case description in Chapter 5.4)

responses of the E-N-companies I investigated. On the one hand, lower level managers interpreted the changes as "driven by the customers" (editor, *Star Print*) and somehow felt the "necessity to react" (ibidem). However, throughout the organization, but particularly at the top management level, organizational members interpreted digitization as "a temporal fad" (CEO of *Star Print* in 2005) for a long time. Furthermore, the E-N-organizations in my sample were struggling with the inner tensions between their inner tendency to follow the environment and the fact that the same environment required actions that contradicted these companies' exclusive, 'book-related' identity. The CEO of *Star Print* commented with resignation:

> *We now have to deal with demands that are initially virulent and do not fall in our business [of a certain type of books]. I would prefer it if product digitization did not exist.*

Interestingly, the interpretation of digitization and the consideration of competitive answers in E-N companies was also characterized by low levels of perceived control, which echoed the overall notion of low levels of self-determination (DeCharms, 1968) and impact (Rotter, 1966). The same CEO stated:

> *We do not know whether we are capable of being part of the digitization wave. Nobody will ask about [...] publishers any more. This is frightening.*

The E-N-firms in my sample did not adopt digitized contents and digitized distribution. Only recently have the two companies started to offer CD-ROMs (which the CEO labels as "rubbish"). However, at both companies, these media are only to supplement the product portfolio in response to requests from customers. The managing editor of *Star Print* explained:

> *We publish books. Everything else needs to go alongside.*

My observations at book fairs corroborated my notion that the top management of E-N companies neglected digital publishing.

In sum, E-N-organizational identities face profound conflicts caused by the growing cognitive distance between the own intra-domain, exclusive focus and the cognitions and beliefs of the stakeholders in that same domain. In particular, once customers start to request changes from the company that contradict pre-existing beliefs and values, and require such companies to implement changes which organizational members feel unable to implement, the company and the environment become increasingly alienated.

> P3-6. Organizations with an exclusive and environment-centric identity are likely to engage in passive responses, including anger, denial, and resignation.

Type 4: Exclusive and self-centric (E-S). Three of my cases belonged to this identity type: *Ars Legendi, Children's Finest,* and *Rocket Book.* As a consequence of their self-centricity, organizations characterized by an E-S identity tend to recognize changes relatively late. Once decision-makers in these organizations acknowledge the change, they respond actively, albeit with domain offense, which can include a defiant retreat into a domain-consistent market niche as recently described by Adner and Snow (2010). Furthermore, of all types, E-S-companies engage most strongly in domain defense. Given the involved conscious focus on, and defense of, the established identity, and the deliberate alienation from the trend, the response of E-S organization could be labeled as 'solipsistic' or 'active encapsulation.'

Similar to I-S-companies, E-S-companies recognize technological discontinuities comparatively late given their relative disregard of external developments. For instance, at *Ars Legendi*, decision-makers limited their market research to the reading of the periodicals and newsletters published by the national industry association and 'keeping their eyes open.' Managers at all levels of *Ars Legendi* regard only three other highly traditional publishers as competitors. Not surprisingly, this company and the other E-S-publishers recognized digitization as something important "as one of the last publishers" (source: archival material), around 2008.

Similar to the E-N-companies, E-S companies tend not to adopt digitization because it is located outside of their constructed identity domain. However, in contrast to E-N companies, E-S companies deliberately *decide* not to adopt the new technology. The production manager of *Ars Legendi* stated:

> If there is one value we need to stick to, it is our quality requirements. [Digitized products] do not meet these quality requirements.

Furthermore their sense of self-determination and 'impact' (Rotter, 1966) caused all E-S companies to remain hopeful and defiant. The CEO of *Ars Legendi* noted:

> *We will see who will die and who will survive. However, I am ready*
> *to fight.*

At the same time, E-S-companies were those firms in my sample that most aggressively attempted to engage in domain offense. For instance, respondents stated that, in response to digitization, they attempted to strengthen their core business by hiring new best-seller authors and by diversifying into new content areas.

Domain creation activities of E-S-publishers were rudimentary. Only in 2010/2011, did the three companies start to offer some non-core-catalogue content as ebooks and apps. All digitized content production was outsourced to contractors. None of the companies provided a marketing budget for ebooks. Finally, documents and interview statements indicated that E-S-firms did not hire new employees for launching digitized products.

> P3-7. Companies with an exclusive and self-centric identity are likely to refrain from adopting a discontinuous technology and to engage in late domain offense and defense.

Incongruent/hybrid cases. As already mentioned above, three cases either changed their identity over time or had ideographically hybrid (Corley et al., 2006) identities. These cases were especially insightful, as they allowed me to scrutinize whether my emerging theory would be replicated by extraordinary cases. As shown in Table 15, all three cases confer to the model; I thus label them as theoretical replications. As an initial I-N company, *Books & More* reacted early, yet somehow rigidly to digitization by launching ebooks as early as 2002. In 2003, however, an identity change from I-N to I-S was triggered by the foundation of a new, central business unit (which now constitutes the 'core' of the company) with a clear self-centric mission and vision. In accordance to my proposition for the I-S reaction pattern, this identity change prompted the organization to commit more operational resources on digitization, overcome established routines, and think about more unique and creative products. As a consequence, in 2011, *Books & More* ranked amongst the most innovative companies in my sample.

My sample also contained two cases with hybrid, ideographic identities (*House of Books, Reader's Publisher*): In both cases, I detected identity discrepancies between top managers and long-tenured employees (E-N) on the one hand and lower hierarchy managers and younger, less-tenured staff (I-N) on the other hand. As a result of the environment-centric locus (according to P3-3), organizational members quickly recognized the environmental turmoil. Top management, however, impeded any active response (in accordance with P3-6). Lower hierarchy managers, however, used their own department budgets to engage in digitization activities. Due to the comparatively low levels of investment, this resulted in early, yet scarcely creative domain creation activities that were highly incongruent amongst the various departments of the organization. In both companies I observed attempts to 'cram' the new technology into the existing market.

5.5.3   Family Influence on Organizational Identity

Besides studying the relationship between organizational identity and adaptation to discontinuous change, investigating the role played by family influence in this context is a major goal of my work. Whilst little is known about what determines the identity of an organization (Gioia, Price, Hamilton, & Thomas, 2010), most scholars agree on the pivotal, influential role played by the founder's values (Fligstein, 1987; Hannan, Baron, Hsu, & Koçak, 2006; Stinchcombe, 1975) and the organizational history (King, Felin, & Whetten, 2010). Prior empirical work provided evidence that maintaining the founder's vision and values over time throughout the company's history is more pronounced for family than for non-family businesses (Ogbonna & Harris, 2001). Moreover, a recent quantitative study conducted by Zachary, McKenny, Short, and Payne (2011) shows how the market orientation of family firms differs from that of their non-family owned counterparts leading to the conclusion that identity of family influenced firms is different from that of non-family influenced firms.

Eight out of the eleven family firms in my sample actively emphasized their status as a family business on the websites as well as during the personal interviews and perceived this idiosyncratic ownership type as a key component of their organizational identity. The three family-owned and influenced companies not referring to the owner-family as a central, distinctive, and enduring element for their organization have all undergone major changes in the recent past: *Rocket Book* and *Reader's Finest* were sold to another family business several years ago. Interestingly, both organizations stayed tight-lipped about that change of ownership and, for instance, neither communicated that radical organizational change in press releases (according to historic website analysis) nor on their organizational self-description on the website. *Book 2020* is a similar yet somewhat different case: The company still belongs to its founding family, yet the family member that was formerly active as CEO for long decades and thereby substantially influenced the company's strategy and culture retrieved from the business operations several years ago.

Despite the commonality of the owner-family as a central, distinctive, and enduring organizational component (Albert & Whetten, 1985; Zachary et al., 2011), the organizational identities of the respective family firms substantially diverged. Referring to the four types of organizational identity presented in the previous chapters of this thesis, one can see that family businesses are homogeneously distributed over all fields of the matrix (see Table 15, two I-N companies, two I-S companies, two E-N companies, three E-S companies, two E-N/I-N hybrids). In contrast to the family firms, the three non-family businesses *Books & More*, *Secret Books*, and *Superbooks* concentrated on identity types with an inclusive as opposed to exclusive focus (two I-S companies, one I-N company). Hence, non-family firms are apparently less inclined to exclusive *foci* of identity. As such, non-family businesses, although they vary tremendously in the speed of their responses to discontinuous change, seem to be likely to ultimately respond with active adoption of discontinuous technologies. In contrast, the responses of family influenced firms to discontinuous change seem to be particularly contingent upon the focus of the identity. In times of discontinuous change, family businesses with exclusive *foci* are likely to be reluctant to adopt the new innovation, and either react passively, exhibiting denial and resignation (E-N types) or actively by domain offense and defense activities (E-S types). Hence, I propose

> P3-8a/b. Family influenced firms are more likely to (a) possess an exclusive focus of identity compared to their non-family influenced counterparts, and (b) consequently refrain from adopting the discontinuous technology.

Extant theory offers some first explanations for the observed phenomenon of exclusive *foci* in family firms. Resource scarcity or 'parsimony' (Carney, 2005) might constrain family firms to concentrate on few and selected market segments and/or products (Miller et al., 2010). Adhering to the highest quality standards whilst at the same time satisfying customers to the highest level achievable, is a core concern for many family business owners who care about their reputation and status (see literature on SEW, e.g., Berrone et al., forthcoming). Over time, such business paradigms might reinforce themselves, ultimately becoming an inextricable part of the organizational identity.

Before concluding this chapter, I will briefly comment on potential factors causing the divergence in (family) firm's organizational identities. While early literature on organizational adaptation viewed industrial association and founder's values as core determinants of identity, Gioia and colleagues (2010) came to the conclusion that more complex patterns are responsible for identity formation. Organizational members of only three of the sampled firms (*Book 2020*, *Peter's Publishing House*, and *Rocket Book*) related back to the founder when referring to the 'essence' of their company. Due to the variance in the values of these companies' founders, the identity types of the respective organizations differed from one another. Companies with owner-families whose primary goal was to maintain financial independence from external shareholders (such as *Arthur & Sons*, *Star Print* or *TopPress*) seemed to be particularly inclined to environment-centric identities, as such organizational self-understanding seemingly best allowed the fulfillment of those goals. Organizations that had experienced and successfully mastered significant changes during their history (such as *Yellow Books* and *Arthur & Sons*) are probably more inclusively rather than exclusively focused, as they were forced to build up what Hatum and Pettigrew label as a flexible organizational identity (2006). Although, due to the slightly deviant focus of this study, I cannot comprehensively conclude the determinants of organizational identity, this work contributes to recent scholarly notions that a variety of factors affects organizational identity (Gioia et al., 2010). In particular, results from my study highlight the effect of the owner's legacy (Ogbonna & Harris, 2001), organizational history and intertwining of identity with the owners' goals.

5.6 Concluding Remarks

5.6.1 Summary and Contribution

My initial goal was to enhance understanding of which dimensions of organizational identity determine organizational response to discontinuous technological change, and how. The overall result of my exploratory, grounded research is a framework that arranges organizational identity into two dimensions and proposes how these dimensions individually, and in combination, lead to specific patterns of incumbent response (see Figure 15). Moreover, this framework denotes the tendency of family-influenced firms to exhibit more exclusive identity *foci* than non-family firms.

The inductive multi-case research strategy has been highly beneficial to my inquiry. In particular, it allowed me to inductively dissect identity types that are particularly salient in times of discontinuous technological change and validate their robustness, as well as that of their effects, in the light of diverse circumstances of each of the cases. Note that my mid-range theory also explains cases that deviate from my model at the first sight. The primarily passive response of the two cases with idiographic hybrid identities (*House of Books* and *Reader's Publisher*) is congruent with the model if one adopts the perspective of upper echelons theory (Hambrick & Mason, 1984) that the cognitive dispositions of the dominant coalition in an organization have a stronger effect on the strategic decisions than those of other members of the organization. Similarly, I argue that the early response of *Books & More*, which contradicts P3-5 as this company showed an I-S identity over the majority of the period of investigation, is a theoretical replication of my theory because this publisher had an I-N identity during the genesis of digital publishing prior to 2003.

My study contributes to research on organizational adaptation by providing a nuanced and empirically based account of the micro-mechanisms that link organizational identity with incumbent response to technological discontinuities. Recent advances have highlighted that organizations are myopic to discontinuities emerging outside of their identity domain and feel little incentive, and lack the capabilities, to respond adequately (Livengood & Reger, 2010; Tripsas, 2009). Stemming from rich evidence from multiple cases, I am able to tell a more fine-grained story. It is not only the focus of the domain that determines organizational response; the notion of 'What is our relationship with the environment? Do we shape it, or does it shape us?' that is deeply embedded in organizations, also affects whether or not an organization is attentive to radical change, and how it responds.

Figure 15: Framework of Types of Organizational Identity and Their Influence on Organizational Adaptation to Discontinuous Change

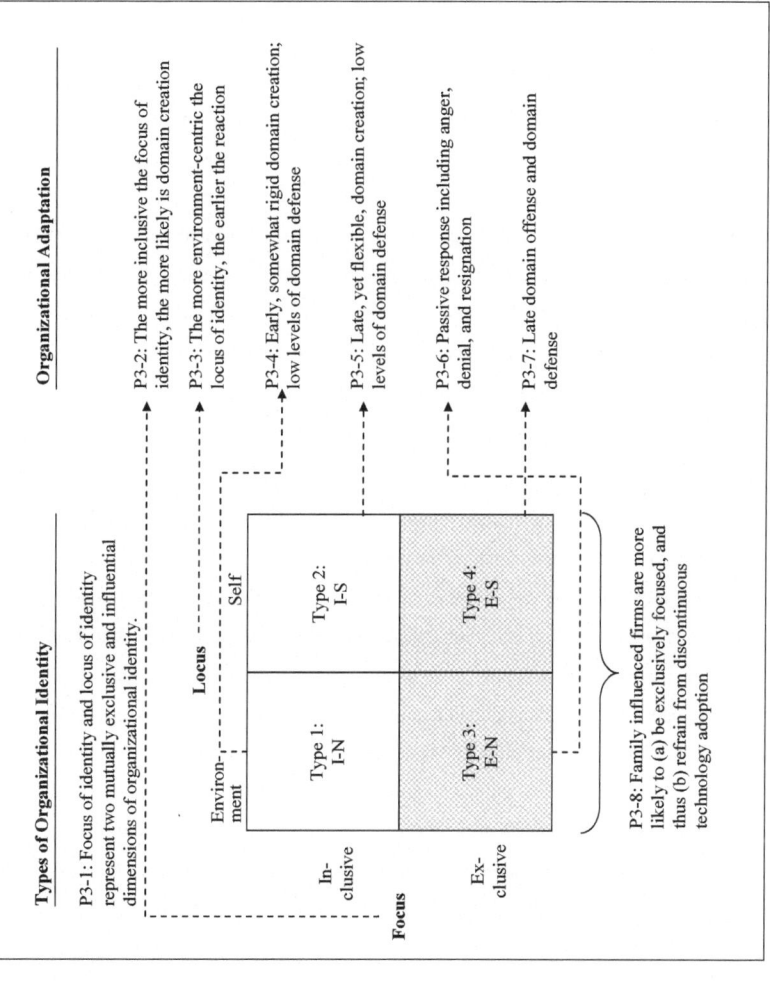

As such, my main contribution is a new explanation of the heterogeneity of incumbent response. Prior theory assumes that incumbents in a competitive field develop 'homogeneous macrocultures' (Abrahamson & Fombrun, 1994) as a consequence of various mimetic effects (DiMaggio & Powell, 1983; Porac et al., 1989). Thus, standard theory inherently predicts that incumbents within one competitive field collectively respond in an inert way when faced with an 'identity-challenging' (Tripsas, 2009) innovation. However, real life evidence is anomalous to this prediction (Gilbert, 2005). My study provides two main insights: First, I show that identities within a group of objectively comparable organizations differ substantially with respect to how they define their identity domain; and second, even if the boundaries of incumbents' identity domains (their *foci*) coincide, the notion of locus of identity explains why these organizations vary in their responses, at least with regard to speed and organizational flexibility.

My findings also run counter to other premises of standard adaptation theory. Most importantly, I add a new facet to research on the role of resource dependence (Pfeffer & Salancik, 1978) in times of discontinuous technological change. Disruptive innovation theory predicts that resource dependence from external constituents, such as customers or investors, is highly influential as it coerces incumbents into pursuing continuous ('sustaining') innovations and reject discontinuous ('disruptive') innovations (Christensen & Bower, 1996). I show that resource dependence is a social construction which is highly dependent on the inherent disposition of an organization: Ceteris paribus, self-centric organizations perceive higher levels of 'self-determination' (DeCharms, 1968) and, thus, lower levels of 'enacted' resource dependence than environment-centric organizations. I then demonstrate that, contrary to Christensen and Bower (1996), E-S-companies reject discontinuous technological change despite their *low* (perceived) resource dependence, and vice-versa, I-N-companies aggressively commit resources despite *high* levels of resource dependence[109].

In addition, I contribute to organizational identity research. Comparable to other researchers (e.g., Fauchart & Gruber, 2011) who have recently attempted to taxonomize identity dimensions of market actors, I develop a typology of organizations facing discontinuous technological change. In this, I combine and extend existing typologies, particularly Miles and Snow's (1978) framework and Albert and Whetten's (1985) dimensions, which have provided much insight but have previously been only loosely related to theory on discontinuous technological change. In particular, my study is unique in

---

[109] In Germany, at the time of this study, ebooks were not a mainstream phenomenon yet, as several consultancy studies (e.g., by PricewaterhouseCoopers (Müller & Spiegel, 2010)) indicate.

introducing the notion of organizational perceptions of locus of control (Brockner et al., 2004; Rotter, 1966) as an important facet of organizational identity. Contrary to other studies, which portray organizational control perception as a part of the cognitive framing of temporal concrete events (Gilbert, 2005), I envisage control perception as a fundamental and long-lasting organizational scheme. In other words, my findings imply that not only individuals but also organizations differ in their tendencies to regard themselves either as 'origins' who give momentum to change or as 'pawns' who merely respond to the environment (DeCharms, 1968). These insights also carry practical implications, as managers who are knowledgeable of their companies' identities are likely to be better able to lead their companies through times of radical change. Overall, my study provides a more fine-grained understanding of organizational identity and its impact on organizational change. Thus, I follow Livengood and Reger's call to empirically "explore [...] identity domain[s] and [their] implications for firm actions and reactions" (2010: 61).

Besides the contributions to literature on organizational identity and adaptation to discontinuous change, my work also amends our understanding of the relationship between family influence and adaptation to discontinuous change, by providing a mediator that explains why some family influenced firms are reluctant to adopt a new innovation at all: In Chapter 3, I compared the adaptation patterns of family influenced and non-family influenced firms but restricted my theorizing to firms that adopted the new technology. In my first empirical study, presented in Chapter 4, I included two non-adopters (*Kiddies* and *Play & More*), however, did not compare the behavior of family and non-family firms. The empirical piece of research presented in this chapter shows that (a) family firms differ from non-family firms regarding their focus of identity, and (b) focus is an important factor determining whether family firms engage in technology adoption or not. Correspondingly, this research may contribute to clarifying the incongruent findings about family influenced businesses' responses to radical change (Hatum et al., 2010), some of which suggest that family firms are particularly flexible and responsive due to the effects of stewardship (Zahra et al., 2008), while others submit the opposite (Basly, 2007). Family influenced firms' focus and locus of identity may provide some clues to reconcile these inconsistencies.

5.6.2 Limitations, Further Research, and Conclusion

As with any empirical research, this study comes along with several limitations, most of them inherent in the inductive research design. Most important is the question whether my findings are affected by retrospective bias of the interviewees. As described earlier, I aimed to mitigate this potential risk by triangulating the ex-post accounts of my informants with real-time information gathered from, for instance, press releases, annual reports, and historic websites. Moreover, interviewing multiple informants per case also contributed to alleviating such biases, and hence increased internal validity (Eisenhardt, 1989). Nevertheless, replications of my study in the same, or related, settings will be important for further scrutinizing the robustness of my data.

Furthermore, replication studies, in particular when combined with extensions, are a promising research approach to investigate potential alternative explanations for the findings provided in this study. An alternative explanation might be that the organizations varied in strength of identity (Labianca et al., 2001) and therefore differed in the activity level and the type of response. However, as I describe in the method section, I controlled for the level of identity strength, with no indication for a systematic influence. Size could be another driver of response variance. However, as I find larger and smaller enterprises across the dimensions, I assume that size is not a major driver of the independent construct I investigate. I also controlled for the effects of market segments, with no model-contradicting findings (although, interestingly, there are more fiction book publishers among the self-centric organizations than among the environment-centric organizations). Finally, the CEO personality could play an important part in the context of my study. CEOs' dispositions can significantly affect organizational behavior (Hambrick & Mason, 1984) and I therefore assessed variance in CEO personality (within and across organizations) based on own impressions and the accounts of respondents from the same company and industry experts. Although my assessment was not based on systematic psychometric scales, my impressions showed remarkably little effect of variance in executives' personalities on the identities of their organizations. Moreover, organizational identity was apparently unaffected by recent CEO changes, as they occurred for four of my sampled companies.[110] Furthermore the fact whether publishing houses own and run their own printing presses or not did not influence whether their focus was tied to print products or not. Nevertheless, future research should test my findings and thereby control for various organizational attributes and external contingencies.

Subsequent inductive case-based and deductive large-scale analyses should test whether my findings can be generalized to other cultural and industrial settings. For large scale testing it will be crucial to develop robust measurement scales, to which my study and

---

[110] Some companies, in particular inclusive family influenced businesses, however, used CEO succession as a measure to trigger and foster change within the organization.

the extant literature provide manifold insight. Large scale testing will be particularly beneficial so as to scrutinize the differences between family influenced and non-family influenced firms, as my findings regarding the effect of ownership structure are somewhat limited, particularly due to the low number of non family firms in my sample.

Another fruitful avenue of future research refers to the link between organizational identity and owners' values and goals. Extant research has so far mostly neglected the important interplay between those two constructs that are likely to affect and cross-pollinate each other, particularly in family firms where owners exert huge influence (Fiss & Zajac, 2004).

Moreover, it would be promising to link the results of this study to other, related areas of research, such as market orientation (e.g., Kohli & Jaworski, 1990; Narver & Slater, 1990; Pelham & Wilson, 1996). While locus of identity, as defined in this study, refers to the internal or external (e.g., customer-focused) sources of legitimization, market orientation denotes how responsive an organization is towards customers (Jaworski & Kohli, 1993). Moreover, similar to organizational identity, market orientation is related to an organization's culture (Narver & Slater, 1990). Integrating the two streams of research—organizational identity and market orientation—is likely to enhance the understanding and predictability of both theoretical perspectives.

To conclude, my aim was to gain detailed insight on the elements of organizational identity. My most important contribution is highlighting the importance of a more nuanced, multi-dimensional, and, at the same time, specific view on organizational identity and its effect on organizations.

6 Summary

6.1 Synthesis of Results

This thesis started with the question, how family influence affects incumbent organizations' adaptation patterns to discontinuous technological change (see Chapter 1.2). My conceptual and empirical findings show that family influence does indeed impact organizations' responses to discontinuous technological changes, regarding their type of adaptation, as well as speed, intensity, stamina, and flexibility of the adoption of the new technology. Due to various interactions of the prevalent manifestations of family influence as well as the characteristic goals and intentions of such organizations, family businesses show an idiosyncratic response pattern. As such, my research shows that investigating the effect of family influence in the context of discontinuous change seems pivotal to increasing the predictability of organizational behavior and industrial change. While detailed results of the family-adaptation and identity-adaptation relationships are outlined in Chapters 3.8, 3.9, 4.4, and 5.5, I will provide a short summary of the most important findings as follows.

In my theorizing, I first proposed that family influenced firms will adopt discontinuous technologies faster than less or non-family influenced firms. Specifically, I argued that such faster responses were due to low levels of formalization, low resource dependence on capital providers, and low levels of political resistance, which are characteristics resulting from family influence (Chapter 3). My qualitative findings (Chapter 4.4) corroborated this theorizing: The family CEOs of the companies I studied frequently engaged in intensive screening of their environment by making effective use of their weak ties, which, in turn, lead to fast awareness of the discontinuity. Furthermore, my research shows that the match between the technological discontinuity and the respective predominant dimension of SEW a family firm strives to maintain impacts the duration of the interpretation phase as well as the ultimate way of implementing the innovation. Family influenced businesses decide faster on how to respond to discontinuous changes because executives in such organizations rely on their intuition rather than on facts from large datasets, which are unavailable at the outset of a discontinuous technology. Furthermore, such firms implement responses more rapidly due to high levels of command, which entail low levels of political resistance. Empirical evidence from my second series of case studies (Chapter 5.5) renders this picture more nuanced by showing that the speed of awareness is also driven by the locus of identity, i.e. whether a company would rather legitimize its actions via internal or external sources. However, this identity dimension is not affected by family influence.

Second, I argued that family influenced firms react with less intensity than less or non family influenced firms. I rationalized this behavior by the 'resource dependence paradox:' Although family influenced firms in principle are less bound to external capital providers, this does not relax their resource rigidity, which goes counter to the predictions of disruptive innovation theory (Christensen & Bower, 1996). Family firms' striving to *maintain* financial independence limits their investment capacity. Empirical evidence from my first series of case studies (Chapter 4.4) grounds this view. Anecdotal evidence indicates that (financial) risk aversion is also a driver for scarce investments. Again, the level of investments depends on the match of the new technology with the company's dominant SEW dimension. For example, online retailing was substantially at odds with *Retail 2000*'s striving to maintain its reputation as a locally embedded provider of goods. Hence, resource investment of this company was initially sparse; only when the company had elaborated a 'compromising strategy'—a business model based on online order and collection of goods from physical stores—did the organization start to invest resources heavily in the new project.

Third, I argued that family firms' low levels of formalization and independence from external capital providers enable them to continuously invest in the new technology. The empirical data from the retailing industry shows that family influenced firms have more stamina and maintain their investment even after (temporary) setbacks—given sufficient CEO commitment to the strategic move. Furthermore, my data shows that open-mindedness of the family CEOs lead decision-makers to reverse some of their decisions on abandonment of investment and enabled them to re-engage in the adoption process.

Fourth, I suggested based on extant literature that family firms will face greater difficulties to adopt new, non-paradigmatic routines due to long tenures of employees, managers, and CEOs and due to these individuals' embeddedness in the firm's history and tradition. Empirical data from my first study provided equivocal results: While routine flexibility of several firms did suffer from conservative, long-term employees with narrow cognitive frames and rigid mental models, this was not true for all companies: For instance, employees of *Alpha Star* and *Toys 2000* significantly contributed to the fast, flexible, and ultimately successful commercialization of the new technology. In the respective subsection of the results of study 1 (Chapter 4.4.4), I argued that this routine flexibility is predominantly rooted in the match of employees' non-economic utility functions with the discontinuous technology. In the second empirical study, I investigated the routine flexibility and type of reaction, building on an organizational identity perspective. I found substantial variance within my sample, depending on whether firms are characterized by an exclusive or inclusive focus. While there is no a priori connection between family influence and focus of identity, the family influenced firms within my sample tended to show exclusive *foci* of identity more frequently than non-family influenced firms. I concluded that, as a consequence, family firms more often than non-family firms refrain from adopting the discontinuous technology at all and, for instance, retreat into niches or engage in domain offense and defense activities. Put differently, the adaptation patterns of family influenced firms are more extreme than those of non-family influenced firms: Family firms either react quickly—or not at all.

Lastly, I theoretically deducted how the openness of the dominant coalition will moderate the relationship between family influence and adaptation to discontinuous change. This impact of characteristics of key decision-makers on organizational adaptation is, to a certain extent, echoed in my empirical findings, which show that the level of a CEO's environmental scanning impacts the speed of awareness: Open CEOs are more likely to engage in environmental scanning for new business opportunities than non-open CEOs. Furthermore, in cases where decision-makers have, once during the 'era of ferment,' decided to abandon the new domain, organizations lead by open-minded dominant coalitions are more likely to reverse the decision and re-engage in investments in the new technology at a later point of time.

264

Figure 16: Summary of Findings

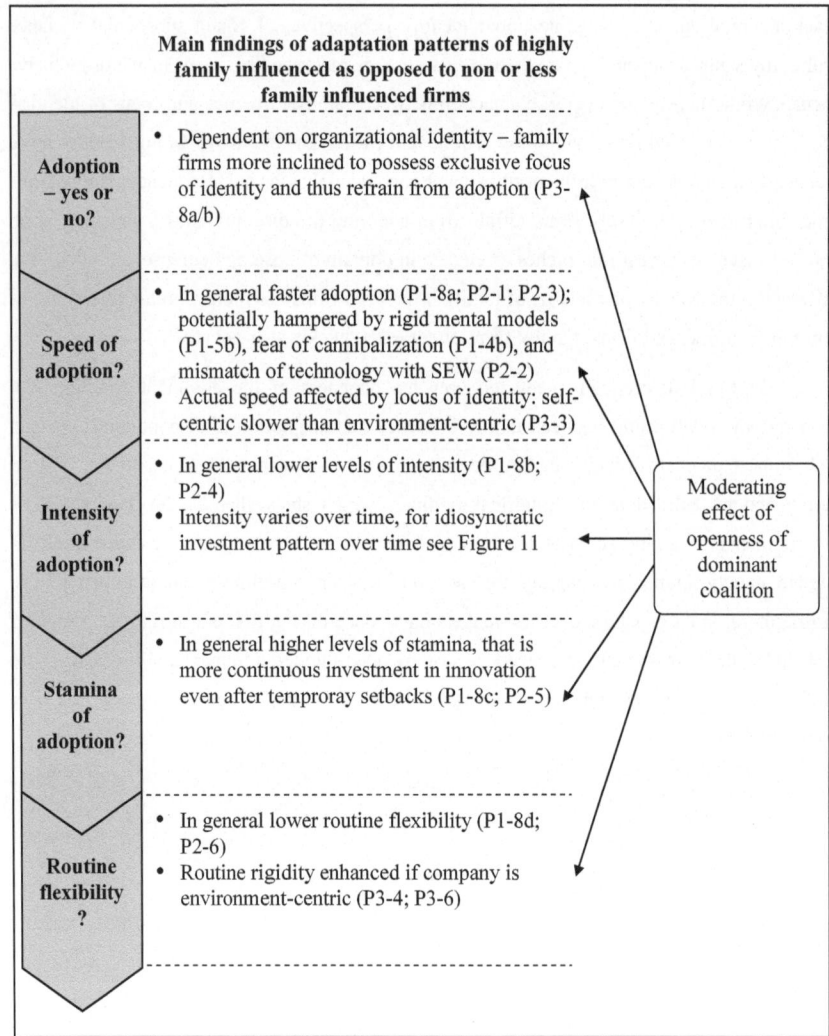

6.2 Theoretical and Practical Implications

My results advance at least three different fields of literature: research on discontinuous change, research on family businesses, and research on organizational identity. While detailed implications are outlined in the specific subchapters (Chapters 3.11.2, 4.5, and 5.6.1), I will use the following passages to provide a holistic overview of those contributions.

First and most importantly, I contribute to literature on organizational adaptation to discontinuous change by providing a more fine-grained picture of how and why organizations do or do not adopt discontinuous technologies. Research on organizational adaptation has so far neglected the pivotal role family owners and managers play for their respective organizations' response patterns to discontinuous technologies. For instance, in one of the most recent and most comprehensive literature reviews on adaptation to discontinuous change, Ahuja and colleagues (Ahuja, Lampert, & Tandon, 2008) do not refer to owner influence in general and family influence in particular at all. By meticulously investigating the links between family influence and determinants of adaptation I showed that organizations' adaptation patterns show variance depending on their respective ownership structures.

Moreover, I showed that the innovator's dilemma that is portrayed in previous, well-received literature (Christensen, 1997) is less relevant for family influenced businesses. Instead, in family influenced businesses a different dilemma arises, impeding adequate adaptation: Although family firms are, in principle, less constrained by disincentives to invest in radical innovations due to their independence from external stakeholders, such firms are likely to refrain from committing aggressively to discontinuous change. The dilemma family businesses face is that they are likely not to invest heavily in the radical technology in order to retain socioemotional wealth even though, in the long run, compromising short-term socioemotional wealth is likely to preserve long-term socioemotional wealth. In turn, they lack the motivation to obtain the money required for necessary long-term investments.

In addition, whereas previous studies have emphasized the impeding role of organizational identity on adaptation to discontinuous technological change (Livengood & Reger, 2010; Tripsas, 2009), my work highlights that specific types of organizational identity *foster* adaptation to such shifts. In particular, organizations with 'inclusive' identities, that is, organizations that focus on similarities of established identity domains rather than boundaries, are more adaptive than firms with 'exclusive' identities. In addition, my empirical work has identified an important barrier to adaptation that has not yet been explicitly addressed by researchers and which I have labeled 'enacted resource dependence.' 'Enacted resource dependence' refers to organizations' perceived, socially constructed rather than objective reliance on firm external forces, such as anticipated customer demands. This considered dependence affects the resource allocation in the respective organizations. As such, my doctoral thesis belongs to an emerging stream of research investigating the role of enactment (Smircich & Stubbart, 1985), identities (Tripsas, 2009), cognition (Eggers & Kaplan, 2009; Kaplan & Tripsas, 2008), and 'self' (Obodaru, 2012) in shaping strategic behavior.

Second, I enhance understanding of the strategic behavior of family influenced companies. Recently, a meta-analysis by O'Boyle Jr., Pollack and Rutherford (2012) revealed that, after decades of debates on whether family influenced firms are a superior or inferior organizational form compared to non-family influenced companies (e.g., Anderson & Reeb, 2003; Chrisman et al., 2004), no a priori performance difference between those two forms of governance can be detected. As a consequence, O'Boyle Jr. et al. (2012) emphasized the need for more nuanced, context-specific analyses on family advantages and disadvantages. By investigating the behavior of family influenced companies in the specific situation of discontinuous technological changes, I followed the call of these authors and took one further step to better understanding the conditions in which family influence constitutes a burden or a benefit and why.

My findings show that family influenced firms are neither better nor worse equipped to adapt to discontinuous technologies. In fact, family influenced firms possess numerous advantages such as low levels of formalization and political resistance allowing for fast reactions and the possibility to persistently pursue long-term goals due to their independence; however there is also a flip side. Striving for independence lowers investment opportunities. Similarly, high emotional attachment to the firm can hamper or hinder adequate adaptation of family businesses. Moreover, such firms' reluctance to involve outsiders and their unwillingness to install decoupled organizational structures with high levels of autonomous responsibility further exacerbates successful adaptation.

Furthermore, my empirical studies provide new insight on SEW and organizational identity of family influenced firms. So far, researchers have made great advances in identifying how high levels of SEW in general affect strategic behavior (Berrone et al., 2010; Chrisman & Patel, 2012; Gómez-Mejía et al., 2011a; Gómez-Mejía et al., 2010; Gómez-Mejía et al., 2007; Zellweger et al., forthcoming-b). However, to the best of my knowledge, none of the previous studies has investigated the interplay of the various dimensions of SEW. With my empirical findings, I provide first insight into the fact that the importance of each of the four SEW dimensions, which have been identified by prior research, varies among family firms; this variation, in turn, exerts substantial influence on strategic behavior. Moreover my results show that organizational identities of family firms differ from those of non-family firms.

Third, I contribute to research on organizational identity. First of all, I follow the call of Corley et al. (2006) to empirically identify which components of organizational identity are important for the explanation of real-world phenomena. I show that, in the context of adapting to discontinuous technologies, two dimensions of organizational identity become salient: focus of identity (which can be either exclusive or inclusive) and locus of identity (which can be either environment- or self-centric). Moreover, my work empirically emphasizes the important role organizational identity plays in shaping firms' strategic behavior and explains the respective links. The contributions of my work to the three research areas are summarized in Figure 17.

Implications from my study, as outlined in detail in the respective subchapters, are of paramount relevance in practice—for both owners and managers. Owners can learn from this study that they—consciously or even unconsciously—affect their respective organizations' adaptation patterns by providing the necessary structures, setting goals, and shaping the organizational identity. It is important for members of owner-families to know about the consequences of such influence. Managers, particularly those working in family influenced firms, however, need to know about the specific 'innovator's dilemma' they face, depending on the organization's ownership structure. Applying remedies—that have been identified by prior research but are not matched with the specific context of ownership and identity the company is embedded in—to such a dilemma may be futile at best and detrimental at worst.

Figure 17: Contributions of this Dissertation

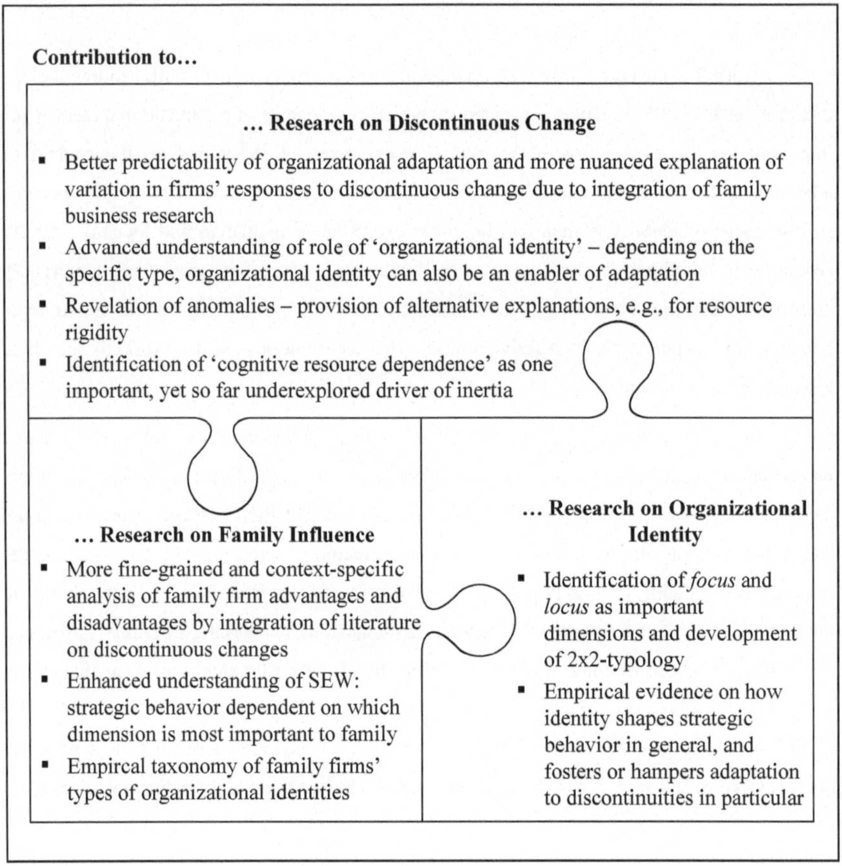

Contribution to...

... Research on Discontinuous Change

- Better predictability of organizational adaptation and more nuanced explanation of variation in firms' responses to discontinuous change due to integration of family business research
- Advanced understanding of role of 'organizational identity' – depending on the specific type, organizational identity can also be an enabler of adaptation
- Revelation of anomalies – provision of alternative explanations, e.g., for resource rigidity
- Identification of 'cognitive resource dependence' as one important, yet so far underexplored driver of inertia

... Research on Family Influence

- More fine-grained and context-specific analysis of family firm advantages and disadvantages by integration of literature on discontinuous changes
- Enhanced understanding of SEW: strategic behavior dependent on which dimension is most important to family
- Empircal taxonomy of family firms' types of organizational identities

... Research on Organizational Identity

- Identification of *focus* and *locus* as important dimensions and development of 2x2-typology
- Empirical evidence on how identity shapes strategic behavior in general, and fosters or hampers adaptation to discontinuities in particular

For instance, family firm owners must be aware of the detrimental consequences that long tenures of employees, managers, and CEOs can have in times of discontinuous change. Whilst such characteristics can result in competitive advantages in continuous times, for example, because employees are specifically loyal and well trained, such a core competence easily becomes a core disadvantage in discontinuous times (Leonard-Barton, 1992). Owners and CEOs can, at least partly, overcome these challenges by putting more weight on external opinion. Externals can and should be included not only in the interpretation and decision-making processes, but also during the implementation phase of technology adoption. For instance, companies can hire external staff experienced in the new technology, acquire firms possessing relevant complementary assets, mandate external consultants, or work in close joint ventures (in which employees strive to build up the competences themselves). Moreover, owners and CEOs need to learn to cede part of their responsibility to (external) managers in order to best balance exploitation and exploration.

In addition, members of family influenced businesses, in particular family CEOs, are advised to review their personal network with whom they usually exchange ideas and rely on for decision-making. Who is part of this network? How broad is the network? How well are these contacts informed about ongoing trends and opportunities? In sum, is this network capable of adverting the CEO to an emerging discontinuous technology? Based on the outcome of this review, the CEO might consider altering his or her established patterns of gathering information.

Furthermore, owners and CEOs need to know and understand the actual and desired dominant dimension of SEW (family influenced firms only) and type of organizational identity (all organizations independent of ownership structure). Dominant dimensions of SEW can be either preservation of family ties, preservation of family influence and control, preservation of reputation and status or preservation of affects and emotions. Organizational identity dimensions differ along the two dimensions 'focus' (inclusive/exclusive) and 'locus' (environment-centric/self-centric). My empirical findings showed that the dimensions of SEW as well as organizational identity substantially affect *how* a company reacts to discontinuous technologies. As such, it is important for organizations to create clarity, transparency, and an agreement within the organization about what the predominant SEW dimension/the type of organizational identity looks like. Moreover, organizational members, in particular the dominant coalition, need to have a common understanding whether (and if so, how) this predominant SEW dimension/type of organizational identity is intended to change in the future, and what the potential strategic consequences of these characteristics are.

Lastly, this study has specific implications for managers of non-family influenced firms. Prior research has already identified 'remedies' such as external influence and ambidexterity (Gilbert, 2005; Hill & Rothaermel, 2003) against incumbent inertia when faced with discontinuous technological changes. This study contributes to this list of valuable measures by illuminating some aspects of how managers of non-family businesses can learn from family influenced organizations. This refers, for instance, to the constructive handling of (temporary) failures within the organization, the idea of 'patient capital,' and also the effects of an organizational culture shaped by a sense of community. Figure 18 summarizes the implications.

From a methodological perspective, this thesis demonstrated a novel way of gathering longitudinal data on non-publically listed, family firms without retrospective bias. 'Waybackmachine' offers free access for researchers to access primary data on firm aspects such as identity and image from today ranging back to the mid-1990s by building on content analysis.

Figure 18: Implications for Owners and Managers

Owners & Managers of Family Influenced Firms

- Review your personal network!—is it broad enough and does it contain the 'right' people to become aware of change?
- Know about the rigidity hazards of long CEO, manager and employee tenures and overcome these by actively involving outsiders!
- Know about your current and desired SEW dimension and the implications of this choice!
- Know about remedies of inertia identified by general management literature (such as structural ambidexterity) but check applicability before implementation!

Managers of Non-Family Influenced Firms

- Learn from family businesses and their way of handling failures!
- Revise the amount of freedom given to core decision makers and employees!

All Owners & Managers

- Become aware of the current and desired organizational identity type of your company!
- Take measures to actively shape the organiztional identity, if required!

6.3 Limitations and Avenues for Further Research

As with any scientific work, this study comes along with several limitations, most of which open up fruitful avenues for further research.

First of all, it is important to re-emphasize, that the focus of my thesis is on building solid and rich theory based on extant literature and empirical, qualitative findings. In this vein, my intention was not to prove or falsify the derived propositions in this work. I strongly encourage researchers to build on my considerations to empirically test my propositions and thus advance the still nascent research in this promising field. Whereas research has recently made substantial advances in operationalizing 'family influence,' e.g., by utilizing the F-PEC (Astrachan et al., 2002) or the FIBER (Berrone et al., forthcoming) scale, research on measuring organizational identity is still premature. Content analysis of primary material such as press releases, letters to the shareholders, and website content have been identified as a promising path to assess such constructs, particularly in the field of family business research (Berrone et al., forthcoming; Micelotta & Raynard, 2011).

Second, it is a promising avenue to go one step further and link adaptation performance, i.e. relaxed routine and resource rigidity, to financial performance. This thesis does not comment on performance of incumbent (family) firms. The choice of recent innovations as opposed to past innovations distant in time is owed to the tradeoff between rich and reliable data with little retrospective bias on the one side and the ability to assess adaptation performance on the other side. Although, the adaptation dimensions discussed in this thesis are important determinants of success as revealed by prior research, these relationships are not necessarily linear (see Chapter 2.1.2.1 for details). Empirical evidence from my case studies emphasizes, in line with extant literature, the importance of the four dimensions, speed, intensity, stamina, and flexibility[111]. However, as most of the markets I investigated are still in turmoil[112], it is not possible to draw solid conclusions on the companies' performance and adaptation success at the time of writing this study.

Third, referring to the conceptual part of this paper (Chapter 3), there might be other factors that moderate the core mechanisms explored. Such moderators, identified as important by other researchers, could include relationship conflicts among family members, as well as the influence of the age or stage of the company. The age/stage of the company has been attributed to different yet ambiguous effects such as increased SEW (Zellweger et al.,

---

[111] In the toy and gaming industry, which is currently in an era of incremental innovations, the two non-adopters of my sample (*Kiddies*, *Play & More*) experience economic problems, whereas *Toys 2000*, a company that adopted the discontinuity in a fast, intense and flexible way outperformed its peers (see case details in Chapter 4.3).

[112] In the smart metering and publishing industry, a dominant design had not yet emerged at the time of this study. The retailing industry had just terminated the era of ferment at the time of this study, with the reaction of two companies from the sample (*Retail 2000*, *White & Blue*) taking place roughly only one year before writing this thesis.

forthcoming-b) or decreased attachment of the family (Gómez-Mejía et al., 2007). . For the sake of parsimony, I deliberately did not include further moderators in my theorizing. However, sharpening my theory by incorporating further contingencies seems to be a promising route for subsequent inquiry.

Fourth, as with all qualitative research, there is a risk of alternative explanations not considered in the respective study. I aimed to rule out potential alternatives such as size of the company (study 1), CEO personality and experience (study 2), or strength of identity (study 2) by measures outlined in extant literature (Yin, 1994). However, I naturally cannot definitively exclude the existence of such errors and hence call for replication of my study.

Fifth, this study has been conducted in a specific national context, focusing on German firms. While this approach allows for thorough comparison of the sampled organizations, it disregards cultural factors that might become important when transferring the results of this study to organizations in different geographic areas (Hofstede, 2001). For instance, the importance of 'customers,' as perceived by the organizational members, might differ from country to country (Hall & Hall, 1990), thus influencing the locus of organizations' identities. Moreover, differences across countries in power distance and long-term orientation as identified by Hofstede (2001) might affect the specifics of how the 4Cs manifest themselves in family firms, and ultimately affect those organizations' strategic behavior.

Further promising avenues for future research open up when broadening the focus of this work: Family owners are an important, yet not the only, group of owners worth investigating. Indeed, it might be interesting to analyze how other groups of owners, such as institutional owners, entrepreneurial founder-owners, businesses privately owned by several individuals without familial ties, or state-owned companies, affect organizational adaptation patterns when challenged with discontinuous technologies.

From a family business research perspective, this thesis invites researchers to join me in investigating the implications of family influence on strategic behavior in specific contextual settings on a very granular level. This will ultimately result in a fine-grained understanding of context specific family firm advantages and disadvantages. Moreover it appears fruitful to dig deeper into the various dimensions of socioemotional wealth, and the question of how they are weighed against each other as well as the resulting implications for firm behavior. Although researchers like Gomez-Mejia, Berrone, Chrisman, and Zellweger have already made great advances in the development of the family firm idiosyncratic concept 'socioemotional wealth,' much remains to be done. For instance, the question of what causes specific socioemotional wealth dimensions to emerge as predominant (as shown in Chapter 4.4.1.2) still remains unclear.

Along the same lines, I encourage organizational researchers to replicate my approach in order to investigate the relevant dimensions of organizational identity as well as their impact on organizational strategic behavior.

6.4   Conclusion

Owners, in particular family owners, strongly affect firm behavior. In the case of discontinuous changes that disrupt an industrial environment, family firms show an idiosyncratic response pattern: They are able to adopt the technology faster and more persistently; however their reaction is less intense and less flexible. The drivers of these differences between family and non family firms are manifold and rooted in several structural, strategic, and emotional aspects tracing back to family firms idiosyncratic characteristics of striving for *continuity*, high levels of *command*, sense of *community* and deep *connections*. Empirical evidence further revealed that socioemotional wealth, particularly the dimension most pronounced for the respective firm, strongly influences the adaptation behavior.

Besides this direct impact of family influence on organizational adaptation to discontinuous change, there is an indirect influence stemming from organizational identity. The history of a family firm as well as the ever-present family owners shaping the company can also exert influence on the organizational identity. Other than portrayed in previous literature (Livengood & Reger, 2010; Tripsas, 2009), I do not find organizational identity to be a definitive inhibitor of change. The more nuanced observations presented in this thesis show that organizations vary in their focus and locus of identity. Depending on the combination of these two dimensions, organizations either react early or late, actively or passively, with domain creation, offense or abandonment to the changing environment. Family firms are particularly inclined to build up organizational identities entailing passive or active non-adoption.

Despite some limitations, as inherent in any scientific work, this thesis shows that organizational adaptation to discontinuous technological change is a much more fine-grained and complex phenomenon than frequently portrayed. Somewhat similar, this work also reveals that family influence is 'neither black nor white,' but an important determinant that shapes strategic behavior in a context-sensitive manner. By providing manifold point-by-point propositions on the interplay of the focal constructs 'organizational adaptation,' 'family influence' and 'organizational identity' I hope to have opened promising and fruitful avenues for further research.

A. Appendix

A-1. Literature Review

Authors	Year	Journal	Title	Type of Change*	Content
Chang, Wu, & Wong	2010	British Journal of Management	Family control and stock market reaction to innovation announcements	Internal change (innovation)	Stock market reacts more negatively to innovation announcements of firms with high as compared to low family control
Ogbonna & Harris	2001	British Journal of Management	The founder's legacy: Hangover or inheritance?	Adaptability to external change	Investigates determinants as to whether the legacy of the founder inhibits or fosters strategic adaptability
Corbetta & Salvato	2004	Entrepreneurship Theory & Practice	Self-serving or self-actualizing? Models of man and agency costs in different types of family firms: A commentary on 'comparing the agency costs of family and non-family firms: Conceptual issues and exploratory evidence"	Internal change (innovation)	Favored model of man results in emphasis on innovation
Ensley	2006	Entrepreneurship Theory & Practice	Family businesses can outcompete: As long as they are willing to question the chosen path	Internal change (strategy)	Commentary on the (stable) strategy present for family businesses with long CEO tenures and depending on the level of task conflict
Habbershon	2006	Journal of Management Studies	Commentary: A framework for managing the familiness and agency advantages in family firms	Internal change (entrepreneurship)	Investigates entrepreneurial virtues in family owned businesses, based on agency view

Author	Year	Journal	Title	Category	Description
Hoy	2006	Entrepreneurship Theory & Practice	The complicating factor of life cycles in corporate venturing	Internal change (innovation)	Commentary on the necessity for family firms to renew through innovation and thereby ensure survivability
Morck & Yeung	2003	Entrepreneurship Theory & Practice	Agency problems in large family business groups	Internal change (innovation)	Investigates agency problems leading to "creative self destruction." Proposes that family firms "quash innovation in one firm to protect its obsolete investment in another" (abstract)
Sharma	2005	Entrepreneurship Theory & Practice	Strategic divestments in family firms: Role of family structure and community culture	Adaptability to external change	Resource based framework suggesting that path dependencies and emotional attachment create inertia regarding divestments in family firms
Zahra, Hayton, Neubaum, Dibrell, & Craig	2008	Entrepreneurship Theory & Practice	Culture of family commitment and strategic flexibility: The moderating effect of stewardship	Adaptability to external (discontinuous) change	Finds empirical evidence that culture of commitment positively affects strategic flexibility. This association is associated by stewardship-oriented culture
Björnberg & Nigel	2007	Family Business Review	The family climate scales - Development of a new measure for use in family business research	Adaptability to external change	Authors develop a scale for measuring the adaptability of families
Blanco-Mazagatos Quevedo-Puente, & Castrillo	2007	Family Business Review	The trade-off between financial resources and agency costs in the family business: An exploratory study	Internal changes (succession)	Investigates changes between first and subsequent generations regarding sources of value and resources

Author	Year	Journal	Title	Change category	Description
Braun & Sharma	2007	Family Business Review	Should the CEO also be chair of the board? An empirical examination of family-controlled public firms	*Not directly related to change*	Examines the influence of CEO duality on firm performance
Cater & Schwab	2008	Family Business Review	Turnaround strategies in established small family firms	Adaptability to external change (crisis)	Case study investigating the turnaround strategies in family firms facing organizational crisis
Chirico & Salvato	2008	Family Business Review	Knowledge integration and dynamic organizational adaptation in family firms	Adaptability to external change	Authors theorize on antecedents of knowledge integration which is a cornerstone for dynamic capabilities
Craig & Moores	2005	Family Business Review	Balanced scorecards to drive the strategic planning of family firms	*Not directly related to change*	Authors add category 'familiness' to existing balanced score card perspectives
Craig & Moores	2006	Family Business Review	A 10-year longitudinal investigation of strategy, systems, and environment on innovation in family firms	Adaptability to external change	Provides evidence that family firms place importance on innovation and are able to adjust
Craig & Dibrell	2006	Family Business Review	The natural environment, innovation, and firm performance: A comparative study	Adaptability to external change (policies)	Family firms are "better able to facilitate environmentally friendly firm policies associated with improved firm innovation" (abstract)
Craig & Moores	2010	Family Business Review	Championing family business issues to influence public policy: Evidence from Australia	Internal change	Investigates attention gaining from policy makers

				Adaptability to external change	
Distelberg & Sorenson	2009	Family Business Review	Updating systems concepts in family businesses	Adaptability to external change	Authors theorize that ownership centrality is associated with adaptability in a reverse U-shaped form. Owners' ability to manage boundaries between systems increases adaptability.
Fernandez & Nieto	2005	Family Business Review	Internationalization strategy of small- and medium-sized family businesses: some influential factors	Adaptability to external change	Provides evidence for family owners' reluctance to change in general and to internationalize in particular
Gilding	2000	Family Business Review	Family business and family change: Individual autonomy, democratization, and the new family business institutions	Internal change (within family system)	Investigates changes within family system such as autonomy and democratization
Gulbrandsen	2005	Family Business Review	Flexibility in Norwegian family-owned enterprises	Internal change (managerial practices)	Investigates flexibility regarding adoption of new managerial practices: Family-CEO as a prohibitor of flexibility
Haberman & Danes	2007	Family Business Review	Father-daughter and father-son family business management transfer comparison: Family FIRO model application	Internal change (succession)	Investigates changes due to succession
Hall, Melin & Nordqvist	2001	Family Business Review	Entrepreneurship as radical change in family business: Exploring the role of cultural patterns	Internal change (entrepreneurship)	Case based work examining the relationship between organizational culture and entrepreneurial processes

Author	Year	Journal	Title	Category	Description
Hatum & Pettigrew	2004	Family Business Review	Adaptation under environmental turmoil: Organizational flexibility in family-owned firms	Adaptability to (discontinuous) external change (environmental turmoil)	Investigates adaptability of Argentinean family firms when faced with environmental turmoil. Identifies a series of antecedents of adaptability.
Leaptrott	2005	Family Business Review	An institutional theory view of the family business	Adaptability to external change	Investigates the organizational forces exerted on family business
Lee	2006	Family Business Review	Impact of family relationships on attitudes of the second generation in family business	Internal change	Investigates the influence on family adaptability, on commitment and satisfaction variables
Litz & Kleysen	2001	Family Business Review	Your old men shall dream dreams, your young men shall see visions: Toward a theory of family firm innovation with help from the Brubeck family	Internal change	Illuminates the role of intergenerational innovation building on a single case study
Muske & Fitzgerald	2006	Family Business Review	A panel study of copreneurs in business: Who enters, continues and exits?	Internal change (connections)	Investigates changes in copreneurial business relationships
Naldi, Nordqvist, Sjöberg, & Wiklund	2007	Family Business Review	Entrepreneurial orientation, risk taking, and performance in family firms	Internal change	Risk taking is a distinct dimension of entrepreneurial orientation and positively associated with innovativeness. Family firms are more risk-averse than non-family firms
Parada, Nordqvist, & Gimeno	2010	Family Business Review	Institutionalizing the family business: The role of professional associations in fostering a change of values	Adaptability to external change	Examines institutional pressures exerted by professional associations

Author	Year	Journal	Title	Category	Description
Sund & Smyrnios	2005	Family Business Review	Striving for happiness and its impact on family stability: an exploration of the Aristotelian conception of happiness	Adaptability to external change	Economic changes cause instabilities within nuclear families and increase complexity
Vago	2004	Family Business Review	Integrated change management: Challenges for family business clients and consultants	Adaptability to external change	Describes how change management can help family firms to adapt to changing environments
Zahra	2005	Family Business Review	Entrepreneurial risk taking in family firms	Internal change (innovation)	Empirical evidence of positive association of family ownership/involvement and risk taking
Aldrich & Cliff	2003	Journal of Business Venturing	The pervasive effects of family on entrepreneurship: Toward a family embeddedness perspective	Internal change (venture creation)	Investigates how characteristics of the entrepreneurs' family systems affect the venture creation process
Winter, Danes, Koh, Fredericks, Paul	2004	Journal of Business Venturing	Tracking family businesses and their owners over time: Panel attrition, manager departure and business demise	Internal change (operations)	Investigates continuity and changes in business operations and owner-management
Miller, Steier, Le-Breton-Miller	2003	Journal of Business Venturing	Lost in time: Intergenerational succession, change, and failure in family business	Internal change (succession)	Investigates failed successions in family firms and identifies antecedents.
Dieleman & Sachs	2008	Journal of Management Studies	Coevolution of institutions and corporations in emerging economies: How the Salim group morphed into an institution of Suharto's crony regime	Internal change (affecting environment)	Investigates whether family groups in emerging countries can influence and change their environments

Eddleston, Kellermanns, & Sarathy	2008	Journal of Management Studies	Resource configuration in family firms: Linking resources, strategic planning and technological opportunities to performance	*Not directly related to change*	Apply resource based view lens to show that family firms can benefit from developing technological assets
Tsui-Auch	2004	Journal of Management Studies	The professionally managed family-ruled enterprise: Ethnic Chinese business in Singapore	Adaptability to external change (crisis)	Investigates change and continuity in the management structure of Asian family businesses
Zahra	2010	Journal of Management Studies	Harvesting family firms' organizational social capital: a relational perspective	Adaptability to (discontinuous) external change	Organizational social capital allows family firms to better connect with new ventures which are an important "vanguard of radical change"
Chung & Luo	2008	Organization Science	Human agents, contexts, and institutional change: The decline of family in the leadership of business groups	Internal change (succession)	Focuses on change-agents and provides evidence that second generation family leaders are more likely to reduce family presence
Tsui-Auch & Lee	2003	Organization Studies	The state matters: Management Models of Singaporean Chinese and Korean Business Groups	Internal change (structures)	Investigates changing management structures in family firms of selected Asian countries on their way to professionalization

A-2. Semi-structured Interview Guides

In this part of the appendix, I provide my semi-structured interview guides. The surveys shown here are targeted towards family owners. When interviewing non-family managers or employees, I basically built on the same framework but adapted certain questions. The interview guides were originally created (and used) in German. I translated the guides for the purpose of showing them in this thesis.

The interview guides are rough templates only. Depending on the specific context of the firms, I had to alter, add, or omit questions. Also the order of the questions frequently varied, in particular because, by responding to one question in detail, informants frequently had already answered some of the other questions included in the interview guide.

The interviews started with a short introduction of the interviewers and the request to record the interview. Given the interviewee's agreement on recording (that was the case in all interviews), I explained to my informant how I would use the data. Moreover, I explained the nature of this interview ('semi-structured') to my informants and encouraged them to provide any information that they believed to be important to me, even if it was not explicitly asked by my questions.

i. Study on Family Influence—Organizational Adaptation

Part 1: Basic Information & Business.
Main intention of this part:
1. Collecting basic information on interviewee to set statements in their correct contexts (van Maanen, 1979a)
2. Gathering information on aspects potentially responsible for alternative explanations (Yin, 1994)
3. Understanding relevant structural (Hannan & Freeman, 1984; Levitt & March, 1988), strategic (Bower, 1970; Burgelman, 1983) and cultural (Barney, 1986) characteristics of the company
4. Collecting information on the organization's historical changes (Hatum & Pettigrew, 2004)

Questions:

Could you please give a short overview of your educational/professional background and your position?

Depending on quality of basic information retrieved before the interview: How many employees work for your firm? What are your main products? Are you internationally active? How?

What does the organizational structure of your company look like?

Do rules or (written) codes exist about firm governance?

How does a typical strategic decision-making process look? (*please illustrate using an example*) What formalized elements exist to standardize this process?

How would you describe the culture of your company?

What is the strategic direction your company is following? Has this changed over time?

What is your competitive advantage? Has this changed over time?

What have your company's major changes during the last decades looked like?

Part 2: Family Influence

Main intentions of this part:

1. Assessing the level of family influence among various dimensions (Astrachan et al., 2002; Chua et al., 1999)

2. Understanding the family's vision and financial as well as non-financial goals(Berrone et al., 2010; Distelberg & Sorenson, 2009; Gómez-Mejía et al., 2007; Zellweger et al., forthcoming-a)

Questions:

Would you consider your business a family business? Why?

How many members of the family and how many non-family members are active in top-management positions? What are their positions?

What generation is running the business right now?

How much of the business is owned by the family/by you?

How are successors involved?

How does the family influence decision-making in the firm (e.g., strategic decision, human relations decisions, etc.)? Could you please give some examples?

How frequent are conflicts in those decisions? How do you deal with them? Does ONE sole decision maker exist within the family?

What is the most important goal to be achieved by your business? What is the most important value? Have these goals and values changed over time?

What is your/the family's vision for the firm 10 years from now?

Part 3: Discontinuous Change

Main intentions of this part:

1. Assessing the level of the family firm's embeddedness in its environment (Hatum & Pettigrew, 2004) and involvement of externals (Gilbert, 2005)

2. Understanding the informant's framing of the discontinuity (Gilbert, 2005; Kaplan & Tripsas, 2008) as well as the sense and decision-making process when faced with the discontinuity (Weick, 1995a)

3. Understanding potential barriers to adaptation such as high levels of formalization, resource dependence, political resistance, avoidance of cannibalization and rigid mental models (see Hill & Rothaermel, 2003 for a summary)

4. Gathering information on the organization's adaptation speed, intensity, stamina and routine flexibility (e.g., Block & MacMillan, 1985; Christensen & Bower, 1996; Gilbert, 2005; Tushman & Anderson, 1986)

Questions:

How do you/does your organization get informed about changes in your industry and innovations? What are your main sources of information? Have they changed over time?

How important is gathering information on changes to you?

How structured and formalized is the process of gathering information?

In what (industry) associations are you/other family members active?

How do you involve externals in your decision-making and implementation processes?

What do you perceive as most impactful and important changes in your environment in the recent past?

Next, I explained the discontinuity of interest (e.g., online business, electronic games, etc.) to the informant to ensure a common understanding for the following parts of the interview.

What is your personal attitude regarding the discontinuity[113]? How has this attitude changed over time and why?

Please explain to us when you first noticed that this discontinuity was going to emerge. Please provide information on the context (when? where? by whom?).

What are your employees' attitudes regarding the discontinuity? How have these attitudes changed over time? Are there differences among the departments of your company?

How did your customers react to the discontinuity? Have your customer groups changed due to the discontinuity?

Please describe, in detail, the process of deciding for or against adopting the new technology? Who was involved? What was the main motivation? What were the individual steps? When did that happen? What were your decision criteria? Did any conflicts occur?

How did/does the implementation of the new technology look? Did you create a plan before starting? Who is responsible for the successful launch of the new technology? Please describe, in detail, the steps you and your organization took to adopt the discontinuity.

How much money (*relative numbers sufficient!*) did you invest in the innovation?

How many employees are working on the implementation of the new technology? Have you hired any new employees? If yes: How many and with what background?

In which department do those employees work? Did you change your organizational structure?

How did you manage to 'motivate' your employees for these new tasks (and how difficult was this)? Did you install any new incentive systems? What elements of your organization (ways to work etc.) had to change because of the new technology?

---

[113] In the interviews, the word 'discontinuity' was replaced by the actual innovation. This also applies to any other mention of 'discontinuity' in the course of this interview guide.

How do you plan to or do you generate revenues with the new technology? Are the revenue channels different from that of your established technology?

How do you measure, whether your adoption of the innovation is successful? As of today— do you believe it has been successful so far?

How far are you along the way to successfully implementing the new innovation? How much of your revenues do you generate with the new technology? What is your unique selling point compared to competitors?

In retrospect: Which parts of the adoption were successful? Which were not successful? Why?

Part 4: Concluding Questions

General intentions of this part:

1. Stimulating further discussion on family related factors driving the organization's adaptation (n/a)

Questions:

Do you think that you/your organization would have behaved differently if you were not a family business?

Open-ended discussion

ii.   Study on Organizational Identity—Organizational Adaptation

Part 1: Business and Family Influence

Main intention of this part:

1. Collecting basic information on interviewee to set statements in their correct contexts (van Maanen, 1979b)

2. Gathering information on aspects potentially responsible for alternative explanations (Yin, 1994)

3. Collecting information on the organization's historical changes (Hatum & Pettigrew, 2004)

4. Assessing the influence of the owner-family on the business (Astrachan et al., 2002)

Questions:

Could you please give a short overview of your educational/professional background and your position?

Depending on quality of basic information retrieved before the interview: How many employees work for your firm? What are your main products? Are you internationally active? How?

What have your company's major changes during the last decades looked like?

How many members of the family and how many non-family members are active in top-management positions? What are their positions?

How much of the business is owned by the family/by you?

How would you describe the culture of your company? How is this culture influenced by the owner-family?

How does the family influence decision-making in the firm (e.g., strategic decision, human relations decisions, etc.)? Could you please give some examples?

Part 2: Identity

Main intention of this part:

1. Assessing how the organization perceives itself—'who are we as an organization?' (Gioia, Schultz, & Corley, 2000; Pratt & Foreman, 2000)

2. Gathering information on the central, distinctive, and enduring elements of the company (Albert & Whetten, 1985; Gioia & Thomas, 1996))

3. Collecting information on how the environment is perceived by the organization (Kaplan, 2011; Porac et al., 1989)

4. Assessing the identification and external image of the company, as perceived by the interviewee (Cornelissen et al., 2007)

5. Assessing the strength of identity (Martins, 2005)

6. Assessing the owner-family's influence on organizational identity (Dyer & Whetten, 2006; Gómez-Mejía et al., 2007; Zellweger et al., 2010)

<u>Questions:</u>

Imagine I was a stranger meeting you somewhere on the street—How would you describe your company to me in two or three sentences?

What is the most important goal of your business? What is the most important value? Have these goals and values changed over time?

What is your/the family's vision for the business? Has this changed over time?

What is your company's core product? Has this changed over time and if yes, how and why?

What do you think: What will be your core products in 5, 10 or 15 years?

Who are your competitors? Who were your competitors in the past?

What is your unique selling point? What are your company's strengths? Has this changed over time?

How do family members influence and shape the company's identity?

How strongly do you identify yourself with your business?

What do you think: How strongly do your employees identify themselves with the business? Why?

How do your customers see your business?

Part 3: Discontinuous Change

Main intentions of this part:

1. Assessing the level of the family firm's embeddedness in its environment (Hatum & Pettigrew, 2004) and involvement of externals (Gilbert, 2005)

2. Understanding the informant's framing of the discontinuity (Gilbert, 2005; Kaplan, 2008b) as well as the sense and decision-making process when faced with the discontinuity (Weick, 1995a)

3. Understanding potential barriers to adaptation such as high levels of formalization, resource dependence, political resistance, avoidance of cannibalization and rigid mental models (see Hill & Rothaermel, 2003 for a summary)

4. Gathering information on the organization's adaptation speed, intensity, stamina and routine flexibility (e.g., Block & MacMillan, 1985; Christensen & Bower, 1996; Gilbert, 2005; Tushman & Anderson, 1986)

Questions:

How do you/does your organization get informed about changes in your industry and innovations? What are your main sources of information? Have they changed over time?

How important is gathering information on changes to you?

How structured and formalized is the process of gathering information?

In what (industry) associations are you/other family members active?

How do you involve externals in your decision-making and implementation processes?

What do you perceive to be the most impactful and important changes in your environment in the recent past?

Next, I explained the discontinuity (digitization.) to the informant to ensure a common understanding for the subsequent parts of the interview.

What is your personal attitude regarding the discontinuity[114]? How has this attitude changed over time?

Please explain to us, when you first noticed that this discontinuity was going to emerge. Please provide information on the context of becoming aware of the new technology (When? Where? By whom?)

Which are your employees' attitudes regarding the technology? How have these attitudes changed over time? Are there differences among the departments of your company?

What was the role of stakeholders such as librarian customers, end customers, and authors in the adaptation process?

---

[114] In the interviews, the word 'discontinuity' was replaced by the actual innovation (digitization). This also applies to any other mention of 'discontinuity' in the course of this interview guide.

290

Please describe, in detail, the process of deciding for or against adopting the technology? Who was involved? What were the individual steps? When did they happen? What were your decision criteria? Did any conflicts occur?

What was your final decision regarding the adoption of the innovation, and if it was 'yes,' how?

How does the implementation of the new technology look? Did you create a plan before starting? Who is responsible for the successful launch of the new technology? Please describe, in detail, the steps you and your organization took to adopt the new technology.

How much money (*relative values sufficient!*) did you invest in the innovation?

How many employees are working on the implementation of the new technology? Have you hired any new employees? If yes: How many and with what background?

In which department do those employees work? Did you change your organizational structure?

How did you manage to 'motivate' your employees for these new tasks (and how difficult was this)? Did you install any new incentive systems? What elements of your organization (ways to work etc.) had to change because of the new technology?

How do you/do you plan to generate revenues with the discontinuity? Do the revenue channels differ from your established technology?

How do you measure whether your adoption is successful? As of today—do you believe it has been successful so far?

How far along are you in successfully implementing the new technology? How much of your revenue do you generate with the new technology? What is your unique selling point when compared to competitors?

In retrospect: Which parts of the adoption were successful? Which were not successful? Why?

Part 4: Concluding Questions

General intentions of this part:

1. Stimulating further discussion on family related factors that drive the organization's adaptation (n/a)

Questions:
Do you think you/your organization would have behaved differently if you were not a family business?
Open-ended discussion

A-3. Coding Guidelines

In this subsection of the Appendix, I will outline the final coding guidelines[115] used for data analysis throughout the studies. I will present the theoretical constructs and definitions the coding is based on, the different items I used to categorize and cluster the cases, as well as descriptions of the individual items. Moreover, I will illustrate each item by providing exemplary statements made by my informants, which I coded into the respective categories.

For each construct, I indicate whether this coding was relevant for the study examining the direct relationship between family influence and organizational adaptation (Chapter 4; noted as study 1), the relationship between organizational identity, family influence and organizational adaptation (Chapter 5, noted as study 2), or both.

---

[115]Selection of the items was informed by extant literature as well as free nodes that emerged throughout the coding process.

Family Influence (used for both empirical studies)			
Definition and Literature	**Item**	**Description**	**Exemplary Quotes** (from my interviews)
Overlap of the family and the business system (Habbershon & Williams, 1999), manifesting itself in the power the owner-family exerts on the businesses' structure, strategy, and culture (Astrachan et al., 2002)	High level of family influence	High levels of family influence refer to (Miller & Le Breton-Miller, 2005b) • High levels of *continuity*, i.e. much experience accumulated over generations, successors involved in business • High levels of *command*, i.e. family members as key decision makers, large fraction of the firm owned by family • High levels of *community*, i.e. family shapes culture of its business, employees feel attached to business • High levels of *connections*, i.e. close ties to selected external stakeholders An interview statement is coded as high level of family influence if it refers to one of the above-mentioned aspects.	We invite employees celebrating a jubilee to our home and [we cook] for them. (CEO, *White & Blue*)
	Medium level of family influence	*In between 'High' and 'Low.' Interview statements referring to family influence indicating neither an extremely high nor an extremely low level of family influence.*	[CEO as ultimate decision maker striving for] joint decisions [in a] cooperative manner. (CEO, *Retail 2000*)

Low level of family influence	High levels of family influence refer to (Miller & Le Breton-Miller, 2005b) • Low levels of *continuity*, e.g., founder-owned firms with potential family-internal successors not intending to take over the business • Low levels of *command*, i.e. family members do not influence strategic decision-making • Low levels of *community*, i.e. culture not shaped by the owner-family • Low levels of *connections*, i.e. connections to stakeholders assessed on a purely rational basis An interview statement is coded as high level of family influence if it refers to one of the above-mentioned aspects.	*Not available as all sampled firms had medium to high family influence on each dimension.* <u>Example outside my sample:</u> Firm culture of *Siemens* not at all shaped by family that owns 6% of shares* (interviews with Siemens employees). * Scholars typically perceive a 5% hurdle to count a firm as a family firm.

Dominant SEW Dimension (used for study 1)			
Definition and Literature	**Item**	**Description**	**Exemplary Quotes** (from my interviews)
Socioemotional wealth (SEW) denotes an umbrella term encompassing all non-financial elements of a family firm's utility function (Gómez-Mejía et al., 2007). The dominant SEW dimension refers to that part of SEW that is most important to preserve for owner-family.	Familial Ties	Refers to family firms that are most concerned about being altruistic to their family members, for instance by appointing them to honorable positions within the firm or striving to pass on the business in the best possible shape (Gómez-Mejía et al., 2007)	[Familial relationships] hold the value [within the firm]. (CEO, *King's Goods*) I could have continued for ten years. [...] But then, my children [...] would say: [...] 'Why didn't you enter the market?' And this was in principle the motive. (CEO, *King's Goods*)
	Power and Influence of the Family	Refers to family firms that have gained SEW through their current and previous control over the firm and whose main goal is to maintain this level of control for future generations (Zellweger et al., forthcoming-b). Control can thereby refer to percentage of family ownership as well as maintaining board and management positions.	We want to remain a family business. We do not want to give the firm into the hands of strangers. (CEO, *Walter & Colleagues*) A loss of control [over the firm] is out of the question. (CEO, *King's Goods*)
	Status and Reputation of the Family	Refers to family firms that are mainly concerned about the image of their business, as the image of the company frequently redounds to the image of the family (Berrone et al., 2010). Image can, for instance, relate to characteristics of the product (e.g., innovative, high quality), the treatment of stakeholders (e.g., customers, employees) or other firm characteristics (e.g., local roots, sustainable way of production)	Providing 'Heimat' [to employees and customers is our main vision.] (CEO, *White & Blue*) We are dedicated to particular values, among them good relationships to suppliers and employees and a deep local rootedness.(website, *Retail 2000*) External example: After the bankruptcy of bakery *Müller*, the founder's daughter strived to redeem the business from the non-family owner in order to re-establish the former good reputation.

| Affects and Emotional Attachment to the Family Firm | Refers to family firms that are most concerned about the positive emotions related to owning and working for their business (Gómez-Mejía et al., 2007). Respective interview statements relate to family owners' emotional attachment to the firm or the 'joy' of being a member of a family firm. | Enjoy[ment] of working [is a key aspect of the company] (CEO, *Kiddies*). My grandfather manufactured [a special kind of] toys, my father manufactured [a special kind of] toys, and I also manufacture [a special kind of] toys. (CEO, *Kiddies*) |

Determinants of Adaptation (used for study 1)

Note: This final list of determinants of adaptation is based on factors identified by prior research (items 1-7) and complemented by relevant drivers that emerged during the interviews (items 8-12). When coding the interview transcripts using the computer program NVivo, I coded each statement either as 'low,' 'medium,' or 'high' value of the respective determinant. I omitted this differentiation into the categories 'low,' 'medium,' or 'high' here for the sake of space.

Definition and Literature	Item	Description	Exemplary Quotes (from my interviews)
Strategic, organizational, economic and emotional-cognitive factors that either enable or hamper organizational adaptation to discontinuous change (Hill & Rothaermel, 2003; Tripsas, 2009).	Formalization	The degree to which processes and structures are formalized and standardized; the level of bureaucracy (Hannan & Freeman, 1984; Levitt & March, 1988). Refers mainly to search and decision-making processes. Written documents are coded as higher level of formalization than verbal agreements.	Low level: Many things are not codified, not written down here. (CEO, *Alpha Star*) Nothing is written down, everything is in the heads of the individuals. (CEO, *Kiddies*)
	Resource Dependence	The degree to which externals such as capital providers and/or customers influence the adaptation process (Christensen, 1997). Refers to the influence on external stakeholders when making decisions. Statements referring to resource scarcity due to the fact that firm owners were reluctant to become resource dependent on externals were coded as a separate sub-item.	Low level (capital providers): We can react without being considerate of shareholders, the stock exchange or things like that. (CEO, *Retail 2000*) High level (customers): We would have to tell our customers that we are unable to deliver. (CEO, *Walter & Colleagues*) Avoidance of dependence on capital providers: We did not have the capacity, the financial resources to take part. (CEO, *Play & More*)

Political Resistance	The degree to which employees resist adaptation due to self-interests, related to personal levels of power (Pfeffer, 1992). Refers to the behavior of managers and employees after the decision to adapt had been taken (early implementation phase).	I once experienced being dependent on banks. I do not want to experience such a situation ever again. (CEO, *King's Goods*) Low level: The employees back the company as one man. (CEO, *Alpha Star*) When [the CEO] averred, 'We'll do it that way,' then we did it that way. There was no time-consuming questioning. (manager, *Walter & Colleagues*)
Avoidance of Cannibalization	The degree to which managers resist the adaptation due to their fear of salvaging their core products (Chandy & Tellis, 1998). Differentiates between whether decision makers perceive technology as a substitute or a complement; and in the case of the former if they fear cannibalization.	Framing as complementary product: Our physical stores will continue to exist for a long time. (CEO, *King's Goods*) E-commerce will not repress stationary retail. (CEO, *Anything & More*) Low level (of avoidance of cannibalization): [Our core business] will remain [besides e-commerce] as long as it is useful for us. (Director, *Anything & More*) Low level (of avoidance of cannibalization): *Not observed*

		High level:
Rigid Mental Models	The degree to which managers and employees resist adaptation due to narrow cognitive frames or emotional attachment[116] to the established product (Kaplan, 2008b; Tripsas & Gavetti, 2000). Differentiates between mental models of decision makers and staff. Differentiates between mental models impacting interpretation/decision-making and those affecting implementation. Refers to emotional attachment to the old technology as well as framing of the technology in a narrow manner.	'We have always done it like that' is the philosophy of most of our employees.(CEO, *Play & More*) Electronics is not communicative. (CEO, *Play & More*) [Employees] regret time gone by. (CEO, *White & Blue*) Some people that sit at my table still do not want to accept that this is part of future. (CEO, *King's Goods*)
External Influence and Scanning of the Environment	The degree to which the firm involves externals in the interpretation and adaptation process (Gilbert, 2005). This included attending industry fairs, engagement in industry fairs and also 'other' (external) sources of information, hiring external managers, and involving external consultants. Each of these aspects was coded in a separate NVivo node.	Low level: Tackling difficult tasks [is] a core task of managers [not of externals].(CEO, *Retail 2000*) What can consultants tell me? Most of them have failed in running their own business and then become a consultant. (CEO, *Play & More*) Medium level: We work with external consultants on rare occasions (regional CEO, *Powerhouse*) High level:

[116] In this categorization I slightly deviate from the conceptual part of this thesis where I attributed 'emotional attachment' to the determinant that describes avoidance of cannibalization. The rationale for this deviation is that in interviews it is hard if not impossible to discover whether narrow frames are rooted in emotional or cognitive factors (for instance: Are the individuals referred to in the fourth quote unable or unwilling "to accept that this is part of the future"?)

		It is important to have an ear to the ground (CEO, *Alpha Star*) Discussion partner, sparring partners [such as one could meet in associations and unions] are absolutely necessary. (CEO, *Toys 2000*) In the recent past, we've intensified working with external consultants (Manager, *Anything & More*)
Organizational Ambidexterity	The degree to which a firm decouples organizational units to allow for undisturbed exploration of the new technology (Gilbert, 2005). Refers to whether the new technology is developed and commercialized by an independent business unit or integrated in existing units.	Low level: We did not have the time to create a new business unit.(CEO, *Alpha Star*) High level: We decided to deliberately set up a new group 'new media' [...] that reports directly to the CEO.(CEO, *King's Goods*)
Intuitive Decision-making	The degree to which decision makers rely on 'gut feeling' rather than rationale or calculative decisions (Kahneman, 2003). Differentiates between two different ways in which decision makers come to conclusions and how they justify them.	Intuitive way: It was gut feeling (CEO, *King's Goods*) The senior CEO has a gut feeling and says: 'Yes [...] The number guys [conclude]: 'no'. But then, a year later, everyone is happy about the CEO's decisions." (manager, *Walter & Colleagues*) Rational way: Decisions are taken based on factual, objective arguments. (CEO, *Retail 2000*)

Change History	The degree to which a firm experienced changes in the past (Hatum & Pettigrew, 2004). Changes can refer to the firm's geography, customer groups, or product portfolios. Major internal changes are also considered. When coding this item, the elapsed time since the change as well as the severity of changes are considered	Low level: There was no single most impactful change [in the firm's history]. The most impactful changes were always related to private life. I do not believe in revolutionary approaches, they have not occurred so far. It's always old wine in a new skin. (CEO, *White & Blue*) Medium level: During my childhood there was a major turmoil in the business. [...] And then we decided to enter the new business. (CEO, *Retail 2000*) High level: *Not observed. Would relate to family firms such as Tata Group that have significantly changed their product portfolio through the company's history.*
Dealing with Failures	Refers to the way firms deal with temporary failures (Staw et al., 1981). Interview statements were coded as constructive way (continuous reinvestment despite failure), resignation (abandoning activities after failure) or destructive way (punishment of failure).	Constructive way: Every day we ask: 'Have you already made a mistake today?' If the answer is no, we say: 'Go, take more risks.' (Industry expert and product manager of international textile manufacturer) Resignation: It is, what it is. (CEO, *Kiddies*)

		Destructive way:
		Not observed.
Long-term Planning Horizon	Refers to the future time frame considered for strategic planning (Das, 1991). Coding refers to the point in time when companies expect discontinuity to exploit its full potential as well as to grace periods granted to every project before first success is expected.	Long-term planning: Planning horizons for online business are 40 years. (CEO, *King's Goods*) The [increased interest of customers] will not be measurable within two or three years. I believe one has to look thirty years ahead. (CEO, *Walter & Colleagues*) Short-term planning: *Not observed*
Risk aversion	Degree of financial risk, the organization is willing to take (Gómez-Mejia et al., 2007; Morck & Yeung, 2003). Statements were coded as either 'risk averse' or 'risk taking'.	Risk aversion: We refrain from risky invests? (CEO, *Walter & Colleagues*) Willingness to take risk: [This] was a risky step of a company of our size. (CEO, *Toys 2000*)

Organizational Identity (used for study 2)

Definition and Literature	Item	Description	Exemplary Quotes (from my interviews)
Organizational identity refers to the 'Who are we as an organization' and depicts the central, distinctive, and enduring elements of an organization (Albert & Whetten, 1985)	Focus	See Chapter 5.5.1 (Due to the novelty of the dimensions 'focus' and 'locus' and the character of their emergence (rather empirical data-than literature-based) they are introduced in the main body of this thesis.)	
	Locus		

Type of Adaptation to Discontinuous Change (used for both studies)

Definition and Literature	Item	Description	Examples (synthesized from my interviews and information provided by historical websites and secondary sources)
Type of adaptation denotes the way in which organizations ultimately do or do not react to discontinuous technological changes. Such adaptation can either be active (domain creation, domain offense, domain consolidation, domain defense) or passive. Active adaptation patterns include adoption and non-adoption of the new technology (Ford & Baucus, 1987; Zammuto & Cameron, 1985)

The term 'domain' is thereby rooted in population ecology (Hannan & Freeman, 1977), referring to an n-dimensional volume in time and space whose borders are defined through the organizations' resource constraints, capabilities, and environmental pressures. | Domain Creation | Refers to adoption of the new technology (in addition to offering products or services of the established domain).

Activities were coded as domain creation if they included radically new product features and/or generated revenue through radically new channels (at least criterion to be fulfilled) | Offering of online based retailing (as opposed to stationary, e.g., *King's Goods, Anything & More*), electronic toys (as opposed to classical toys, *Toys 2000*), database access on smart phones (as opposed to print books, *Yellow Books*), |

Domain Offense	Refers to activities that aim to increase the volume of the organization's current domain. Activities were coded as domain offense if they related to incremental rather than radical product innovations or if the company tried to grow by expanding its market geographically[117] or enhancing customer loyalty.	Strengthening of stationary retail by expanding to other geographical markets (*Retail 2000, King's Goods*) Strengthening of stationary retail by enhancing customer loyalty through retention programs (*White & Blue*) Offering print books on new topics or written by new author (*Ars Legendi*)
Other Active Reactions	Includes Domain Consolidation/Abandonment denoting retrenchment into a niche or exit of a domain respectively; Domain Defense referring to activities aiming to maintain the legitimacy of the established domain; Domain Substitution, that is the exchange of one domain by another. Activities were coded as other active domain reaction if the decision makers were aware of the change and deliberately took actions to face the challenge, but activities were neither domain creation nor offense.	Domain Consolidation: Focusing on specialized products for a special segment of the customer groups, 'not serving the [threatened] markets anymore' (CEO, *Kiddies*) Domain Defense: You cannot digitize everything. [...] It's such a hard and cold device. (*Reader's Finest*)

---

[117] Ford and Baucus (1987) categorized geographical expansion as a domain creation activity. However, in the context of studying adaptation to discontinuous change, this is not adequate as geographical expansion predominantly refers to 'doing the same thing at a different location' and thus differs fundamentally from technology adoption, categorized as domain creation.

	Passive Reaction	Refers to non-reaction of organizations, based on denial, anger, or resignation. Activities were coded as passive reaction if the organization had not changed its activities after becoming aware of the new technology.	Resignation: I think we should continue working on [our core segments]. If we are not able to do so, we have to close our business (CEO, *Play & More*) Denial Although I see the trend towards e-books, I do not think that our market niche will be affected (CEO, *Star Print*)

Characteristics of Adaptation to Discontinuous Change (Speed used for both studies, remaining items used for study 1)

Definition and Literature	Item	Description	Examples (synthesized from my interviews and information provided by historical websites and secondary sources)
An adequate resource investment pattern—that is appropriate regarding timing and value of investment—as well as routine flexibility are two relevant determinants of adaptation performance (Gilbert, 2005)	Speed	Denotes the time that has elapsed since the emergence of the discontinuous technology until the product is commercialized by the firm (e.g., Tushman & Anderson, 1986). Differentiation between speed of awareness, interpretation (focus: cognitive), decision-making (focus: structural), and implementation. Interview statements were coded according to their speed compared with competitors and triangulated with absolute numbers.	Fast speed: Decisions are taken really, really fast. That is three, maybe four phone calls with the management and advisory board and then we'll do it (Industry expert and family CEO of an international family firm with 10 billion EUR annual sales) One of our competitive advantages is speed. Once we have an idea or a decision, we are extremely good at bringing it to the market quickly. That is what we are good at: implementation—very pragmatically, down-to-earth. (CEO, *Alpha Star*)

		Slow speed: For breakthrough innovations that are completely new, we take our time. We do not need to stress [...] Based on our experience, we take time for extensive feedback and delay the decision. (Industry expert and family CEO of an international family firm with roughly 3 billion EUR annual sales)
Intensity	Denotes to the amount of financial and non-financial resources committed to the new technology (Christensen & Bower, 1996). Codings refer to acquisition of firms, hiring new employees, investment in training of established employees, total number of employees working on the new technology, and purchasing of tools and patents	Low intensity: We started with a very cautious resource investment (CEO, *Alpha Star*) Medium intensity: All employees of the [small] new department dealing with the new technology were externally hired. We bought numerous technical tools (CEO, *Toys 2000*) It was an adequate investment. A medium investment (regional CEO, *Powerhouse*)
Stamina	Denotes the degree of continued investment in the technological discontinuity ((Block & MacMillan, 1985) Codings refer to whether companies continued their investment after temporary failures occurred as well as intentions (for industries that have not yet experienced (temporary) setbacks regarding the discontinuity)	High level: [The first shop was a] shadowy existence [for an extended period of time but was kept due to a] dumb feeling that something is going on and we have to stay tuned to it (CEO, *King's Goods*) Low level: *Not observed*

| Routine flexibility | Refers to the degree to which established processes and structures within the organization are replaced by new, non-paradigmatic ones (Gilbert, 2005).

Routine flexibility was measured by assessing the internal structural changes (e.g., referring to organizational setup), the product related changes (e.g., referring to product features, performance criteria), and business models changes (e.g., price structures, revenue channels) | Low level:

For a long time, we were far too cautious, not revolutionary enough. We ran the business from and 'old-world' perspective, as an 'add-on' to stationary retail. We were not sufficiently embedded into the online world (CEO of *King's Goods* describing the first, unsuccessful attempts to enter the e-commerce market)

High level:

[*Anything & More* started to offer 'native' e-commerce products in 2001] |

A-4. Coding Dictionaries

The following dictionaries, used for coding the focus and locus of organizational identities were originally created in German, and then translated into English for the purpose of this thesis.

i. Focus of Identity

Inclusive Focus	Exclusive Focus
Self-descriptions are mainly based on the following terms:	Self-descriptions are mainly based on the following terms:
(Please note: all words are merely indicators of the organization's identity. Singled out and separated from the context, none of these words constitutes a clear predictor of identity.)	*(Please note: all words are merely indicators of the organization's identity. Singled out and separated from the context, none of these words constitutes a clear predictor of identity.)*
• Communicator	• Producer of books
• Mediator	• Seller of books
• Integrative / Integrated	• Printed books
• Whole spectrum	• Printed magazines
• Holistic	• Publishing House
• Broad	• Publisher
• From … to …	• Limits
• Comprehensive	• Borders
• Provider of services	• Narrow
• Provider of solutions	• Focused
• Knowledge	• Out of our core business
• Free to do something else	• Uncompromising

• No limits	• Differences
• Compromising	• Paper
• Similarities	• Printing
• Service packages	• Cobbler stick to your trade
• We see companies of various industries as competitors	• Boundaries
• Across (industry) borders	• Our main competitors are [names of established publishing houses]
• *Any pairs of opposites, such as 'innovation and tradition'*	

ii. Locus of Identity

Self-Centric	Environment-Centric
Self-descriptions are mainly based on the following terms:	Self-descriptions are mainly based on the following terms:
(Please note: all words are merely indicators of the organization's identity. Singled out and separated from the context, none of these words constitutes a clear predictor of identity.)	*(Please note: all words are merely indicators of the organization's identity. Singled out and separated from the context, none of these words constitutes a clear predictor of identity.)*
• To shape	• To be coerced
• Impact	• To be forced
• To act	• Customers demand
• Creativity	• To react
• Employees	• To follow
• Organizational values	• Market dynamics
• R&D efforts	• Consultants told us
• Marketing campaign	• Market research

• To match	• Strong sales force
• To suit	• To listen
• To drive	• To observe
• To tailor	• To be driven
• Our values	• To adopt
• Our norms	• Utilitarian
• Not a self-selling product	• Pull from the customers
• We do not care what others say	• Market/ environment dictates terms
• We have to drum for the product	• External(ly)
• There's no pull from the customers	
• We have to push the product	
• We want to offer value	
• Enjoyment, delight, fun (of employees)	
• Internal(ly)	

B. References

Abernathy, W. J., & Clark, K. B. 1985. Innovation: Mapping the winds of creative destruction. *Research Policy*, 14(1): 3–22.

Abrahamson, E., & Fombrun, C. J. 1994. Macrocultures: Determinants and consequences. *Academy of Management Review*, 19(4): 728–755.

Adner, R., & Kapoor, R. 2010. Value creation in innovation ecosystems: How the structure of technological interdependence affects firm performance in new technology generations. *Strategic Management Journal*, 31(3): 306–333.

Adner, R., & Snow, D. 2010. Old technology responses to new technology threats: Demand heterogeneity and technology retreats. *Industrial and Corporate Change*, 19(5): 1655–1675.

Ahuja, G., Lampert, C. M., & Tandon, V. 2008. Moving beyond Schumpeter: Management research on the determinants of technological innovation. *Academy of Management Annals*, 2(1): 1-98.

Albert, S., & Whetten, D. A. 1985. Organizational identity. *Research in Organizational Behavior*, 7(1): 263–296.

Aldrich, H. E., & Cliff, J. E. 2003. The pervasive effects of family on entrepreneurship: Toward a family embeddedness perspective. *Journal of Business Venturing*, 18(5): 573–596.

Allen, T. J. 1977. *Managing the flow of technology*. Cambridge, MA: MIT Press.

Allio, M. K. 2004. Family businesses: Their virtues, vices, and strategic path. *Strategy & Leadership*, 32(4): 24–33.

Allouche, J., & Amann, B. 1997. Le retour triomphant du capitalisme familial. *L'expansion: Management Review*, 85: 92–99.

Anderson, P., & Tushman, M. L. 1990. Technological discontinuities and dominant designs: A cyclical model of technological change. *Administrative Science Quarterly*, 35: 604–633.

Anderson, R. C., & Reeb, D. M. 2003. Founding-family ownership and firm performance: Evidence form the S&P 500. *The Journal of Finance*, 58(3): 1301–1328.

Argote, L. 1999. *Organizational learning: Creating, retaining, and transferring knowledge*. New York: Springer.

Aronoff, C. E. 1998. Megatrends in family business. *Family Business Review*, 11(3): 181–185.

Arregle, J.-L., Hitt, M. A., Sirmon, D. G., & Very, P. 2007. The development of organizational social capital: Attributes of family firms. *Journal of Management Studies*, 44(1): 73–95.

Arrow, K. J. 1974. *The limits of organization* (1st ed.). New York: Norton.

Astrachan, J. H. 2010. Strategy in family business: Toward a multidimensional research agenda. *Journal of Family Business Strategy*, 1(1): 6–14.

Astrachan, J. H., Klein, S. B., & Smyrnios, K. X. 2002. The F-PEC scale of family influence: a proposal for solving the family business definition problem. *Family Business Review*, 15(1): 45–58.

312

Athanassiou, N., & Nigh, D. 1999. The impact of U.S. company internationalization on top management team advice networks. A tacit. *Strategic Management Journal*, 20(1): 83–92.

Bantel, K. A., & Jackson, S. 1989. Top management and innovations in banking: Does the composition of the top team make a difference? *Strategic Management Journal*, 10: 107–124.

Barker Iii, V. L., & Mueller, G. C. 2002. CEO characteristics and firm R&D spending. *Management Science*, 48: 782–801.

Barley, S. R. 1990. The alignment of technology and structure through roles and networks. *Administrative Science Quarterly*, 35: 61–103.

Barney, J. B. 1986. Organizational culture: Can it be a source of sustained competitive advantage? *Academy of Management Review*, 11(3): 656–665.

Barney, J. B., Bunderson, J. S., Foreman, P., Gustafson, L. T., Huff, A. S., Martins, L. L., Reger, R. K., Sarason, Y., & Stimpert, J. L. 1998. A strategy conversation on the topic of organization identity. In D. A. Whetten, & P. Godfrey (Eds.), *Identity in organizations: Building theory through conversations*. Thousands Oaks, CA: Sage.

Barr, P. S., & Glynn, M. A. 2004. Cultural variations in strategic issue interpretation: Relating cultural uncertainty avoidance to controllability in discriminating threat and opportunity. *Strategic Management Journal*, 25(1): 59–67.

Barr, P. S., Stimpert, J. L., & Huff, A. S. 1992. Cognitive change, strategic action, and organizational renewal. *Strategic Management Journal*, 13(SI): 15–36.

Basly, S. 2007. The internationalization of family SME: An organizational learning and knowledge development perspective. *Baltic Journal of Management*, 2(2): 154–180.

Bechhofer, F., Elliott, B., & McCrone, D. 1984. Safety in numbers: On the use of multiple interviewers. *Sociology*, 18(1): 97–100.

Beck, L., Janssens, W., Debruyne, M., & Lommelen, T. 2011. A study of the relationships between generation, market orientation, and innovation in family firms. *Family Business Review*, 24(3): 252–272.

Becker, W., & Ulrich, P. 2011. Internationalisierung mittelständischer Familienunternehmen – Gründe, Erscheinungsformen, Fallstudien. In F. Keuper, & H. A. Schunk (Eds.), *Internationalisierung deutscher Unternehmen*. Wiesbaden: Gabler Verlag.

Beehr, T. A., Drexler, J. A., & Faulkner, S. 1997. Working in small family businesses: Empirical comparisons to non-family businesses. *Journal of Organizational Behavior*, 18: 297–312.

Bennedsen, M., Nielsen, K. M., Perez-Gonzalez, F., & Wolfenzon, D. 2007. Inside the family firm: The role of families in succession decisions and performance. *Quarterly Journal of Economics*, 122(2): 647–691.

Benner, M. J. 2007. The incumbent discount: Stock market categories and response to radical technological change. *Academy of Management Review*, 32(3): 703–720.

Benner, M. J. 2008. *Selective myopia: A comparison of analysts' reactions when firms respond to technological change*. Paper presented at the annual meeting of the Academy of Management, Anaheim, CA.

Benner, M. J. 2010. Securities analysts and incumbent response to radical technological change: Evidence from digital photography and internet telephony. *Organization Science*, 21(1): 42–62.

Berrone, P., Cruz, C., & Gómez-Mejía, L. R. forthcoming. Socioemotional wealth in family firms: Theoretical dimensions, assessment approaches, and agenda for future research. *Family Business Review*.

Berrone, P., Cruz, C., Gómez-Mejía, L. R., & Larraza-Kintana, M. 2010. Socioemotional wealth and corporate responses to institutional pressures: Do family-controlled firms pollute less? *Administrative Science Quarterly*, 55(1): 82–113.

Bertrand, M., & Schoar, A. 2006. The role of family in family firms. *Journal of Economic Perspectives*, 20(2): 73–96.

Biggart, N. W. 1977. The creative-destructive process of organizational change: The case of the post office. *Administrative Science Quarterly*, 22(3): 410–426.

Björnberg, Å., & Nicholson, N. 2007. The family climate scales—development of a new measure for use in family business research. *Family Business Review*, 20(3): 229–246.

Blanco-Mazagatos, V., De Quevedo-Puente, E., & Castrillo, L. A. 2007. The trade-off between financial resources and agency costs in the family business: An exploratory study. *Family Business Review*, 20(3): 199–213.

Block, Z., & MacMillan, I. C. 1985. Milestones for successful venture planning. *Harvard Business Review*, 63(5): 184–190.

Bloom, N., & van Reenen, J. 2007. Measuring and explaining management practices across firms and countries. *Quarterly Journal of Economics*, 122(4): 1351–1408.

Bockmühl, S., König, A., Enders, A., Hungenberg, H., & Puck, J. 2011. Intensity, timeliness, and success of incumbent response to technological discontinuities: a synthesis and empirical investigation. *Review of Managerial Science*, 5(4): 265–289.

Bogdan, R., & Biklen, S. K. 1982. *Qualitative research for education*. Boston, MA: Allyn and Bacon.

Bower, J. L. 1970. *Managing the resource allocation process*. Boston, MA: Harvard Univ. Division of Research.

Bower, J. L. 1986. *Managing the resource allocation process*. Boston, MA: Harvard Business School Press.

Bower, J. L., & Christensen, C. M. 1995. Disruptive technologies: Catching the wave. *Harvard Business Review*, 73(1): 43–53.

Bowman, C., & Ambrosini, V. 2000. Value creation versus value capture: Towards a coherent definition of value in strategy. *British Journal of Management*, 11(1): 1–15.

Bowman, C., & Ambrosini, V. 2003. How the resource-based and the dynamic capability views of the firm inform corporate-level strategy. *British Journal of Management*, 14(4): 289–303.

Bowman, E. H., & Hurry, D. 1993. Strategy through the option lens: An integrated view of resource investments and the incremental-choice process. *Academy of Management Review*, 18(4): 760–782.

Brannick, T., & Coghlan, D. 2007. In defense of being "native": The case for insider academic research. *Organizational Research Methods*, 10(1): 59–74.

314

Braun, M., & Sharma, A. 2007. Should the CEO also be chair of the board? An empirical examination of family-controlled public firms. *Family Business Review*, 20(2): 111–126.

Brockner, J. 1992. The escalation of commitment to a failing course of action: Toward theoretical progress. *Academy of Management Review*, 17(1): 39–61.

Brockner, J., Spreitzer, G., Mishra, A., Hochwarter, W., Pepper, L., & Weinberg, J. 2004. Perceived control as an antidote to the negative effects of layoffs on survivors' organizational commitment and job performance. *Administrative Science Quarterly*, 49(1): 76–100.

Burgelman, R. A. 1983. A model of the interaction of strategic behavior, corporate context, and the concept of strategy. *Academy of Management Review*, 8(1): 61–70.

Burgelman, R. A. 1994. Fading memories: A process theory of strategic business exit in dynamic environments. *Administrative Science Quarterly*, 39: 24–56.

Bushee, B. J. 2001. Do institutional investors prefer near-term earnings over long-run value? *Contemporary Accounting Research*, 18(2): 207–246.

BVS, & eurotoys. 2010. *Marktdaten*. Köln: Bundesverband des Spielwaren-Einzelhandels. http://www.toy.de/news/rubrik6.html. Accessed on March 27, 2011.

Cabrera-Suarez, K., Saa-Perez, P., & Garcia-Almeida, D. 2001. The succession process from a resource- and knowledge-based view of the family firm. *Family Business Review*, 14(1): 37–48.

Cannella, A. A., Park, J. H., & Lee, H. U. 2008. Top management team functional background diversity and firm performance: Examining the roles of team member colocation and environmental uncertainty. *Academy of Management Journal*, 51(4): 768–784.

Carlock, R. S., & Ward, J. L. 2001. *Strategic planning for the family business*. Houndmills, Basingstoke, Hampshire, NY: Palgrave.

Carney, M. 2005. Corporate governance and competitive advantage in family-controlled firms. *Entrepreneurship Theory and Practice*, 29(3): 249–265.

Cater, J., & Schwab, A. 2008. Turnaround strategies in established small family firms. *Family Business Review*, 21(1): 31–50.

Chandler, A. D. 1994. *Scale and scope* (3rd ed.). Cambridge, MA: Belknap Press of Harvard University Press.

Chandy, R. K., & Tellis, G. J. 1998. Organizing for radical product innovation: The overlooked role of willingness to cannibalize. *Journal of Marketing Research*, 35(4): 474–487.

Chang, S.-C., Wu, W.-Y., & Wong, Y.-J. 2010. Family control and stock market reactions to innovation announcements. *British Journal of Management*, 21(1): 152–170.

Charmaz, K. 2006. *Constructing grounded theory*. London; Thousand Oaks, CA: Sage Publications.

Chattopadhyay, P., Glick, W. H., & Huber, G. P. 2001. Organizational actions in repsonse to threats and opportunities. *Academy of Management Journal*, 44(5): 937–955.

Chen, M.-J. 1996. Competitor analysis and interfirm rivalry: Toward a theoretical integration. *Academy of Management Review*, 21(1): 100–134.

Chesbrough, H. W. 2001. Assembling the elephant: A review of empirical studies on the impact of technical change upon incumbent firms. In R. A. Burgelman (Ed.), *Research on Technological Innovation, Management and Policy*. 1–36. Greenwich, CT: JAI Press.

Chirico, F., & Salvato, C. 2008. Knowledge integration and dynamic organizational adaptation in family firms. *Family Business Review*, 21(2): 169–181.

Chirico, F., Sirmon, D. G., Sciascia, S., & Mazzola, P. forthcoming. Entrepreneurial orientation, generational involvement and participative strategy: A configurational approach to entrepreneurship in family firms. *Strategic Entrepreneurship Journal*, SI.

Chrisman, J. J., Chua, J. H., & Kellermanns, F. W. 2009. Priorities, resource stocks, and performance in family and nonfamily firms. *Entrepreneurship Theory and Practice*, 33(3): 739–760.

Chrisman, J. J., Chua, J. H., Kellermanns, F. W., & Chang, E. P. C. 2007. Are family managers agents or stewards? An exploratory study in privately held family firms. *Journal of Business Research*, 60(10): 1030–1038.

Chrisman, J. J., Chua, J. H., Kellermanns, F. W., Matherne Iii, C. F., & Debicki, B. J. 2008. Management journals as venues for publication of family business research. *Entrepreneurship Theory and Practice*, 32(5): 927–934.

Chrisman, J. J., Chua, J. H., & Litz, R. A. 2004. Comparing the agency costs of family and non-family firms: Conceptual issues and exploratory evidence. *Entrepreneurship Theory and Practice*, 28(4): 335–354.

Chrisman, J. J., Chua, J. H., & Sharma, P. 2005. Trends and directions in the development of a strategic management theory of the family firm. *Entrepreneurship Theory and Practice*, 29(5): 555–575.

Chrisman, J. J., Chua, J. H., & Steier, L. P. 2003. An introduction to theories of family business. *Journal of Business Venturing*, 18(4): 441–448.

Chrisman, J. J., Kellermanns, F. W., Chan, K. C., & Liano, K. 2010. Intellectual foundations of current research in family business: An identification and review of 25 influential articles. *Family Business Review*, 23(1): 9–26.

Chrisman, J. J., & Patel, P. 2012. Variations in R&D investments of family and non-family firms: Behavioral agency and myopic loss aversion perspectives. *Academy of Management Journal*, 55(4): n/a.

Christensen, C. M. 1997. *The innovator's dilemma*. Boston, MA: Harvard Business School Press.

Christensen, C. M., & Bower, J. L. 1996. Customer power, strategic investment, and the failure of leading firms. *Strategic Management Journal*, 17: 197–218.

Christensen, C. M., & Raynor, M. E. 2003. *The innovator's solution*. Boston, MA: Harvard Business School Press.

Chua, J. H., Chrisman, J. J., & Bergiel, E. B. 2009. An agency theoretic analysis of the professionalized family firm. *Entrepreneurship Theory and Practice*, 33(2): 355–372.

Chua, J. H., Chrisman, J. J., & Sharma, P. 1999. Defining the family business by behavior. *Entrepreneurship Theory and Practice*, 23(4): 19–39.

316

Chung, C.-N., & Luo, X. 2008. Human agents: Contexts, and institutional change: The decline of family in the leadership of business groups. *Organization Science*, 19(1): 124–142.

Clark, S. M., Gioia, D. A., Ketchen, D. J., & Thomas, J. B. 2010. Transitional identity as a facilitator of organizational identity change during a merger. *Administrative Science Quarterly*, 55(3): 397–438.

Cohen, J. 1960. A coefficient of agreement for nominal scales. *Educational and Psychological Measurement*, 20(1): 37–46.

Cohen, W. M., & Levinthal, D. A. 1989. Innovation and learning: The two faces of R&D. *Economic Journal*, 99(397): 569–596.

Cohen, W. M., & Levinthal, D. A. 1990. Absorptive capacity: A new perspective on learning and innovation. *Administrative Science Quarterly*, 35: 128–152.

Comte, A. 1853. *The Positive Philosophy of Auguste Comte* (H. Martineau, Trans.). New York: Cosimo.

Corbetta, G., & Salvato, C. 2004. Self-serving or self-actualizing? Models of man and agency costs in different types of family firms: A commentary on "Comparing the agency costs of family and non-family firms: Conceputal issues and exploratory evidence". *Entrepreneurship Theory and Practice*, 28(4): 355–362.

Corley, K. G., Harquail, C. V., Pratt, M. G., Glynn, M. A., Fiol, M. C., & Hatch, M. J. 2006. Guiding organizational identity through aged adolescence. *Journal of Management Inquiry*, 15(2): 85–99.

Cornelissen, J. P., Haslam, S. A., & Balmer, J. M. T. 2007. Social identity, organizational identity and corporate identity: Towards an integrated understanding of processes, patternings and products. *British Journal of Management*, 18(S1): 1–16.

Craig, J., & Dibrell, C. 2006. The natural environment, innovation, and firm performance: A comparative study. *Family Business Review*, 19(4): 275–288.

Craig, J., & Moores, K. 2005. Balanced scorecards to drive the strategic planning of family firms. *Family Business Review*, 18(2): 105–122.

Craig, J., & Moores, K. 2010. Championing family business issues to influence public policy: Evidence from Australia. *Family Business Review*, 23(2): 170–180.

Craig, J. B. L., & Moores, K. 2006. A 10-year longitudinal investigation of strategy, systems, and environment on innovation in family firms. *Family Business Review*, 19(1): 1–10.

Cruz, C. C., Gómez-Mejía, L. R., & Becerra, M. 2010. Perceptions of benevolence and the design of agency contracts: CEO-TMT relationships in family firms. *Academy of Management Journal*, 53(1): 69–89.

Cyert, R. M., & March, J. G. 1963. *A behavioral theory of the firm* (1st ed.). Englewood Cliffs, NJ: Prentice-Hall.

Daft, R. L., & Weick, K. E. 1984. Toward a model of organizations as interpretation systems. *Academy of Management Review*, 9(2): 284–295.

Danneels, E. 2002. The dynamics of product innovation and firm competences. *Strategic Management Journal*, 23(12): 1095–1121.

Danneels, E. 2011. Trying to become a different type of company: Dynamic capability at Smith Corona. *Strategic Management Journal*, 32(1): 1-31.

317

Das, T. K. 1991. Time: The hidden dimension in strategic planning. *Long Range Planning*, 24(3): 49–57.

David, R. J., & Han, S.-K. 2004. A systematic assessment of the empirical support for transaction cost economics. *Strategic Management Journal*, 25(1): 39–58.

Davis, J. H., Schoorman, F. D., & Donaldson, L. 1997. Toward a stewardship theory of management. *Academy of Management Review*, 22(1): 20–47.

Davis, P., & Stern, D. 1988. Adaptation, survival, and growth of the family business: An integrated systems perspective. *Family Business Review*, 1(1): 69–84.

Davis, P. S., & Harveston, P. D. 1999. In the founder's shadow: Conflict in the family firm. *Family Business Review*, 12(4): 311–323.

Day, G. S. 1994. The capabilities of market-driven organizations. *Journal of Marketing*, 58(4): 37–52.

deAngelo, H., & deAngelo, L. 1985. Managerial ownership of voting rights: A study of public corporations with dual classes of common stock. *Journal of Financial Economics*, 14: 33–69.

Debicki, B. J., Matherne Iii, C. F., Kellermanns, F. W., & Chrisman, J. J. 2009. Family business research in the new millennium: An overview of the who, the where, the what, and the why. *Family Business Review*, 22(2): 151–166.

Debruyne, M., & Reibstein, D. J. 2005. Competitor see, competitor do: Incumbent entry in new market niches. *Marketing Science*, 24(1): 55–66.

DeCharms, R. 1968. *Personal causation*. New York: Academic Press.

Dehlen, T., & Zellweger, T. M. 2012. *Acquisitions by family firms: The role of socioemotional wealth.* Unpublished Working Paper. St.Gallen: Center for Family Business, University of St.Gallen.

Deshpande, R., & Webster, F. E., Jr. 1989. Organizational culture and marketing: Defining the research agenda. *The Journal of Marketing*, 53(1): 3–15.

DeYoung, C. G., Peterson, J. B., & Higgins, D. M. 2005. Sources of openness/intellect: Cognitive and neuropsychological correlates of the fifth factor of personality. *Journal of Personality*, 73(4): 825–858.

Dieleman, M., & Sachs, W. M. 2008. Coevolution of institutions and corporations in emerging economies: How the Salim group morphed into an institution of Suharto's crony regime. *Journal of Management Studies*, 45(7): 1274–1300.

DiMaggio, P. J., & Powell, W. W. 1983. The iron cage revisited: Institutional isomorphism and collective rationality in organizational fields. *American Sociological Review*, 48(2): 147–160.

Distelberg, B., & Sorenson, R. L. 2009. Updating systems concepts in family businesses: A focus on values, resource flows, and adaptability. *Family Business Review*, 22(1): 65–81.

Dollinger, S., Urban, K., & James, T. 2004. Creativity and openness: Further validation of two creative product measures. *Creativity Research Journal*, 16(1): 35–47.

Donckels, R., & Fröhlich, E. 1991. Are family businesses really different? European experiences from STRATOS. *Family Business Review*, 4(2): 149–160.

318

Dreux Iv, D. r. 1990. Financing family business: Alternatives to selling out or going public. *Family Business Review*, 3(3): 225–243.

Dutton, J. E., & Dukerich, J. M. 1991. Keeping an eye on the mirror: Image and identity in organizational adaptation. *Academy of Management Journal*, 34(3): 517–554.

Dutton, J. E., & Jackson, S. E. 1987. Categorizing strategic issues: Links to organizational action. *Academy of Management Review*, 12(1): 76–90.

Dyer, W. G. 1986. *Cultural change in family firms*. San Francisco, CA; London: Jossey-Bass.

Dyer, W. G. 1988. Culture and continuity in family firms. *Family Business Review*, 1(1): 37–50.

Dyer, W. G. 1989. Integrating Professional Management into a Family Owned Business. *Family Business Review*, 2(3): 221–235.

Dyer, W. G., & Whetten, D. A. 2006. Family Firms and Social Responsibility: Preliminary Evidence from the S&P 500. *Entrepreneurship Theory and Practice*, 30(6): 785–802.

Eddleston, K. A. 2008. Commentary: The prequel to family firm culture and stewardship: The leadership perspective of the founder. *Entrepreneurship Theory and Practice*, 32(6): 1055–1061.

Eddleston, K. A., Kellermanns, F. W., & Sarathy, R. 2008. Resource configuration in family firms: Linking resources, strategic planning and technological opportunities to performance. *Journal of Management Studies*, 45(1): 26–50.

Eggers, J. P., & Kaplan, S. 2009. Cognition and Renewal: Comparing CEO and Organizational Effects on Incumbent Adaptation to Technical Change. *Organization Science*, 20(2): 461–477.

Eisenhardt, K. M. 1989. Building theories from case study research. *Academy of Management Review*, 14(4): 532–550.

Eisenhardt, K. M., & Bourgeois Iii, L. J. 1988. Politics of strategic decision making in high-velocity environments: Toward a midrange theory. *Academy of Management Journal*, 31(4): 737–770.

Eisenhardt, K. M., & Graebner, M. E. 2007. Theory building from cases: Opportunities and challenges. *Academy of Management Journal*, 50(1): 25–32.

Eisenmann, T. R. 2002. The effects of CEO equity ownership and firm diversification on risk taking. *Strategic Management Journal*, 23(6): 513–534.

Elsbach, K. D., & Kramer, R. M. 1996. Members' responses to organizational identity threats: Encountering and countering the Business Week rankings. *Administrative Science Quarterly*, 41: 442–476.

Ensley, M. 2006. Family businesses can out-compete: As long as they are willing to question the chosen path. *Entrepreneurship Theory and Practice*, 30(6): 747-754.

Fama, E. F., & Jensen, M. C. 1983. Separation of ownership and control. *Journal of Law and Economics*, 26(2): 301-325.

Farjoun, M. 2010. Beyond dualism: Stability and change as a duality. *Academy of Management Review*, 35(2): 202–225.

319

Farmer, S. M., Yao, X., & Kung-Mcintyre, K. 2011. The behavioral impact of entrepreneur identity aspiration and prior entrepreneurial experience. *Entrepreneurship Theory and Practice*, 35(2): 245-273.

Fasse, M. 2009. Thomas Diehl: Fränkischer Optimist, *Handelsblatt*. http://www.handelsblatt.com/unternehmen/mittelstand/thomas-diehl-fraenkischer-optimist-seite-all/3210480-all.html. Accessed on April 27, 2012.

Fauchart, E., & Gruber, M. 2011. Darwinians, communitarians, and missionaries: The role of founder identity in entrepreneurship. *Academy of Management Journal*, 54(5): 935–957.

Feldman, M. S., & Pentland, B. T. 2003. Reconceptualizing organizational routines as a source of flexibility and change. *Administrative Science Quarterly*, 48(1): 94–118.

Fernández, Z., & Nieto, M. J. 2005. Internationalization strategy of small and medium-sized family businesses: Some influential factors. *Family Business Review*, 18(1): 77–89.

Finkelstein, S., & Hambrick, D. C. 1990. Top-management-team tenure and organizational outcomes: The moderating role of managerial discretion. *Administrative Science Quarterly*, 35(3): 484–503.

Finkelstein, S., Hambrick, D. C., & Cannella, A. A. 2009. *Strategic leadership, theory and research on executives, top management team, and boards*. Oxford: Oxford University Press.

Fiol, C. M. 2002. Capitalizing on paradox: The role of language in transforming organizational identities. *Organization Science*, 13(6): 653–666.

Fiss, P. C., & Zajac, E. J. 2004. The diffusion of ideas over contested terrain: The (non)adoption of a shareholder value orientation among German firms. *Administrative Science Quarterly*, 49: 501–534.

Fligstein, N. 1987. The intraorganizational power struggle: Rise of finance personnel to top leadership in large corporations, 1919-1979. *American Sociological Review*, 52(1): 44–58.

Ford, J. D., & Baucus, D. A. 1987. Organizational adaptation to performance downturns: An interpretation-based perspective. *Academy of Management Review*, 12(2): 366–380.

Foreman, P., & Whetten, D. A. 2002. Members' identification with multiple-identity organizations. *Organization Science*, 13(6): 618–635.

Foster, R. N. 1985. Timing technological transitions. *Technology in Society*, 7(2–3): 127–141.

Garud, R., & Rappa, M. A. 1994. A socio-cognitive model of technology evolution: The case of cochlear implants. *Organization Science*, 5(3): 344–362.

Gatignon, H., Robertson, T. S., & Fein, A. J. 1997. Incumbent defense strategies against new product entry. *International Journal of Research in Marketing*, 14(2): 163–176.

Gavetti, G., & Rivkin, J. W. 2007. On the origin of strategy: Action and cognition over time. *Organization Science*, 18(3): 420–439.

Geeraerts, G. 1984. The effect of ownership on the organization structure in small firms. *Administrative Science Quarterly*, 29(2): 232–237.

Gersick, K. E., Davis, J. A., McCollom Hampton, M., & Lansberg, I. 1997. *Generation to generation: Life cycles of the family business*. Boston, MA: Harvard Business School Press.

320

Gersick, K. E., Lansberg, I., Desjardins, M., & Dunn, B. 1999. Stages and transitions: Managing change in the family business. *Family Business Review*, 12(4): 287–297.

Ghemawat, P. 1991. *Commitment: The dynamic of strategy*. New York: Simon & Schuster.

Gibson, C. B., & Birkinshaw, J. 2004. The antecedents, consequences, and mediating role of organizational ambidexterity. *Academy of Management Journal*, 47(2): 209–226.

Gilbert, C. G. 2005. Unbundling the structure of inertia: Resource versus routine rigidity. *Academy of Management Journal*, 48(5): 741–763.

Gilbert, C. G. 2006. Change in the presence of residual fit: Can competing frames coexist? *Organization Science*, 17(1): 150–167.

Gilbert, R. J., & Newbery, D. M. G. 1982. Preemptive patenting and the persistence of monopoly. *American Economic Review*, 72(3): 514–526.

Gilbert, R. J., & Newbery, D. M. G. 1984. Uncertain innovation and the persistence of monopoly: Comment. *American Economic Review*, 74(1): 238–242.

Gilding, M. 2000. Family business and family change: Individual autonomy, democratization, and the new family business institutions. *Family Business Review*, 13(3): 239–250.

Gioia, D. A., & Chittipeddi, K. 1991. Sensemaking and sensegiving in strategic change initiation. *Strategic Management Journal*, 12(6): 433–448.

Gioia, D. A., & Pitre, E. 1990. Multiparadigm perspectives on theory building. *Academy of Management Review*, 15(4): 584–602.

Gioia, D. A., Price, K. N., Hamilton, A. L., & Thomas, J. B. 2010. Forging an identity: An insider-outsider study of processes involved in the formation of organizational identity. *Administrative Science Quarterly*, 55: 1–46.

Gioia, D. A., Schultz, M., & Corley, K. G. 2000. Organizational identity, image, and adaptive instability. *Academy of Management Review*, 25(1): 63–81.

Gioia, D. A., & Thomas, J. B. 1996. Identity, image, and issue interpretation: Sensemaking during strategic change in academia. *Administrative Science Quarterly*, 41(3): 370–403.

Glaser, B., & Strauss, A. 1967. *The discovery of grounded theory: Strategies of qualitative research*. London: Wiedenfeld and Nicholson.

Glasmeier, A. 1991. Technological discontinuities and flexible production networks: The case of Switzerland and the world watch industry. *Research Policy*, 20(5): 469–485.

Glynn, M. A. 2000. When cymbals become symbols: Conflict over organizational identity within a symphony orchestra. *Organization Science*, 11(3): 285–298.

Golden-Biddle, K., & Rao, H. 1997. Breaches in the boardroom: Organizational identity and conflicts of commitment in a nonprofit organization. *Organization Science*, 8(6): 593–611.

Golder, P. N., & Tellis, G. J. 1993. Pioneer advantage: Marketing logic or marketing legend? *Journal of Marketing Research*, 30(2): 158–170.

Gómez-Mejía, L. R., Cruz, C., Berrone, P., & Castro, J. 2011a. The bind that ties: Socioemotional wealth preservation in family firms. *Academy of Management Annals*, 5(1): 653–707.

Gómez-Mejía, L. R., Hoskisson, R. E., Makri, M., Sirmon, D. G., & Campbell, J. 2011b. *Innovation and the preservation of socioemotional wealth in family controlled high technology firms.* Unpublished Work. http://www.nd.edu/~cobweb/doc/InnAndThePreOfSoc.pdf. Accessed on April 1, 2012.

Gómez-Mejía, L. R., Makri, M., & Larraza-Kintana, M. 2010. Diversification decisions in family-controlled firms. *Journal of Management Studies*, 47(2): 223–252.

Gómez-Mejía, L. R., Núnez-Nickel, M., & Gutierrez, I. 2001. The role of family ties in agency contracts. *Academy of Management Journal*, 44(1): 81–95.

Gómez-Mejía, L. R., Takács Haynes, K., Núnez-Nickel, M., Jacobson, K. J. L., & Moyano-Fuentes, J. 2007. Socioemotional wealth and business risks in family-controlled firms: Evidence from Spanish olive oil mills. *Administrative Science Quarterly*, 52: 106–137.

Gottschalk, S., & Keese, D. 2011. *Die volkswirtschaftliche Bedeutung der Familienunternehmen.* München: Stiftung Familienunternehmen.

Govindarajan, V., & Kopalle, P. K. 2006. Disruptiveness of innovations: measurement and an assessment of reliability and validity. *Strategic Management Journal*, 27(2): 189–199.

Granovetter, M. S. 1973. The strength of weak ties. *American Journal of Sociology*, 78(6): 1360–1380.

Granovetter, M. S. 1983. The strength of weak ties: A network theory revisited. *Sociological Theory*, 1: 201–233.

Greene, P. G., & Brown, T. E. 1997. Resource needs and the dynamic capitalism typology. *Journal of Business Venturing*, 12: 161–173.

Greiner, L. E. 1997. Evolution and revolution as organizations grow: A company's past has clues for management that are critical to future success. *Family Business Review*, 10(4): 397–409.

Gulbrandsen, T. 2005. Flexibility in Norwegian family-owned enterprises. *Family Business Review*, 18(1): 57–76.

Gunkel, C. 2011. Erfindung des Rollkoffers, *Spiegel Online*. http://einestages.spiegel.de/static/topicalbumbackground/23990/zieh.html. Accessed on April 27, 2012.

Habbershon, T. G. 2006. Commentary: A framework for managing the familiness and agency advantages in family firms. *Entrepreneurship Theory and Practice*, 30(6): 879–886.

Habbershon, T. G., & Williams, M. L. 1999. A resource-based framework for assessing the strategic advantages of family firms. *Family Business Review*, 12(1): 1–25.

Habbershon, T. G., Williams, M. L., & MacMillan, I. C. 2003. A unified systems theory of family firm performance. *Journal of Business Research*, 18(4): 451–465.

Haberman, H., & Danes, S. M. 2007. Father-daughter and father-son family business management transfer comparison: Family FIRO model application. *Family Business Review*, 20(2): 163–184.

Hall, A., Melin, L., & Nordqvist, M. 2001. Entrepreneurship as radical change in the family business: Exploring the role of cultural patterns. *Family Business Review*, 14(3): 193–208.

Hall, E. T., & Hall, M. R. 1990. *Understanding cultural differences: Germans, French and Americans.* Yarmouth, ME: Intercultural Press.

322

Hambrick, D. C., & Fukutomi, G. D. S. 1991. The seasons of a CEO's tenure. *Academy of Management Review*, 16(4): 719–742.

Hambrick, D. C., & Mason, P. A. 1984. Upper echelons: The organization as a reflection of its top managers. *Academy of Management Review*, 9(2): 193–206.

Hannan, M. T., Baron, J. N., Hsu, G., & Koçak, Ö. 2006. Organizational identities and the hazard of change. *Industrial and Corporate Change*, 15(5): 755–784.

Hannan, M. T., & Freeman, J. 1977. The population ecology of organizations. *American Journal of Sociology*, 82(5): 929–964.

Hannan, M. T., & Freeman, J. 1984. Structural inertia and organizational change. *American Sociological Review*, 49(2): 149–164.

Harris, S. G., & Sutton, R. I. 1986. Functions of parting ceremonies in dying organizations. *Academy of Management Journal*, 29(1): 5–30.

Hatum, A., & Pettigrew, A. M. 2004. Adaptation under environmenal turmoil: Organizational flexibility in family-owned firms. *Family Business Review*, 17(3): 237–258.

Hatum, A., & Pettigrew, A. M. 2006. Determinants of organizational flexibility: A study in an emerging economy. *British Journal of Management*, 17(2): 115–137.

Hatum, A., Pettigrew, A. M., & Michelini, J. 2010. Building organizational capabilities to adapt under turmoil. *Journal of Change Management*, 10(3): 257–274.

Haugh, H. M., & McKee, L. 2003. 'It's just like a family'– shared values in the family firm. *Community, Work & Family*, 6(2): 141–158.

Haveman, H. A. 1993. Follow the leader: Mimetic isomorphism and entry into new markets. *Administrative Science Quarterly*, 38(4): 593–627.

HDE. 2012. *Der deutsche Einzelhandel.* Berlin: Handelsverband Deutschland.

Henderson, R. M. 1993. Underinvestment and incompetence as responses to radical innovation: Evidence from the photolithographic alignment equipment industry. *RAND Journal of Economics*, 24(2): 248–270.

Henderson, R. M., & Clark, K. B. 1990. Architectural innovation: The reconfiguration of existing product technologies and the failure of established firms. *Administrative Science Quarterly*, 35(1): 9–30.

Hill, C. W., & Rothaermel, F. T. 2003. The performance of incumbent firms in the face of radical technological innovation. *Academy of Management Review*, 28(2): 257–274.

Hirn, W., & Jensen, S. 2011. Jugendweihe bei Schlecker, *manager magazin*. http://www.manager-magazin.de/magazin/artikel/0,2828,737713,00.html. Accessed on April 27, 2012.

Hofstede, G. 2001. *Culture's Consequences: Comparing Values, Behaviors, Institutions, and Organizations Across Nations* (2nd ed.). Thousand Oaks, CA: Sage Publications.

Hounshell, D. A. 1975. Elisha Gray and the telephone: On the disadvantages of being an expert. *Technology and Culture*, 16(2): 133–161.

Howell, J. M., & Shea, C. M. 2001. Individual differences, environmental scanning, innovation framing, and champion behavior: key predictors of project performance. *Journal of Product Innovation Management*, 18(1): 15–27.

Hoy, F. 2006. The complicating factor of life cycles in corporate venturing. *Entrepreneurship Theory and Practice*, 30(6): 831–836.

Huff, J. O., Huff, A. S., & Thomas, H. 1992. Strategic renewal and the interaction of cumulative stress and inertia. *Strategic Management Journal*, 13: 55–75.

Hulin, C. L., & Roznowski, M. 1985. Organizational technologies: Effects on organizations' characteristics and individual responses. *Research in Organizational Behavior*, 7: 39–85.

Hustedde, R. J., & Pulver, G. C. 1992. Factors affecting equity capital acquisition: The demand side. *Journal of Business Venturing*, 7(5): 363–374.

IfM, 2012. Institut für Mittelstandsforschung, Bonn. http://www.ifm-bonn.org. Accessed on April 27, 2012

Jaeger, M. 2010. Auf Frauen ist Verlass beim Schuhkauf. *Frankfurter Allgemeine Zeitung*. http://www.faz.net/aktuell/rhein-main/region/familienunternehmen-dielmann-auf-frauen-ist-verlass-beim-schuhkauf-11053539.html. Accessed on April 27, 2012.

James, H. 2006. *Family capitalism*. Cambridge, MA: Belknap-Harvard University Press.

Jaworski, B. J., & Kohli, A. K. 1993. Market orientation: Antecedents and consequences. *Journal of Marketing*, 57(3):53–70.

Jelassi, T., & Enders, A. 2008. *Strategies for e-business* (2nd ed.). Harlow (England); New York: FT Prentice Hall.

Jensen, M. C., & Meckling, W. H. 1976. Theory of the firm: Managerial behavior, agency costs and ownership structure. *Journal of Financial Economics*, 3(4): 305–360.

Jick, T. D. 1979. Mixing qualitative and quantitative methods: Triangulation in action. *Administrative Science Quarterly*, 24: 602–611.

Jones, G. R. 1983. Transaction costs, property rights, and organizational culture: An exchange perspective. *Administrative Science Quarterly*, 28(3): 454–467.

Judge, T. A., Erez, A., Bono, J. E., & Thoresen, C. J. 2003. The core self-evaluation scale: Development of a measure. *Personnel Psychology*, 56(2): 303–331.

Kahneman, D. 2003. A perspective on judgment and choice: Mapping bounded rationality. *American Psychologist*, 58(9): 697–720.

Kahneman, D., & Tversky, A. 1979. Prospect theory: An analysis of decision under risk. *Econometrica*, 47(2): 263–291.

Kaplan, S. 2004. *Framing contests: Micro mechanisms of strategy-making in the face of technological change*. Cambridge, MA: MIT Sloan School of Management.

Kaplan, S. 2008a. Cognition, capabilities, and incentives: Assessing firm response to the fiber-optic revolution. *Academy of Management Journal*, 51(4): 672–695.

Kaplan, S. 2008b. Framing contests: Strategy making under uncertainty. *Organization Science*, 19(5): 729–752.

Kaplan, S. 2011. Research in cognition and strategy: Reflections on two decades of progress and a look to the future. *Journal of Management Studies*, 48(3): 665–695.

Kaplan, S., & Tripsas, M. 2008. Thinking about technology: Applying a cognitive lens to technical change. *Research Policy*, 37(5): 790–805.

Karasek, R. A. 1979. Job demands, job decision latitude, and mental strain: Implications for job redesign. *Administrative Science Quarterly*, 24(2): 285–308.

Karra, N., Tracey, P., & Phillips, N. 2006. Altruism and agency in the family firm: Exploring the role of family, kinship, and ethnicity. *Entrepreneurship Theory and Practice*, 30(6): 861–877.

Katz, R. 1982. The effects of group longevity on project communication and performance. *Administrative Science Quarterly*, 27(1): 81–104.

Kellermanns, F. W., Eddleston, K. A., Barnett, T., & Pearson, A. W. 2008. An exploratory study of family member characteristics and involvement: Effects on entrepreneurial behavior in the family firm. *Family Business Review*, 21(1): 1–14.

Kelly, L. M., Athanassiou, N., & Crittenden, W. F. 2000. Founder centrality and strategic behavior in the family-owned firm. *Entrepreneurship: Theory & Practice*, 25(2): 27–42.

Ketokivi, M., & Mantere, S. 2010. Two strategies for inductive reasoning in organizational research. *Academy of Management Review*, 35(2): 315–333.

Kets de Vries, M. F. R. 2006. *The leader on the couch*. Chichester: John Wiley & Sons.

Kets de Vries, M. F. R., & Carlock, R. S. 2007. *Family business on the couch: A psychological perspective*. Chichester, UK: John Wiley & Sons.

King, B. G., Felin, T., & Whetten, D. A. 2010. Finding the organization in organizational theory: A meta-theory of the organization as a social actor. *Organization Science*, 21(1): 290–305.

Klein, S. B. 2008. Internationale Familienunternehmen - Definition und Selbstbild. In C. Rödl (Ed.), *Internationale Familienunternehmen: Recht, Steuern, Bilanzierung, Finanzierung, Nachfolge, Strategien*. München: C.H. Beck.

Kogut, B., & Zander, U. 1996. What firms do? Coordination, identity, and learning. *Organization Science*, 7(5): 502–518.

Kohli, A. K., & Jaworski, B. J. 1990. Market orientation: The construct, research propositions, and managerial implications. *Journal of Marketing*, 54(2): 1–18.

Koiranen, M. 2002. Over 100 years of age but still entrepreneurially active in business: Exploring the values and family characteristics of old Finnish family firms. *Family Business Review*, 15(3): 175–187.

König, A. 2009. *Cognitive framing and incumbent inertia – A replication and extension of the Gilbert model in the German book retailing Industry*. Berlin: Institut für Unternehmensplanung.

König, A., Schulte, M., & Enders, A. forthcoming. Inertia in response to non-paradigmatic change: The case of meta-organizations. *Research Policy*.

Kraatz, M. S. 1998. Learning by association? Interorganizational networks and adaptation to environmental change. *Academy of Management Journal*, 41(6): 621–643.

Krippendorff, K. 2004. *Content analysis* (2 ed.). Thousand Oaks, CA: Sage Publications.

La Porta, R., Lopez-de-Silanes, F., & Shleifer, A. 1999. Corporate ownership around the world. *Journal of Finance*, 54(2): 471–517.

Labianca, G., Fairbank, J. F., Thomas, J. B., Gioia, D. A., & Umphress, E. E. 2001. Emulation in academia: Balancing structure and identity. *Organization Science*, 12(3): 312–330.

Landis, J. R., & Koch, G. G. 1977. The measurement of observer agreement for categorical data. *Biometrics*, 33: 159–174.

Lansberg, I. 1999. *Succeeding generations*. Boston, MA: Harvard Business School Press.

Lansberg, I., & Astrachan, J. H. 1994. Influence of family relationships on succession planning and training: The importance of mediating factors. *Family Business Review*, 7: 39–59.

Lazarus, R. S., & Launier, R. 1978. Stress-related transactions between person and environment. In L. A. Pervin, & M. Lewis (Eds.), *Perspectives in Interactional Psychology*: 287-327. New York: Plenum Press.

Le Breton-Miller, I., & Miller, D. 2009. Agency vs. stewardship in public family firms: A social embeddedness reconciliation. *Entrepreneurship Theory and Practice*, 33(6): 1169–1191.

Leaptrott, J. 2005. An institutional theory view of the family business. *Family Business Review*, 18(3): 215–228.

Lee, J. 2006. Impact of family relationships on attitudes of the second generation in family business. *Family Business Review*, 19(3): 175-191.

Lee, K. S., Lim, G. H., & Lim, W. S. 2003. Family business succession: Appropriation risk and choice of successor. *Academy of Management Review*, 28(4): 657–666.

Leonard-Barton, D. 1992. Core capabilities and core rigidities: A paradox in managing new product development. *Strategic Management Journal*, 13: 111–125.

Lepine, J. A., Colquitt, J. A., & Erez, A. 2000. Adaptability to changing task contexts: Effects of general cognitive ability, conscientiousness, and openness to experience. *Personnel Psychology*, 53: 563–593.

Lester, R. H., & Cannella, A. A. 2006. Interorganizational familiness: How family firms use interlocking directorates to build community-level social capital. *Entrepreneurship Theory and Practice*, 30(6): 755–775.

Levitt, B., & March, J. G. 1988. Organizational learning. *Annual Review of Sociology*, 14: 319–340.

Lichtenthaler, U. 2009. Absorptive capacity, environmental turbulence, and the complementarity of organizational learning processes. *Academy of Management Journal*, 52(4): 822–846.

Lieberman, M. B., & Montgomery, D. B. 1988. First-mover advantages. *Strategic Management Journal*, 9(S1): 41–58.

Lieberman, M. B., & Montgomery, D. B. 1998. First-mover (dis)advantages: Retrospective and link with the resource-based view. *Strategic Management Journal (SI)*, 19(12): 1111–1125.

Lincoln, Y. S., & Guba, E. G. 1985. *Naturalistic inquiry* (1 ed.). Beverly Hills, CA: Sage Publications.

Litz, R. A., & Kleysen, R. F. 2001. Your old men shall dream dreams, your young men shall see visions: Toward a theory of family firm innovation with help from the Brubeck family. *Family Business Review*, 14(4): 335–352.

Livengood, R. S., & Reger, R. K. 2010. That's our turf! Identity domains and competitive dynamics. *Academy of Management Review*, 35(1): 48–66.

Lounsbury, J. W., Smith, R. M., Levy, J. J., Leong, F. T., & Gibson, L. W. 2009. Personality characteristics of business majors as defined by the big five and narrow personality types. *Journal of Education for Business*, 84: 200–205.

Lubatkin, M. H., Ling, Y., & Schulze, W. S. 2007. An organizational justice-based view of self-control and agency costs in family firms. *Journal of Management Studies*, 44(6): 955–971.

Lüscher, L. S., & Lewis, M. W. 2008. Organizational change and managerial sensemaking: Working through paradox. *Academy of Management Journal*, 51(2): 221–240.

Mannermaa, M. 2004. Traps in future thinking - and how to overcome them. In H. F. Didsbury (Ed.), *Thinking creatively in turbulent times*. Bethesda, Maryland: World Future Society.

March, J. G. 1991. Exploration and exploitation in organizational learning. *Organization Science*, 2(1): 71–87.

March, J. G., & Heath, C. 1994. *A primer on decision making: How decision making happens*. New York: Simon & Schuster.

March, J. G., & Simon, H. A. 1958. *Organizations*. New York: Wiley.

Martins, L. L. 2005. A model of the effects of reputational rankings on organizational change. *Organization Science*, 16(6): 701–720.

McDonald, M. L., Khanna, P., & Westphal, J. D. 2008. Getting them to think outside the circle: Corporate governance, CEOs' external advice networks, and firm performance. *Academy of Management Journal*, 51: 453–475.

Memili, E., Eddleston, K. A., Kellermanns, F. W., Zellweger, T. M., & Barnett, T. 2010. The critical path to family firm success through entrepreneurial risk taking and image. *Journal of Family Business Strategy*, 1(4): 200–209.

Merton, R. K. 1957. The role-set: Problems in sociological theory. *The British Journal of Sociology*, 8(2): 106-120.

Meyer, J. W., & Scott, W. R. 1983. *The institutional environment of organizations*. Beverly Hills, CA: Sage Publications.

Micelotta, E. R., & Raynard, M. 2011. Concealing or revealing the family? Corporate brand identity strategies in family firms. *Family Business Review*, 24(3): 197–216.

Miles, M., & Huberman, M. 1994. *Qualitative data analysis* (2nd ed.). Thousand Oaks, CA: Sage Publications.

Miles, R. E., & Snow, C. C. 1978. *Organizational strategy, structure, and process*. New York: McGraw-Hill.

Miller, D., & Friesen, P. H. 1980. Momentum and revolution in organizational adaptation. *Academy of Management Journal*, 23(4): 591–614.

Miller, D., & Le Breton-Miller, I. 2005a. Management insights from great and struggling family businesses. *Long Range Planning*, 38(6): 517–530.

Miller, D., & Le Breton-Miller, I. 2005b. *Managing for the long run*. Boston, MA: Harvard Business School Press.

Miller, D., & Le Breton-Miller, I. 2006. Family governance and firm performance: Agency, stewardship, and capabilities. *Family Business Review*, 19(1): 73–87.

Miller, D., & Le Breton-Miller, I. 2011. Governance, social identity, and entrepreneurial orientation in closely held public companies. *Entrepreneurship Theory and Practice*, 35(5): 1051–1076.

Miller, D., Le Breton-Miller, I., & Lester, R. H. 2010. Family ownership and acquisition behavior in publicly-traded companies. *Strategic Management Journal*, 31: 201–223.

Miller, D., Le Breton-Miller, I., & Scholnick, B. 2008. Stewardship vs. stagnation: An empirical comparison of small family and non-family businesses. *Journal of Management Studies*, 45(1): 51–78.

Miller, D., Steier, L., & Le Breton-Miller, I. 2003. Lost in time: Intergenerational succession, change, and failure in family business. *Journal of Business Venturing*, 18(4): 513–531.

Milton, L. P. 2008. Unleashing the relationship power of family firms: Identity confirmation as a catalyst for performance. *Entrepreneurship Theory and Practice*, 32(6): 1063–1081.

Minichilli, A., Corbetta, G., & MacMillan, I. C. 2010. Top management teams in family-controlled companies: 'Familiness', 'faultlines', and their impact on financial performance. *Journal of Management Studies*, 47(2): 205–222.

Mischel, W. (Ed.). 1977. *The interactions of person and situation*. Hillsdale, NJ: Lawrence Erlbaum Associates.

Mishra, C. S., & McConaughy, D. L. 1999. Founding family control and capital structure: The risk of loss of control and the aversion of debt. *Entrepreneurship Theory and Practice*, 23: 53–64.

Mitchell, W. 1989. Whether and when? Probability and timing of incumbents' entry into emerging industrial subfields. *Administrative Science Quarterly*, 34(2): 208–230.

Morck, R., Wolfenzon, D., & Bernard, Y. 2005. Corporate governance, economic entrenchment, and growth. *Journal of Economic Literature*, 43(3): 655–720.

Morck, R., & Yeung, B. 2003. Agency problems in large family business groups. *Entrepreneurship Theory and Practice*, 27(4): 367–382.

Morgan, G., & Smircich, L. 1980. The case for qualitative research. *Academy of Management Review*, 5(4): 491–500.

Mowery, D. C. 1983. The relationship between intrafirm and contractual forms of industrial reserch in American manufacturing 1900-1940. *Explorations in Economic History*, 20: 351–374.

Müller, C., & Spiegel, S. W. *E-books in Deutschland: Der Beginn einer neuen Gutenberg-Ära?* Frankfurt am Main: PwC.

Muske, G., & Fitzgerald, M. A. 2006. A panel study of copreneurs in business: Who enters, continues, and exits? *Family Business Review*, 19(3): 193-205.

Nadkarni, S., & Herrmann, P. 2010. CEO personality, strategic flexibility, and firm performance: The case of the Indian business process outsourcing industry. *Academy of Management Journal*, 53(5): 1050–1073.

Nadkarni, S., & Narayanan, V. K. 2007. Strategic schemas, strategic flexibility, and firm performance: The moderating role of industry clockspeed. *Strategic Management Journal*, 28(3): 243–270.

Nag, R., Corley, K. G., & Gioia, D. A. 2007. The intersection of organizational identity, knowledge, and practice: Attempting strategic change via knowledge grafting. *Academy of Management Journal*, 50(4): 821–847.

Naldi, L., Nordqvist, M., Sjöberg, K., & Wiklund, J. 2007. Entrepreneurial orientation, risk taking, and performance in family firms. *Family Business Review*, 20(1): 33–47.

Narver, J. C., & Slater, S. F. 1990. The effect of a market orientation on business profitability. *Journal of Marketing*, 54(4): 20–35.

Nelson, R. R., & Winter, S. G. 1982. *An evolutionary theory of economic change*. Harvard, MA: Harvard University Press.

Noble, D. F. 1984. *Forces of production* (1st ed.). New York: Knopf.

Nuttal, W. J., Zhang, T., Hamilton, D. J., & Roques, F. A. 2009. Sociophysics simulations of technology adoption and consumer behavior, *Second International Symposium on Engineering Systems*: 1–12.

O'Boyle Jr, E. H., Pollack, J. M., & Rutherford, M. W. 2012. Exploring the relation between family involvement and firms' financial performance: A meta-analysis of main and moderator effects. *Journal of Business Venturing*, 27(1): 1–18.

O'Reilly, C. A., & Tushman, M. L. 2008. Ambidexterity as a dynamic capability: Resolving the innovator's dilemma. *Research in Organizational Behavior*, 28: 185–206.

Obodaru, O. 2012. The self not taken: How alterntive selves develop and how they influence our professional lives. *Academy of Management Review*, 37(1): 34-57.

Ocasio, W. 1995. The enactment of economic adversity: A reconciliaton of theories of failure-induced change and threat rigidity. In L. L. Cummings, & B. M. Staw (Eds.), *Research in Organizational Behavior*, Vol. 17: 287–331. Greenwich, CT: JAI Press.

Ocasio, W. 1997. Towards an attention-based view of the firm. *Strategic Management Journal*, 18: 187–206.

Ogbonna, E., & Harris, L. C. 2001. The founder's legacy: Hangover or inheritance? *British Journal of Management*, 12(1): 13–31.

Oliver, C. 1997. Sustainable competitive advantage: Combining institutional and resource-based views. *Strategic Management Journal*, 18(9): 697–713.

Parada, M. J., Nordqvist, M., & Gimeno, A. 2010. Institutionalizing the family business: The role of professional associations in fostering a change of values. *Family Business Review*, 23(4): 355–372.

Pearson, A. W., Carr, J. C., & Shaw, J. C. 2008. Toward a theory of familiness: A social capital perspective. *Entrepreneurship Theory and Practice*, 32(6): 949–969.

Pelham, A. M., & Wilson, D. T. 1996. A longitudinal study of the impact of market structure, firm structure, strategy, and market orientation culture on dimensions of small-firm performance. *Journal of the Academy of Marketing Science*: 24, 27–43.

Pervin, A. 1997. A conversation with Henry Mintzberg. *Family Business Review*, 10(2): 185–198.

Petzold, D. 2009. Warum LEGO für Märklin ein Vorbild sein sollte, *Wirtschaftswoche*. http://www.wiwo.de/unternehmen/modellbahnhersteller-warum-lego-fuer-maerklin-ein-vorbild-sein-sollte/5246326.html. Accessed on April 27, 2012.

Pfeffer, J. 1992. *Managing with power: Politics and influence in organizations*. Boston, MA: Harvard University Press.

Pfeffer, J., & Salancik, G. R. 1978. *The external control of organizations*. New York: Harper & Row.

Popper, K. R. 1935. *The logic of scientific discovery*. New York: Routledge.

Popper, K. R. (Ed.). 1994. *Die beiden Grundprobleme der Erkenntnistheorie*. Tuebingen: J. C. B. Mohr.

Porac, J. F., Thomas, H., & Baden-Fuller, C. 1989. Competitive groups as cognitive communities: The case of Scottish knitwear manufacturers. *Journal of Management Studies*, 26(4): 397–416.

Pratt, M. G., & Foreman, P. O. 2000. Classifying managerial responses to multiple organizational identities. *Academy of Management Review*, 25(1): 18–42.

Raisch, S., & Birkinshaw, J. 2008. Organizational ambidexterity: Antecedents, outcomes, and moderators. *Journal of Management*, 34(3): 375–409.

Reay, T. 2009. Family-business meta-identity, institutional pressures, and ability to respond to entrepreneurial opportunities. *Entrepreneurship Theory and Practice*, 33(6): 1265–1270.

Reinganum, J. F. 1983. Uncertain innovation and the persistence of monopoly. *American Economic Review*, 73(4): 741–748.

Reise, N. 2009. Insider giftet gegen Ikea-Gründer. *Spiegel Online*. http://www.spiegel.de/wirtschaft/unternehmen/0,1518,659742,00.html. Accessed on April 27, 2012.

Remenyi, D. 2002. As the first 50 years of computing draw to an end ...: what kind of society do we want? *Journal of Information Technology*, 17(1): 3–7.

Romano, C. A., Tanewski, G. A., & Smyrnios, K. X. 2000. Capital structure decision making: A model for family business. *Journal of Business Venturing*, 16: 285–310.

Ronte, H. 2001. The impact of technology on publishing. *Publishing Research Quarterly*, 16(4): 11–22.

Rosenbloom, R. S. 2000. Leadership, capabilities, and technological change: The transformation of NCR in the electronic era. *Strategic Management Journal*, 21(10-11): 1083–1103.

Rosenbloom, R. S., & Christensen, C. M. 1998. Technological discontinuities, organizational capabilities, and strategic commitments. In G. Dosi, D. J. Teece, & J. Chytry (Eds.), *Technology, organization, and competitiveness: Perspective on industrial and corporate change*: 215–245. New York: Oxford University Press.

Rothaermel, F. T. 2002. Technological discontinuities and interfirm cooperation: what determines a startup's attractiveness as alliance partner? *Engineering Management, IEEE Transactions on*, 49(4): 388–397.

Rotter, J. B. 1966. Generalized expectancies for internal versus external control of reinforcement. *Psychological Monographs: General and Applied*, 80(1): 1–28.

Rouleau, L. 2005. Micro-practices of strategic sensemaking and sensegiving: How middle managers interpret and sell change every day. *Journal of Management Studies*, 42(7): 1413–1441.

Santos, J., Doz, Y., & Williamson, P. 2004. Is your innovation process global? *MIT Sloan Management Review*, 45(4): 31–37.

Schneeberger, C. 2000. Wir sind berühmt dafür, unprofitabel zu sein, *Handelszeitung*. http://www.pme.ch/de/artikelanzeige/artikelanzeige.asp?pkBerichtNr=25346. Accessed on April 27, 2012.

Schulze, W. S., Lubatkin, M. H., & Dino, R. N. 2002. Altruism, agency, and the competitiveness of family firms. *Managerial and Decision Economics*, 23(4/5): 247–259.

Schulze, W. S., Lubatkin, M. H., & Dino, R. N. 2003a. Exploring the agency consequences of ownership dispersion among the directors of private family firms. *Academy of Management Journal*, 46(2): 179–194.

Schulze, W. S., Lubatkin, M. H., & Dino, R. N. 2003b. Toward a theory of agency and altruism in family firms. *Journal of Business Venturing*, 18(4): 473–490.

Schulze, W. S., Lubatkin, M. H., Dino, R. N., & Buchholtz, A. K. 2001. Agency relationships in family firms: Theory and evidence. *Organization Science*, 12(2): 99–116.

Schumpeter, J. A. 1934. *The theory of economic development*. Cambridge, MA: Harvard University Press.

Schumpeter, J. A. 1942. *Capitalism, Socialism, and Democracy*. New York: Harper.

Schwandt, T. A. 1994. Constructivist, interpretivist approaches to human inquiry. In N. K. Denzin, & Y. S. Lincoln (Eds.), *Handbook of qualitative research*: 221-259. Thousand Oaks: Sage, CA.

Scott, S. G., & Lane, V. r. 2000. A stakeholder approach to organizational identity. *Academy of Management Review*, 25(1): 43–62.

Selznick, P. 1957. *Leadership in Administration.* New York: Harper and Row.

Sharma, P. 2004. An overview of the field of family business studies: Current status and directions for the future. *Family Business Review*, 17(1): 1–36.

Sharma, P., Chrisman, J. J., & Chua, J. H. 1997. Strategic management of the family business: Past research and future challenges. *Family Business Review*, 10(1): 1–35.

Sharma, P., & Manikutty, S. 2005. Strategic divestments in family firms: Role of family structure and community culture. *Entrepreneurship Theory and Practice*, 29(3): 293–311.

Sharma, P., & Salvato, C. 2011. Commentary: Exploiting and exploring new opportunities over life cycle stages of family firms. *Entrepreneurship Theory and Practice*, 35(6): 1199–1205.

Sheff, D. 1999. *Game over: How Nintendo conquered the world*. New York: Vintage Books.

Shepherd, D., & Haynie, J. M. 2009. Family business, identity conflict, and an expedited entrepreneurial process: A process of resolving identity conflict. *Entrepreneurship Theory and Practice*, 33(6): 1245–1264.

Short, J. C., Payne, G. T., Brigham, K. H., Lumpkin, G. T., & Broberg, J. C. 2009. Family firms and entrepreneurial orientation in publicly traded firms: A comparative analysis of the S&P 500. *Family Business Review*, 22(1): 9–24.

Siggelkow, N. 2001. Change in the presence of fit: The rise, the fall, and the renaissance of Liz Claiborne. *Academy of Management Journal*, 44(4): 838–857.

Simon, H. A. 1955. A behavioral model of rational choice. *The Quarterly Journal of Economics*, 69(1): 99–118.

Sirmon, D. G., & Hitt, M. A. 2003. Managing resources: Linking unique resources, management, and wealth creation in family firms. *Entrepreneurship Theory and Practice*, 27(4): 339–358.

Smircich, L., & Stubbart, C. 1985. Strategic management in an enacted world. *Academy of Management Review*, 10(4): 724-736.

Stafford, K., Duncan, K. A., Dane, S., & Winter, M. 1999. A research model of sustainable family businesses. *Family Business Review*, 12(3): 197–208.

Staw, B. M. 1981. The escalation of commitment to a course of action. *Academy of Management Review*, 6(4): 577–587.

Staw, B. M., Sandelands, L. E., & Dutton, J. E. 1981. Threat-rigidity effects in organizational behavior: A multilevel analysis. *Administrative Science Quarterly*, 26(4): 501–524.

Steensma, H. K., & Fairbank, J. F. 1999. Internalizing external technology: A model of governance mode choice and an empirical assessment. *The Journal of High Technology Management Research*, 10(1): 1–35.

Stinchcombe, A. L. 1975. A structural analysis of sociology. *American Sociologist*, 10(2): 57–64.

Strauss, A. L., & Corbin, J. M. 1998. *Basics of qualitative research* (2nd ed.). Thousand Oaks, CA: Sage Publications.

Suarez, F. F., & Lanzolla, G. 2007. The role of environmental dynamics in building a first mover advantage theory. *Academy of Management Review*, 32(2): 377–392.

Suarez, F. F., & Rogelio, O. 2005. Environmental change and organizational transformation. *Industrial and Corporate Change*, 14(6): 1017–1041.

Suddaby, R. O. Y. 2006. From the editors: What grounded theory is not. *Academy of Management Journal*, 49(4): 633–642.

Sull, D. N., Tedlow, R. S., & Rosenbloom, R. S. 1997. Managerial commitments and technological change in the U.S. tire industry. *Industrial and Corporate Change*, 6: 461–501.

Sund, L.-G., & Smyrnios, K. X. 2005. Striving for happiness and its impact on family stability: An exploration of the Aristotelian conception of happiness. *Family Business Review*, 18(2): 155–170.

Sutherland, J. 1975. *Systems: Analysis, administration, and architecture*. New York: Van Nostrand Reinhold Company.

Sutton, R. I., & Staw, B. M. 1995. What theory is not. *Administrative Science Quarterly*, 40(3): 371–384.

Szymanski, D. M., Troy, L. C., & Bharadwaj, S. G. 1995. Order of entry and business performance: An empirical synthesis and reexamination. *Journal of Marketing*, 59(4): 17–33.

Tagiuri, R., & Davis, J. 1996. Bivalent attributes of the family firm. *Family Business Review*, 9(2): 199–208.

Tan, W.-L., & Fock, S. T. 2001. Coping with growth transitions: The case of Chinese family businesses in Singapore. *Family Business Review*, 14(2): 123–139.

Teece, D. J. 2006. Reflections on "profiting from innovation". *Research Policy*, 35(8): 1131–1146.

Teece, D. J., Pisano, G., & Shuen, A. 1997. Dynamic capabilities and strategic management. *Strategic Management Journal*, 18(7): 509–533.

Tellis, G. J. 2006. Disruptive dechnology or visionary leadership? *Journal of Product Innovation Management*, 23(1): 34–38.

Thiede, M. 2009. Der Tanz um den Hirsch, *Süddeutsche Zeitung*. http://www.sueddeutsche.de/wirtschaft/sz-serie-familienunternehmen-der-tanz-um-den-hirsch-1.168488. Accessed on April 27, 2012.

Thomas, J. B., Clark, S. M., & Gioia, D. A. 1993. Strategic sensemaking and organizational performance linkages among scanning, interpretation, action and outcomes. *Academy of Management Journal*, 36(2): 239–270.

Thomas, J. B., & McDaniel, R. R. 1990. Interpreting strategic issues: Effects of strategy and the information-processing structure of top management teams. *Academy of Management Journal*, 33(2): 286–306.

Tilton, J. E. 1971. *International diffusion of technology: The case of semiconductors*. Washington, DC: Brookings Institution.

Tranfield, D., Denyer, D., & Smart, P. 2003. Towards a methodology for developing evidence-informed management knowledge by means of systematic review. *British Journal of Management*, 14(3): 207–222.

Triebelhorn, M. 2012. Der tiefe Fall der Familie Erb. *Neue Züricher Zeitung*. http://www.nzz.ch/aktuell/zuerich/stadt_region/der_tiefe_fall_der_familie_erb_1.1444 1068.html. Accessed on April 27, 2012.

Tripsas, M. 2009. Technology, identity, and inertia through the lens of "the digital photography company". *Organization Science*, 20(2): 441–460.

Tripsas, M., & Gavetti, G. 2000. Capabilities, cognition and inertia: Evidence from digital imaging. *Strategic Management Journal*, 21: 1147–1161.

Tsui-Auch, L. S. 2004. The professionally managed family-ruled enterprise: Ethnic Chinese business in Singapore. *Journal of Management Studies*, 41(4): 693–723.

Tsui-Auch, L. S., & Lee, Y.-L. 2003. The state matters: Management models of Singaporean, Chinese and Korean business groups. *Organization Science*, 24(4): 507–534.

Tushman, M. L., & Anderson, P. 1986. Technological discontinuities and organizational environments. *Administrative Science Quarterly*, 31: 439–465.

Tushman, M. L., & O'Reilly, C. A. 1996. Ambidextrous organizations: Managing evolutionary and revolutionary change. *California Management Review*, 36(4): 8–30.

Tushman, M. L., & Rosenkopf, L. 1996. Executive succession, strategic reorientation and performance growth: A longitudinal study in the U.S. cement industry. *Management Science*, 42(7): 939–953.

Tushman, M. L., Smith, W., Westerman, G., & O'Reilly, C. A. 2010. Organizational designs and innovation streams. *Industrial and Corporate Change*, 19(5): 1331–1366.

Tversky, A., & Kahneman, D. 1974. Judgment under uncertainty: Heuristics and biases. *Science*, 185(4157): 1124–1131.

Utterback, J. M., & Abernathy, W. J. 1975. A dynamic model of process and product innovation. *Omega*, 3(6): 639–656.

Vago, M. 2004. Integrated change management (TM): Challenges for family business clients and consultants. *Family Business Review*, 17(1): 71–80.

Vallejo, M. C. 2008. Is the culture of family firms really different? A value-based model for its survival through generations. *Journal of Business Ethics*, 81(2): 261–279.

van Maanen, J. 1979a. The fact of fiction in organizational ethnography. *Administrative Science Quarterly*, 24(4): 539–550.

van Maanen, J. 1979b. Reclaiming qualitative methods for organizational research: A preface. *Administrative Science Quarterly*, 24(4): 520–526.

Vasudeva, G., & Anand, J. 2011. Unpacking absorptive capacity: A study of knowledge utilization from alliance portfolios. *Academy of Management Journal*, 54(3): 611–623.

von der Hagen, A. 2012. Der Störenfried, *Wir Familienunternehmer*: 6–9. http://www.wir-familienunternehmer.eu/archiv/titelthema/der-strenfried-1136/43/40/. Accessed on April 27, 2012.

Ward, J. L. 1997. Growing the family business: Special challenges and best practices. *Family Business Review*, 10(4): 323–337.

Ward, J. L. 2004. *Perpetuating the family business*. Houndmills, Basingstoke, Hampshire; New York: Palgrave Macmillan.

Weick, K. E. 1995a. *Sensemaking in organizations*. Thousand Oaks, CA: Sage Publications.

Weick, K. E. 1995b. What theory is not, theorizing is. *Administrative Science Quarterly*, 40: 385–390.

Weick, K. E. 2001. *Making sense of the organization*. Oxford, UK; Malden, MA: Blackwell Publishers.

Weick, K. E., & Bougon, M. 1986. Organizations as cognitive maps. In H. Sims, & D. A. Gioia (Eds.), *The thinking organization*. San Francisco, CA: Jossey-Brass.

Weiss, H. M., & Ilgen, D. R. 1985. Routinized behavior in organizations. *Journal of Behavioral Economics*, 14: 57–67.

Weissenborn, C. 2010. Piano Morte – Tod eines Klavierbauers, *Handelsblatt*. http://www.handelsblatt.com/unternehmen/mittelstand/schimmel-insolvenz-piano-morte-tod-eines-klavierbauers/3339688.html. Accessed on April 27, 2012.

Whetten, D. A. 1989. What constitutes a theoretical contribution? *Academy of Management Review*, 14(4): 490–495.

Whetten, D. A., & Mackey, A. 2002. A social actor conception of organizational identity and its implications for the study of organizational reputation. *Business and Society*, 41(4): 393–414.

Wiersema, M. F., & Bantel, K. A. 1992. Top management team demography and corporate strategic change. *Academy of Management Journal*, 35(1): 91–121.

Wiklund, J., & Shepherd, D. A. 2009. The effectiveness of alliances and acquisitions: The role of resource combination activities. *Entrepreneurship Theory and Practice*, 33(1): 193–212.

Wildhagen, A. 2011. Krupp feiert sich selbst - als Fossil. *Wirtschaftswoche*. http://www.wiwo.de/unternehmen/industrie/familienunternehmen-krupp-feiert-sich-selbst-als-fossil/5863728.html. Accessed on April 27, 2012.

Williams, K., & O'Reilly, C. A. 1998. Demography and diversity in organizations: A review of forty years of research. *Research in Organizational Behavior*, 20: 77–140.

334

Winter, M., Danes, S. M., Koh, S.-K., Fredericks, K., & Paul, J. J. 2004. Tracking family businesses and their owners over time: Panel attrition, manager departure and business demise. *Journal of Business Venturing*, 19(4): 535–559.

Wiseman, R. M., & Gómez-Mejía, L. R. 1998. A behavioral agency model of managerial risk taking. *Academy of Management Review*, 23(1): 133–153.

Woods, J. A., Dalziel, T., & Barton, S. L. 2012. Escalation of commitment in private family businesses: The influence of outside board members. *Journal of Family Business Strategy*, 3(1): 18-27.

Yin, R. K. 1994. *Case study research* (3 ed.). Thousand Oaks, CA: Sage Publications.

Zachary, M. A., McKenny, A., Short, J. C., & Payne, G. T. 2011. Family business and market orientation: Construct validation and comparative analysis. *Family Business Review*, 24(3): 233–251.

Zahra, S. A. 2005. Entrepreneurial risk taking in family firms. *Family Business Review*, 18(1): 23–40.

Zahra, S. A. 2010. Harvesting family firms' organizational social capital: A relational perspective. *Journal of Management Studies*, 47(2): 345–366.

Zahra, S. A., Hayton, J. C., Neubaum, D. O., Dibrell, C., & Craig, J. 2008. Culture of family commitment and strategic flexibility: The moderating effect of stewardship. *Entrepreneurship Theory and Practice*, 32(6): 1035–1054.

Zammuto, R. E., & Cameron, K. S. 1985. Environmental decline and organizational response. *Research in Organizational Behavior*, 7: 223–262.

Zellweger, T., Nason, R., Nordqvist, M., & Brush, C. forthcoming-a. Why do family firms strive for nonfinancial performance? *Entrepreneurship Theory & Practice*.

Zellweger, T. M., & Astrachan, J. H. 2008. On the emotional value of owning a firm. *Family Business Review*, 21(4): 347–363.

Zellweger, T. M., Eddleston, K. A., & Kellermanns, F. W. 2010. Exploring the concept of familiness: Introducing family firm identity. *Journal of Family Business Strategy*, 1(1): 54–63.

Zellweger, T. M., Kellermanns, F. W., Chrisman, J. J., & Chua, J. H. forthcoming-b. Family control and family firm valuations by family CEOs: The importance of intentions for transgenerational control. *Organization Science*.